Advanced Technologies in Graphene-Based Materials

Advanced Technologies in Graphene-Based Materials

Guest Editors

Justina Gaidukevic
Jurgis Barkauskas

MDPI

Basel • Beijing • Wuhan • Barcelona • Belgrade • Novi Sad • Cluj • Manchester

Guest Editors
Justina Gaidukevic
Vilnius University
Vilnius
Lithuania

Jurgis Barkauskas
Vilnius University
Vilnius
Lithuania

Editorial Office
MDPI AG
Grosspeteranlage 5
4052 Basel, Switzerland

This is a reprint of the Special Issue, published open access by the journal *Crystals* (ISSN 2073-4352), freely accessible at: https://www.mdpi.com/journal/crystals/special_issues/W1R774G0I1.

For citation purposes, cite each article independently as indicated on the article page online and as indicated below:

Lastname, A.A.; Lastname, B.B. Article Title. *Journal Name* **Year**, *Volume Number*, Page Range.

ISBN 978-3-7258-3049-7 (Hbk)
ISBN 978-3-7258-3050-3 (PDF)
https://doi.org/10.3390/books978-3-7258-3050-3

Contents

About the Editors

Justina Gaidukevic

Justina Gaidukevic is a Professor at the Faculty of Chemistry and Geosciences, Institute of Chemistry, Vilnius University, where she also serves as the Head of the Department of Inorganic Chemistry. She holds a Ph.D. in Chemistry, awarded in 2017, focusing on the synthesis and investigation of nano-structured graphene catalysts. With over nine years of experience in the synthesis and characterization of graphene-based materials, her research interests include the functionalization of these materials and their applications in catalysts, electrocatalysts, and electrochemical (bio)sensors. Dr. Gaidukevic is actively involved in several national and EU-funded research projects, including the development of biosensors and the study of carbon nano-structured membranes. She has contributed to 32 scientific publications and presented her findings at 69 international conferences. Her work has been recognized through various awards, including the Vilnius University Rector's Young Scientist Award in 2022. She is a member of the International Society of Electrochemistry and the Lithuanian Chemical Society. Additionally, she serves on the Editorial Board for the Lithuanian journal "*CHEMIJA*".

Jurgis Barkauskas

Jurgis Barkauskas graduated with an M.Sc. degree in Analytical Chemistry from Vilnius University in 1976. He obtained his PhD degree in Analytical Chemistry at the Institute of Inorganic Chemistry, Riga, Latvian Academy of Sciences, in 1984. In 2004, he defended his habilitated doctorate at Vilnius University in the field of Materials Science. He has more than 20 years of experience as a researcher in European and national projects dealing with nanotechnology, bioapplications, and industrial transfer. Jurgis has participated in 20 scientific projects, acting as the coordinator in 7 them. To date, he has contributed to 70 scientific publications in international journals.

crystals MDPI

Editorial

Advanced Technologies in Graphene-Based Materials

Justina Gaidukevic * and Jurgis Barkauskas

Institute of Chemistry, Faculty of Chemistry and Geosciences, Vilnius University, Naugarduko Str. 24, LT-03225 Vilnius, Lithuania; jurgis.barkauskas@chf.vu.lt
* Correspondence: justina.gaidukevic@chf.vu.lt

Graphene-based materials, including single layer graphene, laser-induced graphene, carbon nanotubes, graphene oxide, and reduced graphene oxide, have become cornerstones of modern materials science due to their exceptional properties and wide range of potential applications. These materials are driving innovation in many fields, such as electronics, catalysis, biosensors, and biomedicine, etc. To unlock the full potential of these materials, the continuous development of new synthesis methods, functionalization techniques and comprehensive characterization approaches is crucial. This Special Issue on "Advanced Technologies in Graphene-Based Materials" presents a collection of groundbreaking studies that explore the multifaceted nature of graphene and its derivatives, and contains a mixture of review articles and original contributions.

The latest research on the preparation of graphene and its application in biomedicine has been reviewed by Klaudia Malisz and Beata Świeczko-Żurek [1]. Their review article highlights graphene's unique physiochemical, mechanical, and biological properties, which have sparked significant interest in its use for various applications, including sensors, implantology, gene and drug delivery, tissue engineering, anticancer therapies, and antimicrobial agents. It emphasizes the importance of the biocompatibility of graphene, making it suitable for use in regenerative medicine and drug delivery systems. Additionally, it discusses the most popular methods of graphene production, particularly chemical vapor deposition. The role of graphene as a support for metal hydrides in energy storage applications is comprehensively reviewed by Cezar Comanescu [2]. The article examines recent advances in graphene-supported metal hydrides, focusing on how these composite systems utilize the unique properties of graphene to enhance the performance of metal hydrides in energy storage. Additionally, it provides a brief overview of the applications of these graphene-supported metal hydride nanocomposites, emphasizing their energy storage properties and potential practical benefits.

Several contributions explored the impact of graphene on the mechanical properties of composite materials. José D. Ríos et al. [3] investigated how variations in the concentration of graphene oxide (GO) and the intensity of ultrasonication influenced the fracture behavior of nano-reinforced cement pastes. Their study demonstrated that the incorporation of GO sheets into cement matrices significantly altered their mechanical and fracture characteristics, with outcomes depending on factors such as the pore size and the dimensions of the GO particles. This research highlighted the potential of GO to improve the durability and structural integrity of construction materials. S. M. Nourin Sultana et al. [4] investigated the effects of few-layer graphene (FLG) on electrical conductivity, mechanical reinforcement, and resistance to photodegradation in polyolefin blends. Their findings showed substantial improvements in electrical conductivity and UV protection due to FLG. As little as 1 wt.% of FLG was sufficient to retard the degradation of the polyolefin composites by UV light. Additionally, the incorporation of FLG helped maintain the mechanical properties of the materials under UV exposure, demonstrating its potential as a multifunctional additive in polymer composites. In a related study, S. M. Nourin Sultana and colleagues [5] explored the impact of FLG on the properties of mixed polyolefin waste streams. The study concluded that FLG enhances the mechanical properties of polyolefin waste composites while

Citation: Gaidukevic, J.; Barkauskas, J. Advanced Technologies in Graphene-Based Materials. *Crystals* **2024**, *14*, 769. https://doi.org/10.3390/cryst14090769

Received: 27 August 2024
Accepted: 27 August 2024
Published: 29 August 2024

maintaining processability. Also, they found that adding FLG up to 10 wt.% did not affect the melt flow index, effective torque during extrusion, or the viscosity of the composites.

The ability to tailor the electronic properties of graphene is crucial for its integration into advanced electronic devices. Isaac Appiah Otoo and colleagues [6] explored how different dielectric materials could modulate the carrier concentration in graphene, offering a method to fine-tune its electronic properties for specific applications. Their findings revealed that the sheet resistance of graphene was significantly influenced by the dielectric substrate, while the carrier scattering time remained unchanged. This insight could open new avenues for designing graphene-based electronics with customized electronic characteristics.

Further expanding the versatile applications of graphene, Andrés Serna-Gutiérrez and Nicolás A. Cordero [7] provided computational insights into the mechanical properties of quasi-square graphene nanoflakes. By challenging conventional assumptions about the mechanical behavior of these nanosystems, the authors proposed new models for understanding the strain-induced electronic modifications in graphene, contributing to the emerging field of straintronics. Stefan A. Pitsch and R. Radhakrishnan Sumathi [8] examined the effect of polar faces of SiC on the epitaxial growth of graphene. Their study revealed how different SiC crystallographic faces influenced the growth mechanism and electrical properties of graphene, providing important insights into optimizing the fabrication of high-quality graphene layers.

The unique surface properties of graphene-based materials also make them an excellent candidate for a variety of functionalization strategies aimed at improving its performance in a wide range of applications. Aya Jezzini et al. [9] investigated the stabilization of reduced graphene oxide (rGO) sheets with copper cations in ethanol and revealed promising photocatalytic activity for dye degradation under visible light. This study demonstrated the potential of rGO–copper systems for environmental remediation and chemical sensing applications. Meanwhile, Hanna Bandarenka and her team [10] demonstrated that graphene-coated Ag/ZrO_2 substrates significantly enhanced their surface-enhanced Raman scattering (SERS) activity. They showed that while UV exposure typically suppressed SERS signals due to zirconia's photocatalytic activity, the presence of graphene effectively counteracted this suppression, resulting in an increased SERS signal. By optimizing the UV treatment conditions, they further increased the sensitivity of these substrates for detecting low concentrations of analytes, such as rhodamine 6G molecules, highlighting the role of graphene in advancing SERS-based sensing technologies.

Innovative theoretical studies also featured prominently in this Issue. Mikhail A. Kalinin et al. [11] presented "graphocrown", a novel two-dimensional oxocarbon material. Through computational analysis, the authors revealed the structural stability and potential electronic applications of graphocrown, suggesting that it could serve as a more effective alternative to traditional graphene in certain contexts. This theoretical work paved the way for future experimental investigations and the development of new graphene-like materials with enhanced properties. The study by Lunwei Yang et al. [12] investigated the adsorption and sensing properties of formaldehyde on chemically modified graphene surfaces. By systematically analyzing the effects of different dopants (B, N, O, P, S, Mg, and Al), the authors provided a theoretical basis for the design of superior graphene-based sensors. They also highlighted the importance of chemical modification in tailoring graphene properties for specific sensing applications.

This Special Issue includes several papers that present new approaches to the synthesis of graphene-based materials. Dongyang Wang et al. [13] prepared polydopamine films on SiO_2 substrates under various conditions, focusing on the effects of pH and dopamine solution concentration on the deposition process. They then carbonized and graphitized these films at high temperatures (800 °C or 1000 °C) to produce a graphene-like material. Raman spectroscopy revealed that the resulting graphene-like film exhibited fewer structural defects compared to previous studies, and X-ray photoelectron spectroscopy showed that most of the carbon atoms were incorporated into a cross-linked honeycomb lattice structure. Additionally, the graphene-like material demonstrated high electrical conductivity and satisfactory mechanical elasticity. Aivaras Sartanavičius et al. [14] used a picosecond laser to convert polyimide (PI) into a conductive graphene network. They confirmed this by Raman spectroscopy and sheet resistance measurements. In addition, they investigated the use of laser-induced graphene (LIG) in PI for high-frequency antenna applications. In particular, they focused on the development of a 2.45 GHz patch antenna, which is commonly used for WiFi frequencies. The study showed that, due to their flexibility and conductivity, LIG-PI composites could be effectively used in antenna applications. By varying the laser parameters, they controlled the quality of the LIG and achieved a minimum sheet resistance of 36.6 ohms/sq. The resulting LIG-PI patch antenna showed a wide operating frequency band and comparable performance to conventional materials, although the conductivity needed to be improved for optimal efficiency. They concluded that LIG-PI composites held great promise for creating flexible and sustainable electronic devices, including antennas. However, future research should focus on optimizing the LIG production process and exploring additional substrates to improve performance.

Taken together, the papers in this Special Issue highlight the ongoing progress in the field of graphene-based materials and their potential to transform a wide range of technologies. The studies presented provide a strong foundation for future research and development, with a focus on expanding the capabilities and applications of these remarkable materials.

Author Contributions: Writing—original draft preparation, J.G.; writing—review and editing, J.G. and J.B. All authors have read and agreed to the published version of the manuscript.

Conflicts of Interest: The authors declare no conflict of interest.

References

1. Malisz, K.; Świeczko-Żurek, B. Graphene Production and Biomedical Applications: A Review. *Crystals* **2023**, *13*, 1413. [CrossRef]
2. Comanescu, C. Graphene Supports for Metal Hydride and Energy Storage Applications. *Crystals* **2023**, *13*, 878. [CrossRef]
3. Ríos, J.D.; Leiva, C.A.; Martinez de la Concha, M.P.; Ariza, H. Cifuentes, Influence of Graphene Oxide Concentration and Ultrasonication Energy on Fracture Behavior of Nano-Reinforced Cement Pastes. *Crystals* **2024**, *14*, 707. [CrossRef]
4. Sultana, S.M.N.; Helal, E.; Gutiérrez, G.; David, E.; Moghimian, N.; Demarquette, N.R. The Influence of a Commercial Few-Layer Graphene on Electrical Conductivity, Mechanical Reinforcement and Photodegradation Resistance of Polyolefin Blends. *Crystals* **2024**, *14*, 687. [CrossRef]
5. Sultana, S.M.N.; Helal, E.; Gutiérrez, G.; David, E.; Moghimian, N.; Demarquette, N.R. Effect of Few-Layer Graphene on the Properties of Mixed Polyolefin Waste Stream. *Crystals* **2023**, *13*, 358. [CrossRef]
6. Otoo, I.A.; Saushin, A.; Owusu, S.; Karvinen, P.; Suvanto, S.; Svirko, Y.; Kuzhir, P.; Fedorov, G. Tailoring the Graphene Properties for Electronics by Dielectric Materials. *Crystals* **2024**, *14*, 595. [CrossRef]
7. Serna-Gutiérrez, A.; Cordero, N.A. Mechanical Properties of Small Quasi-Square Graphene Nanoflakes. *Crystals* **2024**, *14*, 314. [CrossRef]
8. Pitsch, S.A.; Sumathi, R.R. Effect of Polar Faces of SiC on the Epitaxial Growth of Graphene: Growth Mechanism and Its Implications for Structural and Electrical Properties. *Crystals* **2023**, *13*, 189. [CrossRef]
9. Jezzini, A.; Davidson, A.; Hamieh, T.; Toufaily, J. Exploring Reduced Graphene Oxide Sheets Stabilized by Cu(II) and Cu(I) Cations in Ethanol. *Crystals* **2024**, *14*, 654. [CrossRef]
10. Bandarenka, H.; Burko, A.; Laputsko, D.; Dronina, L.; Kovalchuk, N.; Podelinska, A.; Shapel, U.; Popov, A.I.; Bocharov, D. Ultraviolet Exposure Improves SERS Activity of Graphene-Coated Ag/ZrO_2 Substrates. *Crystals* **2023**, *13*, 1570. [CrossRef]
11. Kalinin, M.A.; Kriebel, M.; Oshchepkov, A.S.; Sharapa, D.I. Graphocrown—A Novel, Two-Dimensional Oxocarbon: A Theoretical Study. *Crystals* **2023**, *13*, 909. [CrossRef]
12. Yang, L.; Xiao, W.; Wang, J.; Li, X.; Wang, L. Adsorption and Sensing Properties of Formaldehyde on Chemically Modified Graphene Surfaces. *Crystals* **2022**, *12*, 553. [CrossRef]
13. Wang, D.; Wang, Q.; Lin, Z.; Pang, T.; Ding, N. Effect of pH and Concentration of Dopamine Solution on the Deposited Polydopamine Film and the Prepared Graphene-like Material. *Crystals* **2023**, *13*, 607. [CrossRef]
14. Sartanavičius, A.; Žemgulytė, J.; Ragulis, P.; Ratautas, K.; Trusovas, R. Laser-Induced Graphene in Polyimide for Antenna Applications. *Crystals* **2023**, *13*, 1003. [CrossRef]

crystals

MDPI

Review

Graphene Production and Biomedical Applications: A Review

Klaudia Malisz and Beata Świeczko-Żurek *

Department of Biomaterials Technology, Faculty of Mechanical Engineering and Ship Technology, Gdansk University of Technology, Gabriela Narutowicza 11/12, 80-229 Gdansk, Poland; klaudia.malisz@pg.edu.pl
* Correspondence: beata.swieczko-zurek@pg.edu.pl

Abstract: Graphene is a two-dimensional nanomaterial composed of carbon atoms with sp^2 hybrid orbitals. Both graphene and graphene-based composite have gained broad interest among researchers because of their outstanding physiochemical, mechanical, and biological properties. Graphene production techniques are divided into top-down and bottom-up synthesis methods, of which chemical vapor deposition (CVD) is the most popular. The biomedical applications of graphene and its composite include its use in sensors, implantology, and gene and drug delivery. They can be used for tissue engineering, anticancer therapies, and as antimicrobial agents in implant application. The biocompatibility of graphene-based nanomaterials enables their use in the field of biomedicine. This article reviews the properties of graphene, the methods used to produce it, the challenges associated with its use, and the potential applications of this material in biomedicine, regenerative medicine, and drug delivery systems.

Keywords: graphene; biosensors; tissue engineering

Citation: Malisz, K.; Świeczko-Żurek, B. Graphene Production and Biomedical Applications: A Review. *Crystals* **2023**, *13*, 1413. https://doi.org/10.3390/cryst13101413

Academic Editors: Justina Gaidukevic and Jurgis Barkauskas

Received: 28 July 2023
Revised: 21 August 2023
Accepted: 23 August 2023
Published: 23 September 2023

1. Introduction

Carbon nanostructures have attracted a lot of attention, especially single-walled carbon nanotubes, multi-walled carbon nanotubes, graphene, and reduced graphene oxide (rGO) [1]. Graphene is a two-dimensional material composed of carbon (C) atoms with an sp^2 hybrid orbital [2–5], displaying a single π orbital and three σ bonds perpendicular to the plane [6]. It is a honeycomb like structure formed with a single thick planer carbon sheet [7]. Graphene can be wrapped up into zero-dimensional fullerenes, rolled into one-dimensional carbon nanotubes, or piled together with an inter-planar spacing of 0.335 nm to form three-dimensional graphite [6,8]. Due to its outstanding physical, mechanical, electrical, and chemical properties, materials based on graphene have attracted a lot of interest. These include excellent thermal and electrical conductivity, high specific surface area, elastic moduli (about 1 TPa [9]), an intrinsic strength of 130 GPa [5], adaptability to both flat and irregular surfaces, flexibility in chemical and biological functionalization, and simplicity in mass production, [10–15]. Sensors, energy harvesting, and storage devices such as solar cells and supercapacitors can all be improved by the use of graphene, as can lightweight polymer composites, membranes, and actuators. Graphene derivatives could also be used in biomedical applications such as drug delivery systems, gene therapies, photothermal therapies, antibacterial agents, and bioimaging tools [5,10,16–18]. However, scientists still face numerous challenges when attempting to utilize graphene such as cytotoxicity, biodistribution, and immunological responses.

Currently, there are many ways to produce graphene. Those include chemical vapor deposition, the mechanical cleavage of highly ordered pyrolytic graphite (HOPG), the chemical reduction of chemically exfoliated graphene oxide, epitaxial growth, and chemical synthesis [11,14]. Graphene oxide (GO), reduced graphene oxide (rGO), graphene, and graphene quantum dots (GQDs) are all included in the graphene family [19]. The molecular structures are shown in Figure 1 [20]. Chemical modification can reduce GO to rGO [18]. GQDs, also known as nano-graphene (NG), are nanomaterials that can be used in a range of applications. To create NG, top-down and bottom-up techniques are used [18,21]. Furthermore, there is a separate kind of carbon-based nanomaterial called nano-graphene oxide (NGO). NGO is created by top-down methods. Graphene oxide can be oxidized to form nano-graphene oxide, which has hydrophobic properties and the qualities suitable for creating very stable aqueous colloidal suspensions [18].

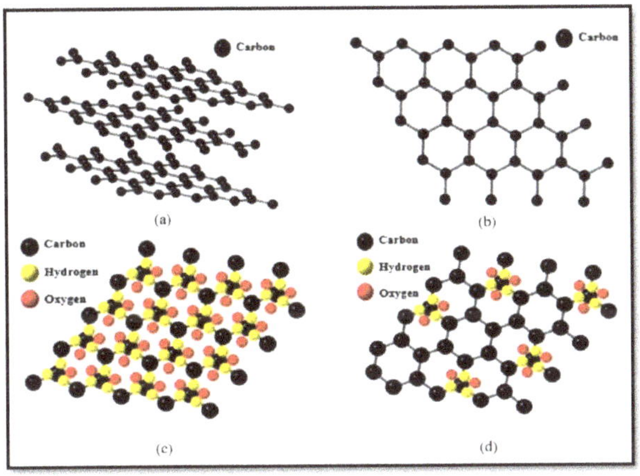

Figure 1. Molecular structure of (**a**) graphite, (**b**) graphene, (**c**) graphene oxide, (**d**) reduced graphene oxide (reprinted from [20], with permission from Elsevier, license number: 5595830853941).

Both GO and rGO are well recognized for their lower dispersion in water but also easy aggregation. This can reduce their surface areas, which would limit their usability and ability to be recycled from treated water. Additionally, graphene composites have significant limitations because of their hydrophobic qualities, which tend to result in the aggregation of sheets, along with curled, folded, and corrugated formations on base materials [19]. However, because of their enormous clinical potential in tissue regeneration therapy, GO and rGO-based scaffolds are particularly promising. The proliferation and differentiation of applied stem cells have been strongly influenced by both GO and rGO, but there are still some challenges to overcome, e.g., cytotoxicity, biodistribution, biotransformation, and immune response [17]. Graphene-reinforced polymer composites are also of great interest. The use of graphene as a reinforcing agent in the polymer matrix improves the composites' properties (for example, adding carbon nanotubes and graphene to metals reduces the coefficient of friction and the wear rate while increasing tensile strength) [5,22,23].

In this review, we will describe the properties of graphene and graphene composites, and the methods used to manufacture them with a particular emphasis on the challenges and potential applications of these materials in biomedicine, tissue engineering, and drug carrier systems.

2. Graphene Production Techniques

Several methods including pyrolysis, epitaxial growth, the chemical and electrochemical exfoliation of graphite, physical vapor deposition, and chemical vapor deposition have been used to produce graphene [6,11,14]. We divide these techniques into top-down (mechanical exfoliation, chemical exfoliation, chemical synthesis) and bottom-up synthesis methods (epitaxial growth, pyrolysis, CVD, etc.) [6]. To obtain graphene, it is possible to use the thermal expansion of graphite/graphite oxide [8,24], which can lead to its formation. However, this process rarely leads to results in the complete exfoliation of graphene to the atomic level of individual sheets of graphene [8]. There are also known methods such as liquid phase exfoliation of graphite employing ultrasonication, chemical exfoliation methods, or the electrochemical exfoliation method [8,25–27]. The disadvantages of graphene obtained by liquid phase exfoliation of graphite employing ultrasonication include the graphene sheet size because it is typically smaller than 1 μm^2, and the graphene yield is too low to be utilized in technological applications. Chemical exfoliation methods involve the oxidation of graphite to graphene oxide and then reduction chemically or thermally. These methods have attracted a lot of interest due to their potentially low cost and easy production, but to obtain the graphene structure, the thermal reduction of graphite oxide is carried out at high temperatures. However, the oxidized forms of graphite oxide cannot be completely removed with a chemical reducing agent, which may cause the degradation of the electronic properties [8].

Among these technologies, CVD appears to be the most promising process for industrial development and scale-up production of graphene. Nearly all of the transition metals included in the periodic table may be used as catalysts for the production of graphene, according to recent developments in the CVD process for growing graphene [28]. Most commonly, CVD of graphene involves injecting a precursor in the gas phase into a reaction chamber. In the reaction chamber at elevated temperature, the precursor reacts with the catalyst, and graphene is produced on the surface of the catalyst. The precursor may be a hydrocarbon (for example, methane or ethylene) or also low-molecular-weight alcohols. The temperatures of growth range from a few hundred degrees Celsius to the melting point of the catalyst metal [29]. Nickel (Ni) and copper (Cu) are two catalysts mostly used to prepare homogeneous and high-quality graphene. Nickel-catalyzed graphene, consisting of a large proportion of few-layer or multilayer, could therefore be widely used as stretchable transparent electrodes. In contrast, graphene grown on Cu is a monolayer with a large area that has great potential in high-mobility field-effect transistors. The catalyst surface is where graphene grains often form during the CVD growth of graphene on metals. During graphene growth, the grains coalesce and eventually lead to the formation of grain boundaries, which are mainly dispersed in the as-grown graphene layer. As a result, CVD-grown graphene is frequently polycrystalline and composed of a patchwork of grains with different sizes and orientations. Two methods can be used to obtain monocrystalline graphene without grain boundaries using chemical vapor deposition. The first method involves complete control of the number of nucleation centers, as reducing the nucleation number to one will finally result in the formation of individual graphene crystals. The second method, on the other hand, involves controlling the orientation of the graphene grains. In the absence of grain boundaries in the stitching areas, grains with aligned crystal lattices would merge neatly, resulting in large-sized graphene crystals [28].

The catalyst-assisted CVD approach is an intriguing synthetic method for producing wafer-sized graphene. When the nucleation and growth of crystallized graphene domains take place in an atmospheric pressure CVD process, the average size of graphene domains raises with increasing temperature and CH_4 partial pressure, but the domain density decreases (but it is independent of the CH_4 partial pressure). In addition, research by Liu L. et al. [30] indicates that the nucleation of graphene domains on Cu depends on the initial annealing temperature [30]. Plasma-enhanced chemical vapor deposition (PECVD) has great potential for graphene production due to its low-temperature growth and fast reaction rate. The high temperature provides the energy needed for graphene nucleation and growth in the thermal CVD process. However, in the PECVD process, the growth of multilayer graphene is dominated mostly by radicals generated in the plasma, so it does not need to use high temperatures. Notwithstanding, the disadvantages of this method include poor controllability and quality of obtained graphene after the graphene is grown. Due to these limitations, this method is not widely used [11,31]. Li N. et al. [11] studied the nucleation and growth of graphene at different temperatures by PECVD. They noted that graphene could not be grown at temperatures under 600 °C. In the temperature range of 650–800 °C, they obtained a high nucleation density of nanoscale graphene. In the temperature range from 850 °C to 900 °C, they formed bigger grains of graphene, and over 950 °C, the coexistence of large and small grains was produced [11].

Electrochemical exfoliation graphene is a popular top-down graphene recovery method. It is composed of oxidizing, expanding, and exfoliating graphite to produce graphene with a lower oxygen content. The process is schematically shown in Figure 2. The electrochemical exfoliation method is an environmentally friendly technique to fabricate graphene and enables production at room temperature [8,32,33]. This technique uses many types of graphite, including graphite foils, rods, plates, and powders, as electrodes in an aqueous or non-aqueous electrolyte, together with an electric current to cause electrode expansion [6]. The surface shows high roughness after electrolytic exfoliation, which is associated with a large surface area. The highest surface area was obtained after 1 h of exfoliation, and after that it declined rapidly [8]. Applying a voltage of 10 V resulted in the formation of multilayer graphene platelets with high defect concentration. Awasthi G. et al. [33] produced multi-layered (i.e., 1–5), high-quality graphene nanoplatelets using low voltage (3–4 V) during the process. For this purpose, they used potassium hydroxide solution as an electrolyte. The stirring of an electrolyte during electrolysis affected the properties of graphene, i.e., the average size of graphene nanoplatelets increased in that case, but also a reduction in the defect concentration was observed with an increase in several layers of C atoms in graphene [33]. To create high-yield solution-processable graphene, Zhang Y. et al. [34] used an electrochemical anode and cathode co-exfoliation technique. They obtained graphene with an ultralow defect and oxygen content on electrodes. The flexible supercapacitor exhibits great electrochemical performance and can be potentially used in wearable electronics [34].

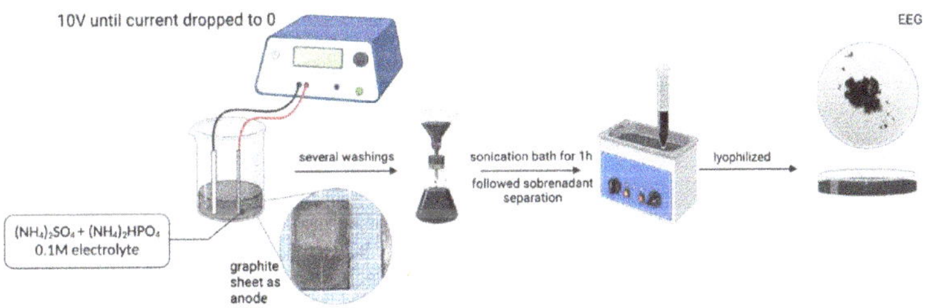

Figure 2. Electrochemical exfoliation method (reprinted from [32], with permission from Elsevier, license number: 5595791108264).

3. Application of Graphene

Graphene has received a lot of attention in a variety of sectors, including energy storage, electrochemical sensors, nanoelectronics, optical biosensors, membranes, composites, drug delivery systems, etc. [5,9,10].

3.1. Biosensors

A biosensor is a device that detects chemical or biological reactions by producing signals proportional to the concentration of an analyte in the reaction. Applications for biosensors include the detection of pollutants, disease-causing microorganisms, and markers that are indicators of disease in bodily fluids like blood, saliva urine, or even sweat. Biosensors are also used in the discovery of new medicines and the monitoring of diseases. A typical biosensor consists of a bioreceptor, transducer, electronics, and display. A bioreceptor is a molecule that recognizes the analyte specifically [35]. A transducer turns biological interactions into physical signals (for example, optical, chemical, electrical, or thermal signals) [36]. Electronics is a part that processes the transduced signal and prepares it for display. On the display, the output signal is presented in numerical, graphical, tabular, or pictorial form, depending on the requirements of the user [35].

Due to its excellent charge transfer, high specific surface area (2620 m^2/g), thermal conductivity of 500 W/mK, optical properties (visible and infrared light transmittance of 98%), and ability to immobilize molecules, graphene has become a widely used material in sensor manufacturing [7,9]. Graphene can be modified, for example, with gold or silver nanoparticles. It has been proven that the addition of nanogold resulted in greater surface area, enhanced electrochemical reactions, and interaction between elements or nanoparticles, as well as simplified charge transfer for better robustness, sensitivity, and specificity [9,37]. Narayanan J. et al. [38] successfully developed an electrochemical immunosensor for the detection of botulinum neurotoxin-E, where glassy carbon electrodes were modified using graphene nanosheets-aryldiazonium salt as a sensing platform and enzyme induced silver nanoparticles, which were deposited on gold nanoparticles as a signal amplifier [38].

Graphene-based electrochemical sensors are used for the detection of cancer biomarkers and neurotransmitters, but also as immunosensors and for the detection various other bioanalytes [39]. Graphene's self-assembling characteristics are easily monitored, allowing it to be used in highly sensitive biosensors for DNA detection. Furthermore, due to their large surface area, superior electrical conductivity, and ability to load analytes, GO-based electrochemical biosensors have shown promising results in detecting cancers [18].

Ansari G. et al. [40] prepared a nanosensor using zinc oxide nanowires (ZnO NWs) and graphene. It can detect changes in hemoglobin content in vivo. ZnO NWs were produced using vapor-liquid–solid fabrication. The graphene layer can improve the sensitivity and response time of the biosensor, while the ZnO NWs can offer a significant surface area for immobilizing the biological molecules. The biosensor's sensitivity is improved by graphene's changed chemical potential. Furthermore, combining these two materials raises the selectivity of this device [40]. Soman G. et al. [41] designed a GO-based molecularly imprinted polymer for selective uric acid detection in blood serum. The main advantages of the manufactured sensor include the possibility of direct analysis of blood samples without requiring any pre-treatment, as well as extraordinary stability and reproducibility. The findings showed the potential of the developed sensor in uric acid analysis [41].

Additionally, Mubarakali A. et al. [42] worked on a sensor to detect glucose. They obtained that using colloidal Cu nanoparticles modified with graphene coated on indium-coated tin oxide glass substrate as a working electrode. The results indicate that the bioelectrode shows adequate stability and repeatability with a shorter response time. Because of those properties, the electrode may be one option for a non-enzymatic glucose biosensor [42]. Li B. et al. [43] also studied biosensors for glucose detection. A flexible glucose oxidase/chitosan/graphene sponge/Prussian blue biosensor for sweat glucose detection was designed (fabrication process is presented in Figure 3). The immobilization of glucose oxidase with chitosan offers glucose oxidase good isolation from the environment and protection, which increases glucose oxidase's stability, reusability, and activity. The gaphene sponge (graphene aerogel) is characterized by a large specific surface area, high porosity, excellent electrical conductivity, and biocompatibility. As an electron transport medium, Prussian blue was electrochemically deposited on the working electrode. The results presented that the biosensor has a high selectivity for glucose and responds well to glucose in human sweat. In addition, the many binding sites of the graphene sponge allow for further composites with other functional materials to improve the flexible sweat glucose biosensors [43].

However, Pareek S. et al. [44] fabricated an electrochemical DNA biosensor for human papillomavirus-16 (HPV-16) detection. Cervical cancer is caused by the HPV-16 virus. GO and silver-coated gold nanoparticles were used to modify an indium tin oxide-coated glass electrode. The results indicate that the biosensor has a high sensitivity for the detection of HPV-16, and it is possible to detect it in the early stage, which can be crucial in developing point-of-care devices [44]. Bao J. et al. [45] developed an electrochemical biosensor for methylated DNA detection. DNA methylation is related to cell proliferation and differentiation. More and more often, there is information that abnormal methylation contributes to the occurrence of diseases, especially in tumorigenesis. It can also be used as biomarkers to predict response to chemotherapy strategies [45,46]. They proposed the biosensor, which included gold electrodes and nanocomposite based on gold nanoparticles, rGO, and graphite carbon nitride. The nanocomposite showed excellent electrochemical properties by the synergistic effects of each component. The proposed biosensor also displayed features of high specificity, stability, and reproducibility. Because of the synergistic effect of each component, the nanocomposite demonstrated good electrochemical characteristics [45].

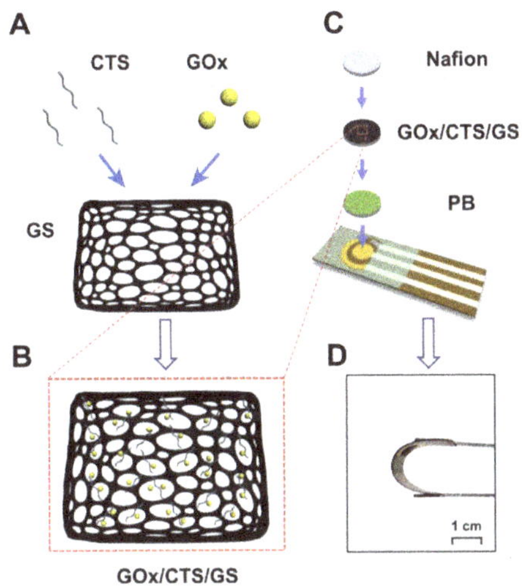

Figure 3. Fabrication process flexible glucose oxidase/chitosan/graphene sponge/Prussian blue biosensor for sweat glucose detection. (**A**) Graphene sponge (GS) mixed with chitosan (CTS) and glucose oxidase (GOx); (**B**) obtained GOx/CTS/GS composite; (**C**) Prussian blue (PB), composite and Nafion solution were added to the gold three-electrode; (**D**) flexible biosensor (reprinted from [43], with permission from Elsevier, license number: 5611890451239).

3.2. Tissue Engineering

Tissue engineering is "an interdisciplinary field that applies the principles of engineering and the life sciences toward the development of biological substitutes that restore, maintain, or improve tissue function" [47]. Tissue engineering is often known as "regenerative medicine". It offers a variety of solutions, from building new organs for transplant surgery to repairing damaged tissues. Tissue engineering, apart from cell culture and stem cell differentiation, also uses biodegradable and biologically safe materials to create structures called "scaffolds" [48]. Due to different design restrictions related to tissue engineering, features including mass transport, scaffold degradation, and biocompatibility must be taken into consideration while designing scaffolds [49]. For successful tissue engineering, it is essential to produce a biocompatible porous scaffold to support the cells, delivering bioactive molecules (for example, growth factors); as a result, they can produce physical and chemical cues that will ultimately determine the cell fate and cellular organization [50]. The cell microenvironment has an impact on the adhesion, migration, differentiation, communication, and proliferation of cells on the surface, biomaterials, or inside the extracellular matrix (ECM). A perfect biomaterial for tissue engineering and regenerative medicine often aims to imitate the corresponding ECM, providing cells with the right microenvironment [51].

The commonly used materials for tissue engineering are presented in Table 1. In addition, 2D nanomaterials have recently attracted great interest. 2D nanomaterials have a thickness of a few nanometers but a large lateral size. The unique energy level structure and optical properties, controllable thickness, simplicity in modification and doping, inherent bioactivity, high biocompatibility, and biodegradability are just a few of the distinctive properties of 2D nanomaterials. Additionally, 2D nanomaterials' large specific surface area makes them good carriers of nanoparticles or medicines [52]. In addition to graphene, scientists in recent years have proposed the following: 2D oxide and hydroxide nanosheets [53], black phosphorus nanoparticles, nanodots, nanosheets [54], non-spherical metal nanomaterials (noble metal or transition metal dichalcogenides) [55], or hexagonal boron nitride (hBN) nanoparticle [56]. Tarhan T. et al. [56] used hexagonal boron nitride nanoparticles, silver nanoparticles, and polydopamine to stimulate wound healing. Silver nanoparticles were chosen because of their anti-inflammatory properties and hBN for induced proliferation and migration of cells. Even the low concentration of the obtained agent enhances wound healing besides it reduces ROS production, promotes wound closure, and reorganizes tube formation in cells [56]. Whereas, non-spherical metal nanoparticles are used in biomedicine, including regenerative medicine and cancer therapy. What is important, the architecture of metal-based nanomaterials influences their interaction with biological systems, i.e., particle geometry is an important factor and can affect their biodistribution, interactions with blood vessels, tumor penetration, transport across endothelial cells, etc. [56,57].

Table 1. Examples of materials used for tissue engineering [49,51,58].

Natural Materials	Synthetic Polymers	Inorganic Material
collagens; hyaluronic acid; chondroitin sulfate; chitosan (CS); silk fibroin	polyethylene glycol (PEG); poly(lactic-co-glycolic acid) (PLGA); polycaprolactone (PCL); ultra-high molecular weight polyethylene (UHMWPE)	hydroxyapatite (HAp); bioglass; calcium phosphate cement (CPC)

Sharifi S. et al. [59] proposed magnesium–zinc–graphene oxide nanocomposite scaffolds for bone tissue engineering. Scaffolds were fabricated by using powder metallurgy method. Because of their biodegradability, adequate mechanical qualities, and required biocompatibility, magnesium (Mg) scaffolds are an appropriate option for bone tissue regeneration in several studies. The addition of Zn to Mg increases its corrosion resistance and mechanical qualities. Furthermore, zinc is an element crucial for the human organism; therefore, the released products should not be toxic to the body. Among the known alloys, Mg-6Zn is interesting because of its corrosion resistance, mechanical properties, and cell compatibility. However, GO indicates a positive effect on the adhesion and differentiation of osteoblasts and stimulates biomineralization [59–62]. The addition of graphene oxide made the mechanical strength and corrosion resistance significantly improved compared to Mg-Zn scaffolds. Additionally, Mg-Zn-GO scaffolds were biocompatible and not cytotoxic in contact with L-929 cells [59].

Chitosan is an amino polysaccharide that is largely found in the cell walls of fungi, plants, insects, and marine invertebrates. Due to its chemical stability, high biocompatibility, antibacterial activity, biodegradability, ability to adsorb proteins, and to accelerate wound healing, CS is used in implantology, as a drug delivery system or scaffold [58,63,64]. Valencia A.M. [65] proposed to use chitosan functionalized GO by a covalent bond (CS-GO). The results show the potential application of the CS-GO compound in tissue engineering because of a low inflammatory response in vivo test and advanced resorption at 60 days of implantation [65]. Nakagawa de Arruda M. et al. [32] designed a few-layer graphene using the electrochemical exfoliation of graphite. It was supplemented with chitosan to form a homogeneous composite with low oxygen content. In addition to increasing surface area and enhancing chemical interactions, chitosan creates homogenous coatings with graphitic materials that improve the percolation of charges and electrical signals, increasing conductivity or capacitance. Because of its higher dispersion and resonance active groups, the graphene composite's electrical conductivity increased. Furthermore, it is environmentally friendly because it uses less water and produces less toxic waste [32].

Hydroxyapatite ($Ca_{10}(PO_4)_6(OH)_2$) is a commonly used material because of its similarity with natural bone in terms of chemistry and structure. It has high biocompatibility, bioactivity, and osteoconductivity. HAp also shows osteoinductive properties and it promotes bone regeneration [66–69]. However, HAp's limited mechanical strength and fracture toughness limit its clinical applicability because they could cause the propagation of cracks and an increase in corrosion rate [66,67,69,70]. To improve mechanical properties, some materials are used as reinforcements of Hap, i.e., polyethylene, Al_2O_3, TiO_2, and carbon nanotubes. Unfortunately, these reinforcements may reduce the biological properties of hydroxyapatite and affect adjacent tissues [66]. However, HAp/GO composite showed improved in vitro osteoblast adhesion and apatite mineralization [71].

Liu Y. et al. [72] proposed HAp and HAp–graphene nanosheet (GN) composites synthesized using a liquid precipitation approach and deposited using vacuum cold spraying. The results demonstrated that the osteoblast cells spread and proliferated more readily on the GN-containing coatings [72]. Ramadas M. et al. [71] proposed hydroxyapatite nanorods grown on a graphene oxide sheet using a hydrothermal process. To assess cytotoxicity at various concentrations of material, they used human skin cancer cells. The nanocomposites presented no cytotoxicity effects on cancer cell lines. Although HAp/GO provides excellent biocompatibility [71]. However, Sánchez-Campos D. et al. [73] prepared a composite containing HAp, GO, and silver nanoparticles using a microwave-assisted hydrothermal method, a modified Hummer's synthesis, and by using dietary quercetin for silver nanoparticles formation and deposition. A dose-dependent toxic response was found in eukaryotic cells. It depended on the presence of GO, but superior cytocompatibility was observed for the samples with HAp. Furthermore, the composite increased the inhibition zone for *S. aureus* and *E. coli* compared to silver nanoparticles. The results of the antibacterial study are presented in Figure 4. These indicate that the material can be applied as a bactericidal agent for tissue replacement and surface coating [73].

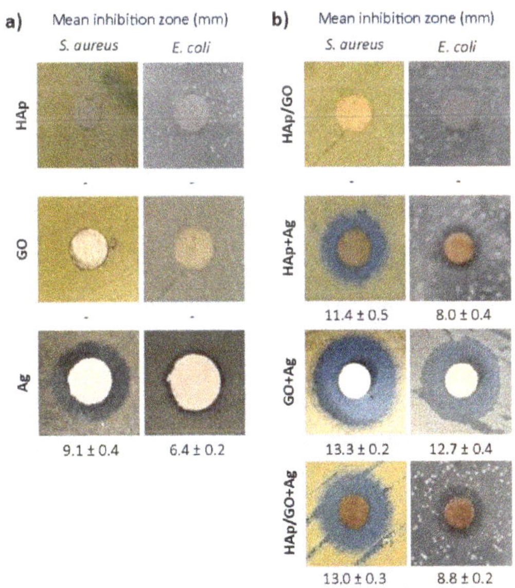

Figure 4. Pictures present inhibition zones (mm) of materials against *S. aureus* and *E. coli*: (**a**) the pure HAp, GO and silver nanoparticles (AgNPs); (**b**) HAp/GO, HAp + Ag, GO + Ag and HAp/GO + Ag composites (reprinted from [73], with permission from Elsevier, license number: 5611920074326).

Han W. et al. [66] prepared graphene oxide/hydroxyapatite composite coatings on a titanium substrate using the electrophoretic deposition method. GO/HAp composite coatings outperformed HAp coatings in terms of adhesion strength and corrosion resistance. An appropriate amount of graphene oxide promoted the proliferation of the mouse fibroblast cells and enhanced the mineralization properties. The results showed that 5 wt%

GO exhibited optimal mechanical properties, bioactivity, corrosion resistance, and good biocompatibility [66]. Fardi S.R. et al. [67] deposited hydroxyapatite–graphene oxide nanocomposites on titanium using an ultrasound-assisted electrophoretic technique. Ti sheets were anodized and then coated with HAp-GO with various compositions (0, 1, and 3 wt% of GO). They improved the mechanical properties with the best reported adhesion strength demonstrated by coating with 1 wt% GO. This coating showed excellent bioactivity and corrosion resistance [67]. Daulbayev C. et al. [74] fabricated a graphene oxide/calcium hydroxyapatite/polycaprolactone composite using an electrospun technique. Graphene oxide/hydroxyapatite composite was dispersed in biodegradable PLC. The results showed good antimicrobial activity against Gram-positive and Gram-negative bacterial strains, which depended on the GO loading. Furthermore, the composite had no cytotoxic effect on the preosteoblast MC3T3-E1 cell line. This indicates that this composite scaffold can be used for potential bone tissue regeneration [74]. Zhao H. et al. [75] fabricated porous graphene oxide/hydroxyapatite composite ceramic scaffolds using digital light processing technology. They demonstrated that adding a small amount of GO (0.1–0.4 wt%) to composite ceramics can improve their mechanical properties. The findings revealed that no scaffolds are cytotoxic. The composite scaffold increased cell adhesion, proliferation, and the expression of osteogenesis-related genes, while GO (0.1–0.2 wt%)/HAp scaffolds demonstrated superior alkaline phosphatase activity and more effective bone mineralization, as well as osteoinductivity [75].

3.3. Drug Delivery

Most conventional therapies and drug delivery systems have disadvantages such as rapid metabolism and excretion of drugs before reaching the target, poor water solubility, non-specificity to the target site, and opposing impacts on normal tissues. Therefore, nanotechnology and nanomaterials are becoming more and more popular in the development of innovative drug transport mechanisms. The efficiency of intelligent drug delivery systems is continually being improved. This is to maximize therapeutic activity and reduce side effects [76–78].

Li W. et al. [79] constructed a rectal delivery system using temperature-sensitive hydroxybutyl chitosan gel cross-linked with GO with pingyangmycin as a model drug. The temperature-sensitive gels are mostly based on polyisopropyl acrylamide, poloxamer, and CS. Because of their good biocompatibility, high permeability, and low drug toxicity, they are frequently used for injectable drug administration, oral drug delivery, rectal drug delivery, or mucosal drug delivery. The obtained material presented an effectively prolonged action time of the drug in vivo with rectal administration, and a sustained release effect [79]. Whereas, pH-responsive chitosan-based hydrogel can also be used as a drug delivery system. Hydrogel was efficiently fabricated by varying amounts of GO for controlled cephradine release. The content of GO proportionally affected the thermal stability. The material was most effective against Escherichia coli and Pseudomonas aeruginosa. Furthermore, pH-responsive swelling (schematic diagram presented in Figure 5) and biodegradability were confirmed in phosphate-buffered saline and proteinase K solutions [80].

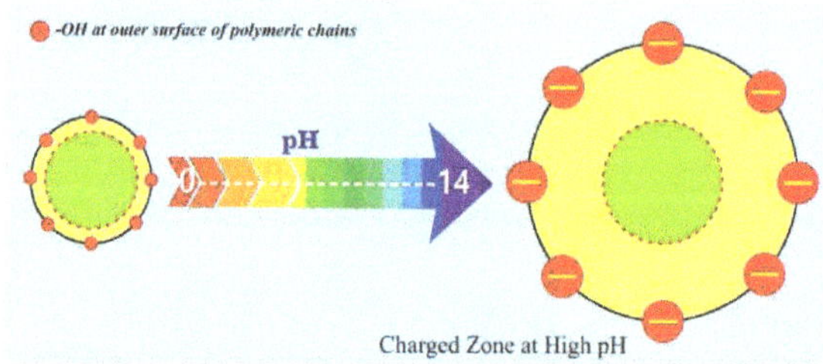

Figure 5. Schematic diagram for swelling of hydrogel due to charge zones at polymeric chains (reprinted from [80], with permission from Elsevier, license number: 5611930020590).

Breast cancer is one of the most common cancers in women and has a high mortality rate. Chemotherapy, immunotherapy, radiotherapy, and surgery are just a few of the methods that have been used to treat cancer, and combining these methods typically results in a more successful treatment [81–83]. However, scientists are still working on improving therapeutic methods. Matiyani M. et al. [84] developed a multi-functionalized GO-based novel drug nanocarrier for the delivery of the chemotherapeutic. They used quercetin and curcumin, which are potent anticancer drugs. The drug delivery system was created by combining graphene oxide with a hydrophilic polymer and metal oxides, zinc oxide and titanium dioxide. The results showed that the nanocarrier exhibits a pH-sensitive release of loaded drugs, besides demonstrating cytotoxicity towards breast cancer cells [84]. Whereas, Rajaei M. et al. [81] synthesized pH-sensitive hydrogel of chitosan/agarose/graphene oxide with glyoxal as the cross-linker. It was prepared using the water-in-oil-in-water emulsification technique. The results indicate that the loading and entrapment efficiencies of the drug are satisfactory. A highly effective and sustained medicine release profile was observed at pH 5.4; within 48 hours, almost the entire content of the drug model was released. In addition, effective cytotoxicity was observed against breast cancer cell lines (MCF-7) [81].

Quantum dots can be used to minimize the size of sheets to less than 100 nm. Moreover, quantum dots show better properties with more active groups on their surface than typical graphene and graphene oxide. They exhibit unique properties, i.e., water solubility, non-toxicity, high biocompatibility, a small size with a large surface area, high drug loading capacity, and better cellular uptake [85–87]. Mohammed-Ahmed H.K. et al. [85] examined graphene oxide quantum dots conjugated with glucosamine and boric acid with and without doxorubicin (DOX) for cancer therapy. The results showed that nanocomposite with boric acid improved the loading and release of DOX, with higher cellular internalization. Seyyedi Zadeh et al. [86] also worked on the production of a drug carrier based on quantum dots. They synthesized a magnetic graphene quantum dots-Fe_3O_4 nanocarrier targeted with folic acid (FA) and loaded with curcumin. pH-dependent drug release was observed for both GQDs-Fe_3O_4 and GQDs-Fe_3O_4-FA. The bare nanocarrier had no cytotoxic effect on normal cells. In addition, folic acid acted as a selective targeting agent in the nanocarriers [86].

4. Conclusions

Single-walled carbon nanotubes, multiwalled carbon nanotubes, graphene, and reduced graphene oxide are just a few of the carbon nanostructures that have drawn a lot of research interest. Graphene is a 2D sheet of sp^2-hybridized carbon atoms of atomic thickness that are organized in a honeycomb crystal lattice. Graphene is attracting a lot of interest due to its unique structural, optical, chemical, thermal, mechanical, electrical, and biological properties. Its derivatives are graphene oxide and reduced graphene oxide and graphene quantum dots. Graphene can be created using a variety of processes, including pyrolysis, epitaxial growth, chemical and electrochemical exfoliation of graphite, physical vapor deposition, and chemical vapor deposition. But the most promising method for industrial development and the production of graphene among those approaches is CVD.

Biomedical applications such as biosensing, medication delivery, regenerative medicine, and diagnostic tools are rapidly expanding. Graphene-based scaffolds are a particularly promising technology and have attracted interest due to their strong clinical potential in tissue regeneration therapies due to their substantial influence on the proliferation and differentiation of used stem cells. However, scientists still have many problems to solve, the biggest challenges are as follows: cytotoxicity, biodistribution, biotransformation, and immune response.

In conclusion, graphene nanomaterials have a bright future in biomedical applications. However, scientists still have many problems to solve, the biggest challenges being cytotoxicity, biotransformation, and immune response.

Author Contributions: Conceptualization, K.M.; writing—original draft preparation, K.M.; writing and editing, B.Ś.-Ż. supervision, B.Ś.-Ż.—substantial assistance in writing the article, methodology. All authors have read and agreed to the published version of the manuscript.

Funding: This research received no external funding.

Data Availability Statement: No new data were created or analyzed in this study. Data sharing is not applicable to this article.

Acknowledgments: We thank Michael Tokarz for checking the correctness of the English language.

Conflicts of Interest: The authors declare no conflict of interest.

References

1. Chakrabarty, N.; Chakraborty, A.K.; Kumar, H. Nickel-Hydroxide-Nanohexagon-Based High-Performance Electrodes for Supercapacitors: A Systematic Investigation on the Influence of Six Different Carbon Nanostructures. *J. Phys. Chem. C* **2019**, *123*, 29104–29115. [CrossRef]
2. Meng, X.; Hou, L.; Jin, H.; Li, W.; Wang, S.; Wang, Z.; An, J.; Wen, C.; Ji, G.; Xu, X.; et al. Study on Corrosion Protection Properties of PANI/ZnO/Zn/Graphene Coating on Aluminum Alloy. *Diam. Relat. Mater.* **2023**, *136*, 110067. [CrossRef]
3. Geim, A.K.; Novoselov, K.S. The Rise of Graphene. *Nat. Mater.* **2007**, *6*, 183–191. [CrossRef] [PubMed]
4. Abergel, D.S.L.; Apalkov, V.; Berashevich, J.; Ziegler, K.; Chakraborty, T. Properties of Graphene: A Theoretical Perspective. *Adv. Phys.* **2010**, *59*, 261–482. [CrossRef]
5. Mohan, V.B.; Lau, K.; Hui, D.; Bhattacharyya, D. Graphene-Based Materials and Their Composites: A Review on Production, Applications and Product Limitations. *Compos. Part B Eng.* **2018**, *142*, 200–220. [CrossRef]
6. Mbayachi, V.B.; Ndayiragije, E.; Sammani, T.; Taj, S.; Mbuta, E.R.; khan, A.U. Graphene Synthesis, Characterization and Its Applications: A Review. *Results Chem.* **2021**, *3*, 100163. [CrossRef]
7. Shahdeo, D.; Roberts, A.; Abbineni, N.; Gandhi, S. Chapter Eight—Graphene Based Sensors. In *Comprehensive Analytical Chemistry*; Hussain, C.M., Ed.; Elsevier: Amsterdam, The Netherlands, 2020; Volume 91, pp. 175–199.
8. Bose, S.; Kuila, T.; Kim, N.H.; Lee, J.H. 4—Graphene Produced by Electrochemical Exfoliation. In *Graphene: Properties, Preparation, Characterisation and Devices*; Skákalová, V., Kaiser, A.B., Eds.; Woodhead Publishing: Sawston, UK, 2014; pp. 81–98.
9. Lin, L.P.; Tan, M.T.T. Biosensors for the Detection of Lung Cancer Biomarkers: A Review on Biomarkers, Transducing Techniques and Recent Graphene-Based Implementations. *Biosens. Bioelectron.* **2023**, *237*, 115492. [CrossRef]
10. Kumar, Y.; Sahoo, S.; Chakraborty, A.K. Mechanical Properties of Graphene, Defective Graphene, Multilayer Graphene and SiC-Graphene Composites: A Molecular Dynamics Study. *Phys. B Condens. Matter* **2021**, *620*, 413250. [CrossRef]
11. Li, N.; Li, D.; Zhen, Z.; Zhang, R.; Mu, R.; Xu, Z.; He, L. Nucleation and Growth of Graphene at Different Temperatures by Plasma Enhanced Chemical Vapor Deposition. Mater. *Today Commun.* **2023**, *36*, 106568. [CrossRef]
12. Kenry; Lee, W.C.; Loh, K.P.; Lim, C.T. When Stem Cells Meet Graphene: Opportunities and Challenges in Regenerative Medicine. *Biomaterials* **2018**, *155*, 236–250. [CrossRef]
13. Sanchez, V.C.; Jachak, A.; Hurt, R.H.; Kane, A.B. Biological Interactions of Graphene-Family Nanomaterials: An Interdisciplinary Review. *Chem. Res. Toxicol.* **2012**, *25*, 15–34. [CrossRef]
14. Wang, Y.; Li, Z.; Wang, J.; Li, J.; Lin, Y. Graphene and Graphene Oxide: Biofunctionalization and Applications in Biotechnology. *Trends Biotechnol.* **2011**, *29*, 205–212. [CrossRef] [PubMed]
15. Balandin, A.A.; Ghosh, S.; Bao, W.; Calizo, I.; Teweldebrhan, D.; Miao, F.; Lau, C.N. Superior Thermal Conductivity of Single-Layer Graphene. *Nano Lett.* **2008**, *8*, 902–907. [CrossRef]
16. Tale, B.; Nemade, K.R.; Tekade, P. V Graphene Based Nano-Composites for Efficient Energy Conversion and Storage in Solar Cells and Supercapacitors: A Review. *Polym. Technol. Mater.* **2021**, *60*, 784–797. [CrossRef]
17. Raslan, A.; Saenz del Burgo, L.; Ciriza, J.; Pedraz, J.L. Graphene Oxide and Reduced Graphene Oxide-Based Scaffolds in Regenerative Medicine. *Int. J. Pharm.* **2020**, *580*, 119226. [CrossRef]
18. Arshad, F.; Nabi, F.; Iqbal, S.; Khan, R.H. Applications of Graphene-Based Electrochemical and Optical Biosensors in Early Detection of Cancer Biomarkers. *Colloids Surf. B Biointerfaces* **2022**, *212*, 112356. [CrossRef]
19. Ndagijimana, P.; Rong, H.; Ndokoye, P.; Mwizerwa, J.P.; Nkinahamira, F.; Luo, S.; Guo, D.; Cui, B. A Review on Activated Carbon/Graphene Composite-Based Materials: Synthesis and Applications. *J. Clean. Prod.* **2023**, *417*, 138006. [CrossRef]
20. Zhu, S.; Zhang, J.; Liu, X.; Li, B.; Wang, X.; Tang, S.; Meng, Q.; Li, Y.; Shi, C.; Hu, R.; et al. Graphene Quantum Dots with Controllable Surface Oxidation, Tunable Fluorescence and up-Conversion Emission. *RSC Adv.* **2012**, *2*, 2717–2720. [CrossRef]
21. Hulagabali, M.M.; Vesmawala, G.R.; Patil, Y.D. Synthesis, Characterization, and Application of Graphene Oxide and Reduced Graphene Oxide and Its Influence on Rheology, Microstructure, and Mechanical Strength of Cement Paste. *J. Build. Eng.* **2023**, *71*, 106586. [CrossRef]
22. Dorri Moghadam, A.; Omrani, E.; Menezes, P.L.; Rohatgi, P.K. Mechanical and Tribological Properties of Self-Lubricating Metal Matrix Nanocomposites Reinforced by Carbon Nanotubes (CNTs) and Graphene—A Review. *Compos. Part B Eng.* **2015**, *77*, 402–420. [CrossRef]

23. Kuilla, T.; Bhadra, S.; Yao, D.; Kim, N.H.; Bose, S.; Lee, J.H. Recent Advances in Graphene Based Polymer Composites. *Prog. Polym. Sci.* **2010**, *35*, 1350–1375. [CrossRef]
24. Schniepp, H.C.; Li, J.-L.; McAllister, M.J.; Sai, H.; Herrera-Alonso, M.; Adamson, D.H.; Prud'homme, R.K.; Car, R.; Saville, D.A.; Aksay, I.A. Functionalized Single Graphene Sheets Derived from Splitting Graphite Oxide. *J. Phys. Chem. B* **2006**, *110*, 8535–8539. [CrossRef] [PubMed]
25. Hernandez, Y.; Nicolosi, V.; Lotya, M.; Blighe, F.M.; Sun, Z.; De, S.; McGovern, I.T.; Holland, B.; Byrne, M.; Gun'Ko, Y.K.; et al. High-Yield Production of Graphene by Liquid-Phase Exfoliation of Graphite. *Nat. Nanotechnol.* **2008**, *3*, 563–568. [CrossRef] [PubMed]
26. Stankovich, S.; Dikin, D.A.; Piner, R.D.; Kohlhaas, K.A.; Kleinhammes, A.; Jia, Y.; Wu, Y.; Nguyen, S.T.; Ruoff, R.S. Synthesis of Graphene-Based Nanosheets via Chemical Reduction of Exfoliated Graphite Oxide. *Carbon N. Y.* **2007**, *45*, 1558–1565. [CrossRef]
27. Liu, F.; Wang, C.; Sui, X.; Riaz, M.A.; Xu, M.; Wei, L.; Chen, Y. Synthesis of Graphene Materials by Electrochemical Exfoliation: Recent Progress and Future Potential. *Carbon Energy* **2019**, *1*, 173–199. [CrossRef]
28. Geng, D.; Wang, H.; Yu, G. Graphene Single Crystals: Size and Morphology Engineering. *Adv. Mater.* **2015**, *27*, 2821–2837. [CrossRef]
29. Frank, O.; Kalbac, M. 2—Chemical Vapor Deposition (CVD) Growth of Graphene Films. In *Graphene: Properties, Preparation, Characterisation and Devices*; Skákalová, V., Kaiser, A.B., Eds.; Woodhead Publishing: Sawston, UK, 2014; pp. 27–49.
30. Liu, L.; Zhou, H.; Cheng, R.; Chen, Y.; Lin, Y.-C.; Qu, Y.; Bai, J.; Ivanov, I.A.; Liu, G.; Huang, Y.; et al. A Systematic Study of Atmospheric Pressure Chemical Vapor Deposition Growth of Large-Area Monolayer Graphene. *J. Mater. Chem.* **2012**, *22*, 1498–1503. [CrossRef]
31. Terasawa, T.; Saiki, K. Growth of Graphene on Cu by Plasma Enhanced Chemical Vapor Deposition. *Carbon N. Y.* **2012**, *50*, 869–874. [CrossRef]
32. de Arruda, M.N.; Palley, B.F.; de Souza, G.F.; Lopez, L.; Pontes, M.A.P.; Fernandes, W.P.; Gandara, M.; Machado, F.B.C.; Ferrão, L.F.D.A.; Gonçalves, E.S. Physicochemical Characterization of a Low-Oxygen Composite Based on Chitosan and Electrochemically Exfoliated Graphene. *Mater. Chem. Phys.* **2023**, *306*, 128049. [CrossRef]
33. Awasthi, G.; Mistry, K.; Jamnapara, N.; Salot, M.; Santhy, K.; Mandal, D.; Chaudhury, S.K. Effect of Stirring on Characteristics of Electrochemically Exfoliated Graphene. *Materialia* **2023**, *30*, 101818. [CrossRef]
34. Zhang, Y.; Xu, Y.; Liu, R.; Niu, Y. Synthesis of High-Quality Graphene by Electrochemical Anodic and Cathodic Co-Exfoliation Method. *Chem. Eng. J.* **2023**, *461*, 141985. [CrossRef]
35. Bhalla, N.; Jolly, P.; Formisano, N.; Estrela, P. Introduction to Biosensors. *Essays Biochem.* **2016**, *60*, 1–8. [CrossRef] [PubMed]
36. Prasad, S.V.S.; Kumar, M.; Arulananth, T.S.; Ravi, B.; Kumar, B.; Kiran Kumar, B. Graphene/ZnO Nanocomposite Based Optical Biosensors. *Mater. Today Proc.* **2023**. [CrossRef]
37. O'Brien, C.; Varty, K.; Ignaszak, A. The Electrochemical Detection of Bioterrorism Agents: A Review of the Detection, Diagnostics, and Implementation of Sensors in Biosafety Programs for Class A Bioweapons. *Microsyst. Nanoeng.* **2021**, *7*, 16. [CrossRef]
38. Narayanan, J.; Sharma, M.K.; Ponmariappan, S.; Sarita; Shaik, M.; Upadhyay, S. Electrochemical Immunosensor for Botulinum Neurotoxin Type-E Using Covalently Ordered Graphene Nanosheets Modified Electrodes and Gold Nanoparticles-Enzyme Conjugate. *Biosens. Bioelectron.* **2015**, *69*, 249–256. [CrossRef]
39. Ashraf, G.; Aziz, A.; Iftikhar, T.; Zhong, Z.-T.; Asif, M.; Chen, W. The Roadmap of Graphene-Based Sensors: Electrochemical Methods for Bioanalytical Applications. *Biosensors* **2022**, *12*, 1183. [CrossRef] [PubMed]
40. Ansari, G.; Pal, A.; Srivastava, A.K.; Verma, G. Detection of Hemoglobin Concentration in Human Blood Samples Using a Zinc Oxide Nanowire and Graphene Layer Heterostructure Based Refractive Index Biosensor. *Opt. Laser Technol.* **2023**, *164*, 109495. [CrossRef]
41. Soman, G.; Molahalli, V.; Hegde, G. Molecularly Imprinted Graphene Based Biosensor as Effective Tool for Electrochemical Sensing of Uric Acid. *Sens. Int.* **2023**, *4*, 100243. [CrossRef]
42. Mubarakali, A.; Gopinath, S.; Parthasarathy, P.; Arun Kumar, U.; Alavudeen Basha, A. Highly Efficient and Sensitive Non-Enzymatic Glucose Biosensor Based on Flower-Shaped CuO-Colloid Nanoparticles Decorated with Graphene-Modified Nanocomposite Electrode. *Measurement* **2023**, *217*, 113145. [CrossRef]
43. Li, B.; Wu, X.; Shi, C.; Dai, Y.; Zhang, J.; Liu, W.; Wu, C.; Zhang, Y.; Huang, X.; Zeng, W. Flexible Enzymatic Biosensor Based on Graphene Sponge for Glucose Detection in Human Sweat. *Surf. Interfaces* **2023**, *36*, 102525. [CrossRef]
44. Pareek, S.; Jain, U.; Bharadwaj, M.; Saxena, K.; Roy, S.; Chauhan, N. An Ultrasensitive Electrochemical DNA Biosensor for Monitoring Human Papillomavirus-16 (HPV-16) Using Graphene Oxide/Ag/Au Nano-Biohybrids. *Anal. Biochem.* **2023**, *663*, 115015. [CrossRef]
45. Bao, J.; Ding, K.; Zhu, Y. An Electrochemical Biosensor for Detecting DNA Methylation Based on AuNPs/RGO/g-C_3N_4 Nanocomposite. *Anal. Biochem.* **2023**, *673*, 115180. [CrossRef]
46. Heyn, H.; Esteller, M. DNA Methylation Profiling in the Clinic: Applications and Challenges. *Nat. Rev. Genet.* **2012**, *13*, 679–692. [CrossRef] [PubMed]
47. Langer, R.; Vacanti, J. Advances in Tissue Engineering. *J. Pediatr. Surg.* **2016**, *51*, 8–12. [CrossRef] [PubMed]
48. Elitok, M.S.; Gunduz, E.; Gurses, H.E.; Gunduz, M. Chapter 20—Tissue Engineering: Towards Development of Regenerative and Transplant Medicine. In *Omics Technologies and Bio-Engineering: Towards Improving Quality of Life*; Barh, D., Azevedo, V., Eds.; Academic Press: Cambridge, MA, USA, 2018; pp. 471–495.

49. Webber, M.J.; Khan, O.F.; Sydlik, S.A.; Tang, B.C.; Langer, R. A Perspective on the Clinical Translation of Scaffolds for Tissue Engineering. *Ann. Biomed. Eng.* **2015**, *43*, 641–656. [CrossRef]
50. Ghosal, K.; Mondal, P.; Bera, S.; Ghosh, S. Graphene Family Nanomaterials- Opportunities and Challenges in Tissue Engineering Applications. *FlatChem* **2021**, *30*, 100315. [CrossRef]
51. Gao, J.; Yu, X.; Wang, X.; He, Y.; Ding, J. Biomaterial–Related Cell Microenvironment in Tissue Engineering and Regenerative Medicine. *Engineering* **2022**, *13*, 31–45. [CrossRef]
52. Cao, Z.; Bian, Y.; Hu, T.; Yang, Y.; Cui, Z.; Wang, T.; Yang, S.; Weng, X.; Liang, R.; Tan, C. Recent Advances in Two-Dimensional Nanomaterials for Bone Tissue Engineering. *J. Mater.* **2023**, *9*, 930–958. [CrossRef]
53. Ma, R.; Sasaki, T. Two-Dimensional Oxide and Hydroxide Nanosheets: Controllable High-Quality Exfoliation, Molecular Assembly, and Exploration of Functionality. *Acc. Chem. Res.* **2015**, *48*, 136–143. [CrossRef]
54. Deng, L.; Xu, Y.; Sun, C.; Yun, B.; Sun, Q.; Zhao, C.; Li, Z. Functionalization of Small Black Phosphorus Nanoparticles for Targeted Imaging and Photothermal Therapy of Cancer. *Sci. Bull.* **2018**, *63*, 917–924. [CrossRef]
55. Makvandi, P.; Zarepour, A.; Zheng, X.; Agarwal, T.; Ghomi, M.; Sartorius, R.; Zare, E.N.; Zarrabi, A.; Wu, A.; Maiti, T.K.; et al. Non-Spherical Nanostructures in Nanomedicine: From Noble Metal Nanorods to Transition Metal Dichalcogenide Nanosheets. *Appl. Mater. Today* **2021**, *24*, 101107. [CrossRef]
56. Tarhan, T.; Şen, Ö.; Ciofani, M.E.; Yılmaz, D.; Çulha, M. Synthesis and Characterization of Silver Nanoparticles Decorated Polydopamine Coated Hexagonal Boron Nitride and Its Effect on Wound Healing. *J. Trace Elem. Med. Biol.* **2021**, *67*, 126774. [CrossRef]
57. Zhu, X.; Vo, C.; Taylor, M.; Smith, B.R. Non-Spherical Micro- and Nanoparticles in Nanomedicine. *Mater. Horiz.* **2019**, *6*, 1094–1121. [CrossRef]
58. Ma, P.; Li, M.; Hu, J.; Li, D.; Tao, Y.; Xu, L.; Zhao, H.; Da, J.; Li, L.; Zhao, G.; et al. New Skin Tissue Engineering Scaffold with Sulfated Silk Fibroin/Chitosan/Hydroxyapatite and Its Application. *Biochem. Biophys. Res. Commun.* **2023**, *640*, 117–124. [CrossRef]
59. Sharifi, S.; Ebrahimian-Hosseinabadi, M.; Dini, G.; Toghyani, S. Magnesium-Zinc-Graphene Oxide Nanocomposite Scaffolds for Bone Tissue Engineering. *Arab. J. Chem.* **2023**, *16*, 104715. [CrossRef]
60. Wen, C.; Zhan, X.; Huang, X.; Xu, F.; Luo, L.; Xia, C. Characterization and Corrosion Properties of Hydroxyapatite/Graphene Oxide Bio-Composite Coating on Magnesium Alloy by One-Step Micro-Arc Oxidation Method. *Surf. Coat. Technol.* **2017**, *317*, 125–133. [CrossRef]
61. Toghyani, S.; Khodaei, M. Fabrication and Characterization of Magnesium Scaffold Using Different Processing Parameters. *Mater. Res. Express* **2018**, *5*, 35407. [CrossRef]
62. Seyedraoufi, Z.S.; Mirdamadi, S. In Vitro Biodegradability and Biocompatibility of Porous Mg-Zn Scaffolds Coated with Nano Hydroxyapatite via Pulse Electrodeposition. *Trans. Nonferrous Met. Soc. China* **2015**, *25*, 4018–4027. [CrossRef]
63. Jugowiec, D.; Łukaszczyk, A.; Cieniek, Ł.; Kot, M.; Reczyńska, K.; Cholewa-Kowalska, K.; Pamuła, E.; Moskalewicz, T. Electrophoretic Deposition and Characterization of Composite Chitosan-Based Coatings Incorporating Bioglass and Sol-Gel Glass Particles on the Ti-13Nb-13Zr Alloy. *Surf. Coat. Technol.* **2017**, *319*, 33–46. [CrossRef]
64. Kumar, S.; Deepak, V.; Kumari, M.; Dutta, P.K. Antibacterial Activity of Diisocyanate-Modified Chitosan for Biomedical Applications. *Int. J. Biol. Macromol.* **2016**, *84*, 349–353. [CrossRef]
65. Valencia, A.M.; Valencia, C.H.; Zuluaga, F.; Grande-Tovar, C.D. Synthesis and Fabrication of Films Including Graphene Oxide Functionalized with Chitosan for Regenerative Medicine Applications. *Heliyon* **2021**, *7*, e07058. [CrossRef] [PubMed]
66. Han, W.; Liu, L.; Yu, B.; Zhu, Q. Electrodeposition of Graphene Oxide-Hydroxyapatite Composite Coating on Titanium Substrate. *Ceram. Int.* **2023**, *49*, 9647–9656. [CrossRef]
67. Fardi, S.R.; khorsand, H.; Askarnia, R.; Pardehkhorram, R.; Adabifiroozjaei, E. Improvement of Biomedical Functionality of Titanium by Ultrasound-Assisted Electrophoretic Deposition of Hydroxyapatite-Graphene Oxide Nanocomposites. *Ceram. Int.* **2020**, *46*, 18297–18307. [CrossRef]
68. Erakovic, S.; Jankovic, A.; Tsui, G.C.P.; Tang, C.-Y.; Miskovic-Stankovic, V.; Stevanovic, T. Novel Bioactive Antimicrobial Lignin Containing Coatings on Titanium Obtained by Electrophoretic Deposition. *Int. J. Mol. Sci.* **2014**, *15*, 12294–12322. [CrossRef] [PubMed]
69. Bensalem, A.; Kucukosman, O.K.; Raszkiewicz, J.; Topkaya, F. Synthesis, Characterization, Bactericidal Activity, and Mechanical Properties of Hydroxyapatite Nano Powders Impregnated with Silver and Zinc Oxide Nanoparticles (Ag-ZnO-Hap). *Ceram. Int.* **2021**, *47*, 21319–21324. [CrossRef]
70. Akao, M.; Aoki, H.; Kato, K. Mechanical Properties of Sintered Hydroxyapatite for Prosthetic Applications. *J. Mater. Sci.* **1981**, *16*, 809–812. [CrossRef]
71. Ramadas, M.; Bharath, G.; Ponpandian, N.; Ballamurugan, A.M. Investigation on Biophysical Properties of Hydroxyapatite/Graphene Oxide (HAp/GO) Based Binary Nanocomposite for Biomedical Applications. *Mater. Chem. Phys.* **2017**, *199*, 179–184. [CrossRef]
72. Liu, Y.; Dang, Z.; Wang, Y.; Huang, J.; Li, H. Hydroxyapatite/Graphene-Nanosheet Composite Coatings Deposited by Vacuum Cold Spraying for Biomedical Applications: Inherited Nanostructures and Enhanced Properties. *Carbon N. Y.* **2014**, *67*, 250–259. [CrossRef]

73. Sánchez-Campos, D.; Peña-Juárez, M.G.; Salado-Leza, D.; Mendoza-Anaya, D.; Rodríguez-Lugo, V.; González-Calderón, J.A.; Pérez, E. A Non-Toxic Synergistic Antibacterial Platform Based on Green Silver Nanoparticles Deposited on Hydroxyapatite/Graphene Oxide Composites. *Mater. Today Commun.* **2023**, *35*, 106097. [CrossRef]
74. Daulbayev, C.; Sultanov, F.; Korobeinyk, A.V.; Yeleuov, M.; Taurbekov, A.; Bakbolat, B.; Umirzakov, A.; Baimenov, A.; Daulbayev, O. Effect of Graphene Oxide/Hydroxyapatite Nanocomposite on Osteogenic Differentiation and Antimicrobial Activity. *Surf. Interfaces* **2022**, *28*, 101683. [CrossRef]
75. Zhao, H.; Xing, H.; Lai, Q.; Zhao, Y.; Chen, Q.; Zou, B. Additive Manufacturing of Graphene Oxide/Hydroxyapatite Bioceramic Scaffolds with Reinforced Osteoinductivity Based on Digital Light Processing Technology. *Mater. Des.* **2022**, *223*, 111231. [CrossRef]
76. Saharan, R.; Paliwal, S.K.; Tiwari, A.; Tiwari, V.; Singh, R.; Beniwal, S.K.; Dahiya, P.; Sagadevan, S. Exploring Graphene and Its Potential in Delivery of Drugs and Biomolecules. *J. Drug Deliv. Sci. Technol.* **2023**, *84*, 104446. [CrossRef]
77. Safari, J.; Zarnegar, Z. Advanced Drug Delivery Systems: Nanotechnology of Health Design A Review. *J. Saudi Chem. Soc.* **2014**, *18*, 85–99. [CrossRef]
78. Goodarzi, S.; Da Ros, T.; Conde, J.; Sefat, F.; Mozafari, M. Fullerene: Biomedical Engineers Get to Revisit an Old Friend. *Mater. Today* **2017**, *20*, 460–480. [CrossRef]
79. Li, X.; Chen, A.; Liu, Y.; Li, L. Preparation and Rectal Administration of Hydroxybutyl Chitosan/Graphene Oxide Composite Thermosensitive Hydrogel. *React. Funct. Polym.* **2023**, *189*, 105608. [CrossRef]
80. Qureshi, M.A.U.R.; Arshad, N.; Rasool, A. Graphene Oxide Reinforced Biopolymeric (Chitosan) Hydrogels for Controlled Cephradine Release. *Int. J. Biol. Macromol.* **2023**, *242*, 124948. [CrossRef] [PubMed]
81. Rajaei, M.; Rashedi, H.; Yazdian, F.; Navaei-Nigjeh, M.; Rahdar, A.; Díez-Pascual, A.M. Chitosan/Agarose/Graphene Oxide Nanohydrogel as Drug Delivery System of 5-Fluorouracil in Breast Cancer Therapy. *J. Drug Deliv. Sci. Technol.* **2023**, *82*, 104307. [CrossRef]
82. Dolatkhah, M.; Hashemzadeh, N.; Barar, J.; Adibkia, K.; Aghanejad, A.; Barzegar-Jalali, M.; Omidi, Y. Graphene-Based Multifunctional Nanosystems for Simultaneous Detection and Treatment of Breast Cancer. *Colloids Surf. B Biointerfaces* **2020**, *193*, 111104. [CrossRef]
83. Siegel, R.L.; Miller, K.D.; Jemal, A. Cancer Statistics, 2017. *CA. Cancer J. Clin.* **2017**, *67*, 7–30. [CrossRef]
84. Matiyani, M.; Rana, A.; Karki, N.; Garwal, K.; Pal, M.; Sahoo, N.G. Development of Multi-Functionalized Graphene Oxide Based Nanocarrier for the Delivery of Poorly Water Soluble Anticancer Drugs. *J. Drug Deliv. Sci. Technol.* **2023**, *83*, 104412. [CrossRef]
85. Mohammed-Ahmed, H.K.; Nakipoglu, M.; Tezcaner, A.; Keskin, D.; Evis, Z. Functionalization of Graphene Oxide Quantum Dots for Anticancer Drug Delivery. *J. Drug Deliv. Sci. Technol.* **2023**, *80*, 104199. [CrossRef]
86. Seyyedi Zadeh, E.; Ghanbari, N.; Salehi, Z.; Derakhti, S.; Amoabediny, G.; Akbari, M.; Asadi Tokmedash, M. Smart PH-Responsive Magnetic Graphene Quantum Dots Nanocarriers for Anticancer Drug Delivery of Curcumin. *Mater. Chem. Phys.* **2023**, *297*, 127336. [CrossRef]
87. Iannazzo, D.; Pistone, A.; Salamò, M.; Galvagno, S.; Romeo, R.; Giofré, S.V.; Branca, C.; Visalli, G.; Di Pietro, A. Graphene Quantum Dots for Cancer Targeted Drug Delivery. *Int. J. Pharm.* **2017**, *518*, 185–192. [CrossRef] [PubMed]

crystals

Review

Graphene Supports for Metal Hydride and Energy Storage Applications

Cezar Comanescu

National Institute of Materials Physics, Atomistilor Str. 405 A, 077125 Magurele, Romania;
cezar.comanescu@infim.ro

Abstract: Energy production, distribution, and storage remain paramount to a variety of applications that reflect on our daily lives, from renewable energy systems, to electric vehicles and consumer electronics. Hydrogen is the sole element promising high energy, emission-free, and sustainable energy, and metal hydrides in particular have been investigated as promising materials for this purpose. While offering the highest gravimetric and volumetric hydrogen storage capacity of all known materials, metal hydrides are plagued by some serious deficiencies, such as poor kinetics, high activation energies that lead to high operating temperatures, poor recyclability, and/or stability, while environmental considerations related to the treatment of end-of-life fuel disposal are also of concern. A strategy to overcome these limitations is offered by nanotechnology, namely embedding reactive hydride compounds in nanosized supports such as graphene. Graphene is a 2D carbon material featuring unique mechanical, thermal, and electronic properties, which all recommend its use as the support for metal hydrides. With its high surface area, excellent mechanical strength, and thermal conductivity parameters, graphene can serve as the support for simple and complex hydrides as well as RHC (reactive hydride composites), producing nanocomposites with very attractive hydrogen storage properties.

Keywords: graphene; energy storage; hydrogen; metal hydrides; composites; nanoconfinement; supercapacitor; batteries

Citation: Comanescu, C. Graphene Supports for Metal Hydride and Energy Storage Applications. *Crystals* **2023**, *13*, 878. https://doi.org/10.3390/cryst13060878

Academic Editors: Justina Gaidukevic and Jurgis Barkauskas

Received: 15 May 2023
Revised: 24 May 2023
Accepted: 25 May 2023
Published: 27 May 2023

1. Introduction

Energy storage is a key driver and supporter of the everyday needs of society. Within this context, metal hydrides are promising systems with the ability to store and release hydrogen gas, the sole element promising a sustainable, emission-free future [1–9]. While there are many binary and complex hydrides known, only those belonging to lightweight metals have practical implications due to their high gravimetric and volumetric H_2 content, in line with DOE's current targets. However, their slow reaction kinetics, possible side reactions, and poor recyclability have prompted researchers to find mitigation strategies to overcome these shortcomings. The recent advances in SSHS (solid-state hydrogen storage) materials have been summarized, and they entail both physisorption and chemisorption of hydrogen [10]. In contrast to conventional systems based on high-pressure gas phase or low-temperature liquid-state hydrogen, SSHS systems have the tangible potential to offer high hydrogen storage capacity, sustainable performance, recyclability with little loss in storage capacity, as well as manageable production costs [1–10].

Among recent advances, nanoconfinement in a properly chosen support has shown important improvements regarding stability, thermodynamic parameters, and overall kinetics, oftentimes affording reversible systems where pristine hydrides showed little or no reversibility. For instance, MgH_2, which arguably is the metal hydride most investigated to date, features too high energy barriers ($\Delta H = -75$ kJ/mol, $T_{des} = 573$ K) to be considered in its neat form in hydrogen storage tanks as such. Additionally, the dehydrogenated material (Mg) is a very reactive metal ($\varepsilon_{ox} = +2.37$ V), susceptible to oxidation by even minute oxygen traces, which would lead to MgO and, upon moisture exposure, to $Mg(OH)_2$; this way, the reversibility of the cycle Mg/MgH_2 would be lost. Additionally, bare MgH_2 is also plagued by particle agglomeration during cycling, which can lead to a continuous decrease in H_2 storage capacity and even slower kinetics. These downsides observed for the model system Mg/MgH_2 have prompted scientists to further tune the thermodynamic parameters by altering the morphological features of MgH_2. Nanostructuring, alloying, and compounding with additives or catalysts are among the most utilized mitigating strategies.

Confinement in nano-sized supports would afford better particle size control upon absorption/desorption a/d cycling, along with increased stability due to the confinement matrix (typically silica or carbon). Furthermore, additional hydride may support interactions that further decrease total system energy. While many types of carbo materials have been investigated as supports (OMC-ordered mesoporous carbon, AC activated carbon, carbon nanotubes CNTs, etc.), graphene has emerged as a two-dimensional carbon material with unique electronic, mechanical, and thermal properties that can act as an efficient support material for metal hydrides. Several properties are essential to accomplishing this task: its large surface area (in excess of 2500 m^2/g), the layered structure forbidding hydride NP from agglomeration, high electrical conductivity, and excellent mechanical strength. The inclusion of hydride materials into nanoporous scaffolds is also an important aspect; while ball milling (BM) seems to be the prevalent method, the incipient wetness/infiltration method and in situ generation of hydride materials inside the matrix are gaining momentum, especially due to the better dispersion and polydispersity of NPs produced by the latter methods. With a higher surface-to-volume ratio, hydride NPs are expected to behave differently in hydrogenation studies, with lower enthalpies compared to bulk and enhanced kinetics due in part to more efficient H_2 diffusion in the thin hydride layer of a few nm [1–3]. Solvent infiltration/evaporation is more commonly used in the case of complex hydride solutions, given their higher solubility in ethereal solutions such as MTBE and similar inert ethers [2–6].

The current review aims to discuss the recent advances in graphene-supported metal hydrides for energy storage applications, covering the main features of composite systems based on graphene (G, GO, and related) and metal hydrides (LiH, MgH_2, $LiBH_4$, $LiAlH_4$, $NaBH_4$, $NaAlH_4$, $Mg(BH_4)_2$, Mg_2NiH_4, $Mn(BH_4)_2$, and others) [11]. The applications of the generated nanocomposites have also been briefly described, with an emphasis on the energy storage properties of said composites. The conclusions and future outlook directions identify some areas where improvement strategies may yield enhanced energy storage properties of hydride/graphene composites, as well as raising awareness regarding the proper handling and processing of end-of-life composite energy materials, a prerequisite for a sustainable future built around a hydrogen economy.

2. Overview of Metal Hydrides and Graphene Supports

In an effort to reduce CO_2 emissions and the subsequent greenhouse effect, H_2 production can be achieved from renewable sources (biomass, water) by electrolysis, thermolysis, or photocatalytic splitting [11]. However, the lightweight and limited space requirements for a viable vehicular fuel have shifted attention towards solid-state hydrogen materials, and metal hydrides are an excellent example of such materials. However, metal hydrides do not cumulatively meet the requirements for high gravimetric content (~6.5 wt.%, ideally), fast and reliable recycling behavior (over 1000 cycles without significant degradation), and moderate thermodynamic parameters (enthalpies around 30 kJ/mol) that would allow operation under the expected range (−40 . . . 100 °C). Some of these shortcomings can be alleviated by the inclusion of the active hydrogen-storing compound in a nanoporous matrix such as graphene [12]. Graphene is a 2D material with very appealing properties, highlighting its potential use as support for various reactive species, including metals and metal hydrides [13]. By embedding hydride species into graphene supports, valuable nanocomposites can be obtained with direct use for energy storage applications (Figure 1) [11–13].

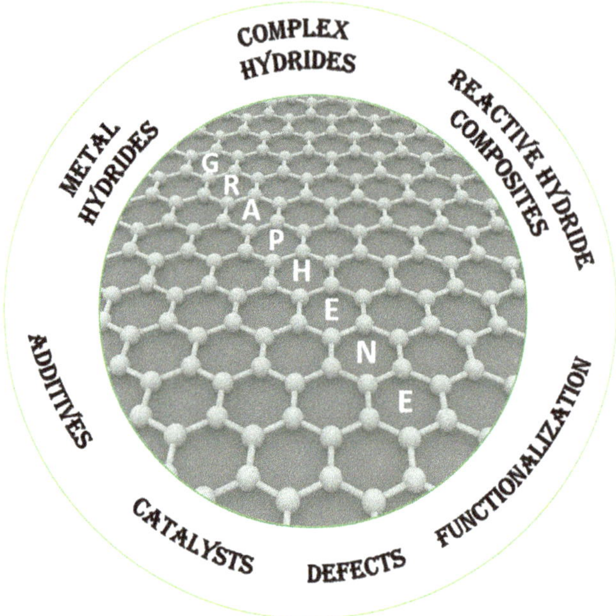

Figure 1. Graphene sits at the core of various improvement strategies for better and more effective embedding of active hydrogen storage materials.

Engineering MH_x@G composites has been no easy task, but a vast majority of researchers have turned their attention towards the Mg/MgH_2 system, in part due to its favorable thermodynamic properties (ΔH = 65.8–75.2 kJ/mol), high H_2 storage capacity (7.7 wt.%, 109 g H_2/L), and most important, due to the reversible nature of the hydrogenation/dehydrogenation process [14–16]. Assessing the support contribution to H_2 storage is equally importantly, and functionalization strategies have yielded interesting results in ball-milled Mg-B-electrochemically synthesized reduced graphene oxide (erGO) with enhanced H-uptake kinetics [17]. B-doped graphene has also been investigated by density-functional theory DFT studies [18]. Tan et al. have shown, using principal calculations, that the adsorption energy of H_2 ... B-doped G can be tuned, enhancing the interaction by positively charging the support. BC5, for instance, can store up to 5.3 wt.% H_2. More significantly, the study sheds light on the heteroatom doping of graphene; B-doped G forms easier than N-doped G (5.6 eV/atom for the former vs. 8.0 eV/atom for the latter) [18]. Combining metal doping (Ni) with B-doped carbon-based materials, Wang et al. studied and compared various forms of carbon with those of 3D graphene-based materials, obtaining composites with surfaces functionalized with O-containing groups that further enhanced the spillover effect with a positive effect on H_2 storage properties [19]. Edge-functionalization of G for kinetic control of HSM (hydrogen storage materials) has also been reported [20].

Among various strategies to improve the behavior of hydrogen storage materials at the nanoscale, nanoconfinement and nanostructuring were shown to be very effective [2,21,22]. Nanostructured metal hydrides MH_x showed a tangible improvement in H_2 storage properties [2,21], while nanostructures of type hydride @ C in particular were proven to be a model system for the future solid-state hydrogen storage fuel, as evidenced by DFT computation by Shevlin and Guo in 2008 [22]. DFT simulations have played a pivotal role in the expansion of the hydrogen storage field, the above-mentioned resource comparing a few of the most important hydrides: MgH_2, $LiNH_2$, $LiBH_4$, and derived reactive hydride composites RHCs, as well as their interaction with various carbon systems [22].

2.1. Metal-Decorated Graphene

An important strategy for tuning graphene for energy storage applications has been to decorate the support with metals or clusters of metals. Among the investigated materials are metallized siligraphene nanosheets (SiG) with varying light metal decorations (Li, Na, K, Mg, Ca, Sc, Ti) [23], metal-decorated graphene (Li, Na, Mg)/G, DFT study [24], or K in K @ B-substituted G [25]. Given the potential of AlH_3 for hydrogen storage, Al-doping has also been explored in Al/G composites [26,27], Al_n clusters supported by coronene and graphene G (DFT study) [27], and Al/Si –SLG (SLG, single layer graphene + Si +Al) [28].

Palladium is well known for its high affinity for H_2, and many studies have been devoted to the theoretical modelling of this interaction; Pd-decorated N-doped G, DFT study [29]; Pd_n @ G (n = 1–4) in BC3 variant [30], and Cu- and Pd-decorated G, DFT study [31]. Titanium and its clusters have also been investigated: Ti_3 clusters [32]; Ti_4—decorated B/N-doped G [33]; Ti_4 & Ni_4–doped G nanoplatelets [34]; and Ti–Al subnanoclusters on G [35].

Most reports, however, focus on Mg-doped porous carbonaceous materials, such as Mg@ G flake nanocomposites (H_2 generation from H_2O) [36], Mg@graphite for comparison purposes [37], Mg@rGO layers [38,39], Mg@Heteroatom–doped G [40], Mg@B–doped G [41], and Mg/defected GO [42]. Additionally, various alloys have also been studied for graphene supports: Mg alloy @rGO–V_2O_3 [43], rGO–EC@AB5 hybrid material (EC = ethyl cellulose, AB5 = $La(Ni_{0.95}Fe_{0.05})_5$) (LNF) [44], or $MmNi_{3.55}Co_{0.75}Mn_{0.4}Al_{0.3}$/G nanoplatelets (Mm denotes mischmetals) [45].

2.2. Mechanistic Insight and Kinetics of $H_2 \ldots$ Support Interaction

Pristine graphene can chemically absorb H_2 and its theoretical storage capacity is 7.7 wt.%; the hydrogenated graphene (graphane, $(CH)_n$, a sp^3 hybridized analog of graphene) releases $H_2(g)$ at ~400 °C, with an Ea = 158 kJ/mol (1.64 eV) [46]. While an intriguing material in its pristine form, its thermodynamic parameters make it less feasible for scaling up processes aimed at vehicular applications; however, it is worth noting the similarity of the activation energy deduced for graphane and that of metal hydrides.

The fundamental understanding of the adsorption/desorption mechanism of H_2 in graphene is paramount to developing new materials aimed at this task; a pertinent comparison between physisorption and chemisorption on graphene was reported in 2011, where the physical limitations of G (5 wt.% H_2 storage) were correlated to the entropy contribution $T\Delta S$ and the large van der Waals distance between two H_2 molecules (0.3 nm), further preventing the increase in gravimetric storage capacity of pristine graphene [7]. The interaction $H_2 \ldots$ G was studied by DFT in single- and double-vacancy graphene by Wu et al., with direct implications for the behavior of defected graphene during hydrogenation studies [47]. The mechanism of H_2 interaction with Al-doped porous graphene has been reported by Ao et al., showing by DFT that Al/G can store up to 10.5 wt.% H_2, with a relatively low H_2 adsorption energy of −1.11 to −0.41 eV/H_2, which would potentially allow hydrogen absorption/desorption near room temperature conditions, in agreement with the findings from the analysis of atomic charges, electronic distribution, and density of states (DOS) of the system [26]. The enhanced interaction was potentially due to the polarization of the adsorbed H_2 molecules.

Akilan et al. have studied by DFT the adsorption of H_2 molecules on B/N-doped defected (5-8-5, 55-77, 555-777 and 5555-6-7777 defects) graphene sheets [48]. The N-atom addition (donor behavior, n-type semiconductor) increases the delocalized electrons, while the B atoms (acceptor, p-type semiconductor) increase the localized electrons in the considered system. The most efficient adsorption was modelled when the H_2 molecule approached the sheet in a perpendicular direction (−80 meV), while the least efficient interaction was observed in a parallel orientation (−9 meV), while the delocalized electron density was higher on the fusion points of the pentagonal and hexagonal rings and would therefore enhance H_2 adsorption [48]. Another supporting DFT study of H_2 storage on TM-doped defected graphene (TM = transition metal) revealed that in the case of TM = Sc, the 555-777/Sc structure doped with Sc showed the maximum H_2 capacity, with H_2 binding energies in the range 0.2–0.4 eV/H_2 [49].

The advances regarding TM-loaded Mg-based alloys/G have been reviewed recently [8]. A few important points are attributed to graphene: It can inhibit grain growth, thus aiding the overall cyclability of the composite, and it can (co-)catalyze the hydrogenation process, in which the electron transfer between Mg and C plays a key role [8].

The cyclic behavior of metal hydrides can be affected by issues related to grain growth. This can be partly overcome with the formation of G layers encapsulating MgH_2 to prevent grain growth [50]. In this report, Lototskyy et al. used various carbon sources (graphite, AC, MWCNTs, etc.) and showed that the formation of graphene sheets during high-energy reactive ball milling in hydrogen (HRBM) is responsible for the encapsulation of MgH_2, noting an increase in a/d cycling behavior along with a more reduced size of the MgH_2 crystallites (40–125 nm vs. 180 nm in pristine form) [50]. The catalytic role of graphene nanoplatelets (GNP) over H_2 storage kinetics in Mg has been studied by Ruse et al. [51]. The enhancement of more than an order of magnitude was attributed to GNP properties (size, thickness, defect density, and specific surface area), and these can be further tuned to alter H_2 storage kinetics in Mg–GNP nanocomposites [51]. A carbon-neutral, reversible, and sustainable process that produces H_2 is the formate-bicarbonate system, where graphene has also served as a support of Pd and Ru metals [52].

2.3. Manufacturing Techniques

Several techniques have been utilized to introduce metal catalysts into graphene, synthesizing (nano)composites containing graphene and carbonaceous materials [9]. While ball milling and its variants remain a key technique, other options have been explored: Electrostatic layer-by-layer self-assembled G/MWCNTs [53], Uranium U-decorated G (H_2 and D_2 adsorption) [54], and plasma-assisted milling in Mg@FLG composites (few-layer graphene nanosheets) [55]. Given the remarkable properties of 2D graphene on hydride storing materials, the synthesis of 2D MgH_2 has also been proposed in DFT studies [56].

3. Applications of Graphene-Based Hydride Nanocomposites

Graphene-supported metal hydrides have potential applications in hydrogen storage, battery electrodes, and fuel cells, among others. These are mainly the contributions of the embedded metal hydrides that have been utilized in energy storage devices such as batteries and fuel cells. By embedding metal hydrides in high-surface-area graphene, the obtained nanocomposites can further enhance the performance and efficiency of pristine hydrides, potentially yielding competitive novel energy storage devices with improved cycling stability and durability [2,21].

3.1. Batteries—Battery Electrodes

Graphene-supported metal hydrides have also been studied as electrode materials in batteries and fuel cells. The addition of graphene to metal hydride electrodes can improve their cycling stability and reduce their electrode polarization. This improves the efficiency and lifespan of the batteries and fuel cells. The high surface area of graphene also enhances the electrochemical performance of the electrodes by providing more active sites for electrochemical reactions.

Composites such as MH_x@G have been utilized for batteries [57–59]. Mixed-valence oxides (Co_3O_4) and hydroxides ($Ni(OH)_2$) have also been explored in Co_3O_4 nanocubes on N-doped G for Ni-metal hydride batteries [60] and $Ni(OH)_2$-coated N-doped G aerogel as replacement NiCd and NiMH batteries [61]. Various anode materials based on graphene have recently emerged, such as MgH_2 self-assembled on G as anode material for Li-ion batteries [62].

3.2. (Super)capacitors, Electrochemical Storage

Supercapacitors are another energy storage device that can benefit from the addition of graphene-supported metal hydrides in MH_x@G composites [57,58]. Supercapacitors are electrochemical devices that can store energy by electrostatically adsorbing ions at the surface of the electrode [53]. The improvement of using Li-ion (120–170 Wh/Kg) vs. Ni-metal hydride (40–100 Wh/Kg) has to be considered when designing new storage devices [58]. The high surface area and conductivity of graphene can enhance the electrochemical performance of supercapacitors. The addition of metal hydride nanoparticles to the graphene electrodes can further enhance the energy storage capacity and stability of supercapacitors. Other materials based on graphene have also been investigated, for instance Fe_3O_4 @ G composite for high-performance Li-ion hybrid capacitors [63]. Electrochemical storage properties have also been studied on heteroatom-substituted materials, such as N-doped C [64], but also on $La_{0.7}Mg_{0.3}(Ni_{0.85}Co_{0.15})_{3.5}$@Ni/Graphene [65].

3.3. Solar Cells and Portable Electronic Devices

Graphene-supported metal hydrides have also been studied for their application in solar cells. The addition of metal hydrides to the graphene electrodes in solar cells can improve their photovoltaic performance by enhancing charge separation and transport properties. The high conductivity and large surface area of graphene also facilitate electron transfer and reduce recombination losses in the solar cells. In addition to the above applications, graphene-supported metal hydrides have also been studied for their potential use in energy storage for portable electronic devices, power grid stabilization, and energy harvesting. The unique properties of graphene and metal hydrides can be tailored to meet the specific requirements of these applications and enhance their performance [2].

3.4. Energy Storage

One of the most promising applications of graphene-supported metal hydrides is in hydrogen storage for fuel cell vehicles. Fuel cell vehicles are gaining popularity due to their high efficiency and low emissions. However, the limited availability of hydrogen storage materials has hindered their widespread adoption. This aspect will be discussed in more detail in the following section (Section 4) [2,21].

4. Hydrogen Storage Properties of Composites MH_x@G

Metal hydrides have shown promise as hydrogen storage materials, but their low hydrogen storage capacity and slow kinetics have limited their application. The addition of graphene as a support material can enhance the hydrogen storage capacity and kinetics of metal hydrides, making them a promising candidate for hydrogen storage in fuel cell vehicles.

4.1. Binary Hydrides—The Case of MgH_2

While light metal hydrides show great promise as H_2-storing materials, the literature and scientific research are still focused on MgH_2 due to its advantageous gravimetric content, activation energies for desorption/adsorption, and multiple synthesis methods. The following subsections explore the H_2 storage properties of non-catalyzed and catalyzed nanocomposites such as MgH_2@G.

4.1.1. Non-Catalyzed Support

A great variety of reports in the literature focus on the Mg/MgH_2 system embedded in various supports, including graphene [1] and GLM (graphene-like materials) (Figure 2) [16], although some other binary hydrides (layer-by-layer lithiated composites such as LiH/G for instance) have also been reported [66]. For instance, MgH_2 NPs nanoparticles have been embedded in GNS (graphene nanosheets) to yield MgH_2 NPs@GNS composites [67], MgH_2 NPs nanoparticles uniformly-grown@G [68].

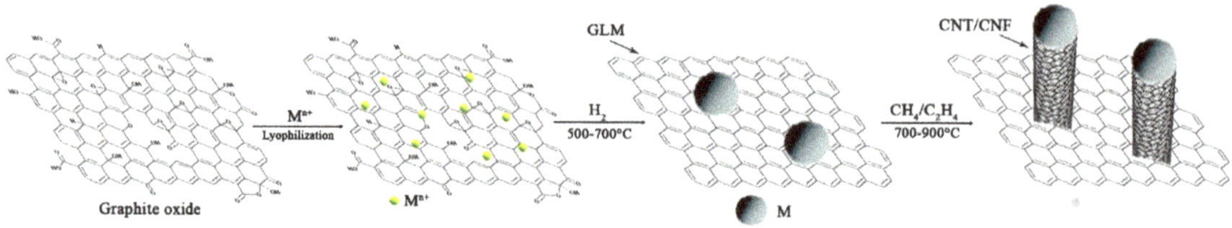

Figure 2. Synthesis of metal-carbon-graphene composites. Reprinted from ref. [16] under Creative Commons Attribution (CC BY) license.

The mechanism of MgH_2@G composite synthesis by reactive mechanical grinding (under 2 MPa H_2) has been studied by Jang et al., who concluded that G played a catalytic role in H_2 absorption, considering the absorption kinetics and PCT (pressure-composition-temperature) curves [69]. The composite MgH_2–5 wt.% graphene could store 3.69 wt.% H_2 at 423 K, while increasing the graphene amount in MgH_2–5 wt.% graphene allowed 5.08 wt.% at the same temperature (423 K) [69]. Comparing the PCI isotherms for the two samples, it seemed that MgH_2–10 wt.% graphene displays a lower activation energy Ea and better cyclic stability compared to MgH_2–5 wt.% graphene [69].

Another strategy to enhance the stability of MgH_2 during cycling was to create Mg/rGO multilaminates to produce Mg/MgH_2@rGO during cycling. These composites exhibited a very high 6.5 wt.% reversible H_2 storage and a capacity of 0.105 Kg H_2/L [70]. When MgH_2 interacts with 5 wt.% GO incorporated in MgH_2 during hydrogenation studies, it also causes a reduction in the porous support (GO→rGO), as highlighted by a study by Pukazhselvan et al. (Figure 3) [71].

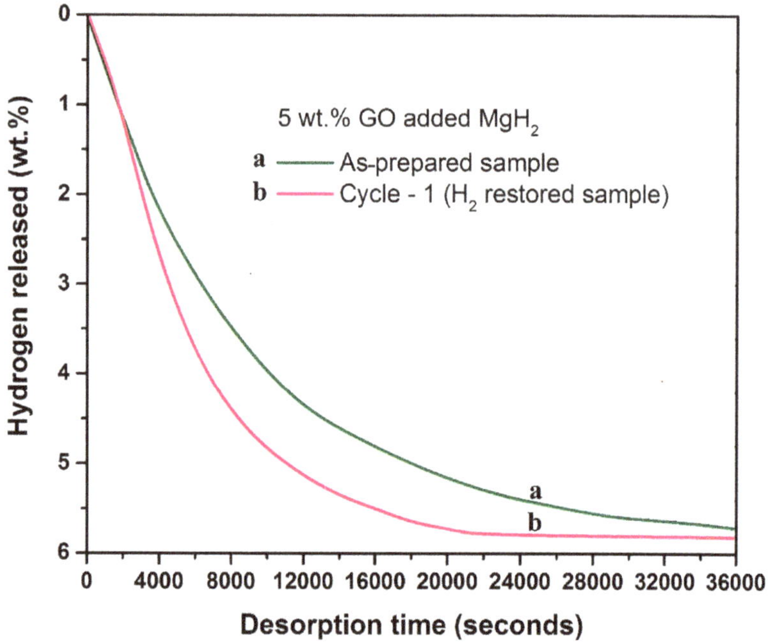

Figure 3. Dehydrogenation kinetics profiles obtained for 5 wt.% GO incorporated MgH_2. Profile "a" corresponds to the as-prepared sample, and profile "b" corresponds to the hydrogen-restored sample. Dehydrogenation temperature/pressure: 315 °C/1 bar. Reprinted from reference [71] under Creative Commons Attribution (CC BY) license.

The reduction occurs even during composite preparation, namely during mechanical milling, and it affects the dehydrogenation temperature of MgH_2 (a reduction of 60 °C vs. pristine) by lowering the dehydrogenation activation energy (by 20 kJ/mol). Considering the affinity of Mg for oxygen, this could also be used as a method to obtain MgO-decked rGO composites [71]. Various carbon scaffolds have been used for MgH_2, including but not limited to graphite, graphene derivatives, and carbon fiber [72]. Combined experimental and DFT studies on the helical form of graphene nanofibers (HGNF) by Singh et al. [73] revealed that the extra carbon sheet edges exposed would further improve MgH_2 dehydrogenation compared to G-catalyzed MgH_2, and the experimental results showed a 45 °C improvement due to HGNF use in MgH_2 + 7 wt.% HGNF nanocomposites. The DFT results supported this finding by revealing a weakening of Mg-H bonds by interaction with zigzag and armchair type HGNF, forming C–Mg–H and C–H bonds [73]. Combined DFT and experimental computations have also been performed on other MgH_2@G variants, which confirmed once again the catalytic role of G in reducing the dehydrogenation enthalpy and dehydrogenation activation energy of MgH_2 [74].

Other approaches tackle graphene support functionalization, as is the case of MgH_2–FG (FG = fluorographene nanosheets). Using fluorographene, the nanocomposite MgH_2–FG absorbed 6.0 wt.% within 5 min, and desorbed 5.9 wt.% within 50 min, at 300 °C [75].

4.1.2. Catalyzed Support

Various catalysts have been utilized to improve the behavior of hydride@graphene composites. For instance, MgH_2 catalyzed by TiH_2@Gr (desorption onset at 204 °C), with comparative improvements relative to the catalytic activity of Ti@Gr, TiO_2@Gr, and TiH_2@Gr [76]. Other Ti-based salts such as $TiCl_3$ have been used in MgH_2 catalyzed by $TiCl_3$ and GO-based porous carbon (GC) in composites such as MgH_2/GC-$TiCl_3$, obtained by isothermal hydrogenation of $MgBu_2$/GC, first to yield MgH_2/GC and then ball milling with $TiCl_3$ to form the final composite with about 78% MgH_2 [77]. The MgH_2/GC-$TiCl_3$ composite can release reversibly up to 7.6 wt.% H_2 within 9 min at 300 °C, a result superior to even the MgH_2-GC-$TiCl_3$ mixture (weight ratio of 1:1:0.08, produced by direct ball milling for 5 h) (Figure 4) [77]. This highlights the advantages of the stepwise synthesis procedure outlined above.

Among the metals showing the most striking improvements in hydrogenation studies, Ni has a distinct role. For instance, MgH_2 has shown improvements in dehydrogenation behavior when catalyzed by Ni@N-doped C spheres [78], or by NiCo–functionalized G [79]. MgH_2–Ni@NCS composites produced by Wang et al. were obtained by hydriding combustion and high energy ball milling Ni@NCS into Mg [78]. These composites showed good stability even after 10 a/d cycles, and moderate H_2 storage parameters (a: 5. wt.%, d: 4.3 wt.% H_2 within 8 min at 623 K; a: 4.2 wt.% in 60 min at 373 K (Figure 5)) [78].

When NiCo bimetallic NPs (4–6 nm) were used to functionalize graphene oxide supports, NiCo/rGO were obtained via a one-pot synthesis and produced energy storing nanocomposites MgH_2–NiCo/rGO with excellent cycling stability, featuring reduced Ea = 105 kJ/mol, absorbing 6.1 wt.% H_2 within 350 s at 300 °C under 0.9 MPa H_2 [79]. The metal sources were the corresponding metal chlorides ($NiCl_2.6H_2O$ and $CoCl_2.6H_2O$) that underwent sonication with GO/EG, reflux, and reduction with $N_2H_4.H_2O$/NaOH [79].

Other support functions have also been pursued, such as MgH_2@PTh core-shell NPs on G (PTh = polythiophene) for use in Li-ion batteries [59], or MgH_2@NiPc (NiPc = nickel phthalocyanine)/G [80].

A summary of the main parameters in other doped-graphene supports for binary hydrides or alloys is given in Table 1.

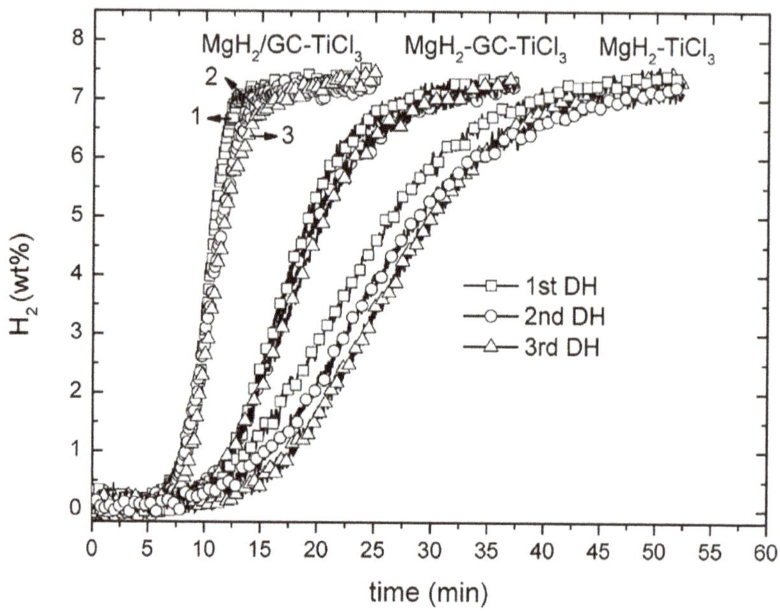

Figure 4. Cyclic dehydriding (DH) curves of the MgH_2-$TiCl_3$, MgH_2/GC-$TiCl_3$ and MgH_2-GC-$TiCl_3$ composite samples at 300 C under 0.1 bar H_2. Reprinted with permission from Ref. [77].

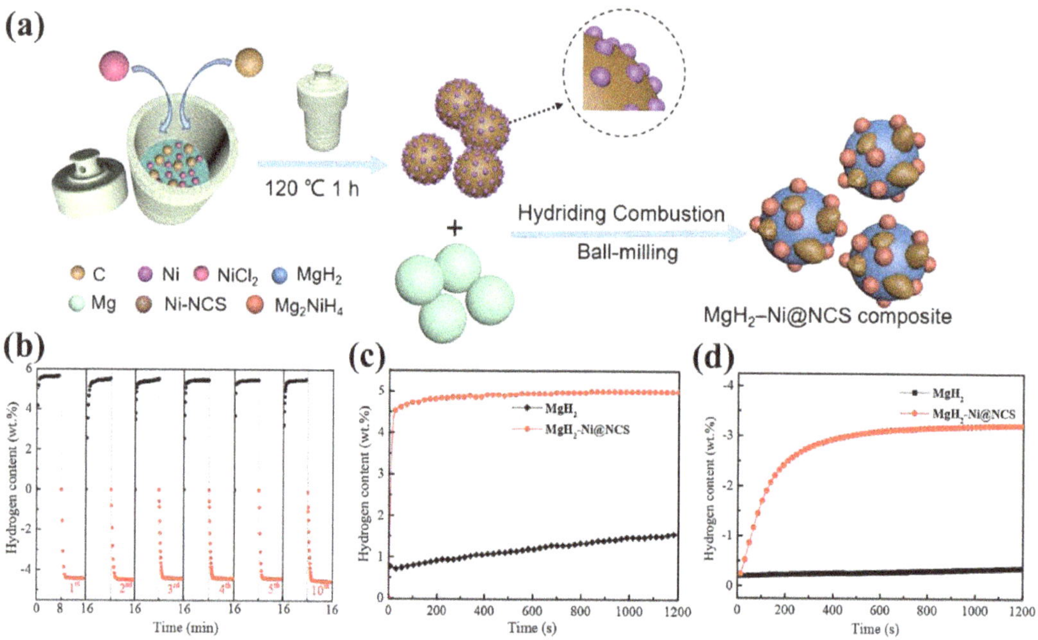

Figure 5. (**a**) Preparation process of the MgH_2–Ni@NCS composite. (**b**) Absorption and desorption cycling of MgH_2–Ni@NCS composite at 623 K; (**c**) Isothermal hydrogenation curves of the MgH_2–Ni@NCS composite and milled MgH_2 at 423 K; (**d**) Isothermal dehydrogenation curves of the MgH_2–Ni@NCS composite and milled MgH_2 at 598 K. Reprinted with permission from Ref. [78].

Table 1. Doped-graphene supports for Mg/MgH_2 and related alloys.

Hydrogen Storing Species	Graphene Dopant	Nanocomposite	Observations	Ref.
Mg	5 wt.% NiS	Mg—5 wt.% NiS/rGO (by HPMR, hydrogen plasma metal reaction)	ball milling Mg and NiS/rGO, the NiS-catalyzed support was prepared by co-reduction of Ni^{2+}—impregnated GO support; des: 3.7 wt.% H_2 in 10 min and 4.5 wt.% H_2 in 60 min; $E_{a,\ abs}$ = 44.47 kJ/mol; $E_{a,\ des}$ = 63.02 kJ/mol	[81]
MgH_2	Ni	MgH_2@Ni/G	(MgH_2 + 10 wt.%G + 10 wt.%Ni) composites were investigated from a combined DFT and experimental viewpoint; des. onset at 339.5 °C	[82]
MgH_2	Ni	MgH_2@Ni–Gn	nickel-decorated graphene support (Ni–Gn) afforded abs: 6.28 wt.%, 100 min, 373 K, and des: 5.73 wt.%, 1800 s, 523 K	[83]
MgH_2	ZrCo	MgH_2@10 wt.% ZrCo/G	The ZrCo dopant was present as 2D nanosheets; des: 6.3 wt.% H_2, 5 min, 300 °C, $E_{a,des}$ = 90.4 kJ/mol; abs: 4.4 wt.% H_2, 10 min, 120 °C, 3 MPa H_2, $E_{a,ads}$ = 57.6 kJ/mol.	[84]
MgH_2	KOH	MgH_2 + KOH/graphene	$KMgH_3$, MgO formed in-situ; Ea = 109.89 kJ/mol; 5.43 wt.% H_2 reversibly. Graphene provides more H diffusion channels and better disperses the catalysts	[85]
Alloys HSAs	-	(LaCeY)$(NiMnCoAl)_5$@rGO	HSAs = hydrogen storage alloys; retention rate 51.25% at a discharge current density 3000 mA/g	[86]
Mg–Al alloy	Ni	$Mg_{90}Al_{10}$-x (80 wt.%Ni@Gn)	Mg–Al alloy in graphene–supported Ni; x = 0,4, 8, 12 wt.%; for x = 8, a: 5.11 wt.% in 400 s, 523 K; d: 5.81 wt.%, 1800 s, 573 K	[87]
Mg–Al alloy	TiF_3	Mg–Al alloy–TiF_3@G	Mg–Al-M (M = G, TiF_3, and TiF_3@G) composites; Ea = 139.8 kJ/mol for TiF_3@G support; Mg–Al–TiF_3@G released 5.41 wt.% at 350 °C	[88]
Mg–Al alloy	Y_2O_3	Mg–Al alloy—Y_2O_3@G	Mg-Al-Y_2O_3@rGO reversibly store/release H_2 at 250 °C; with 5 wt.% of Y_2O_3@rGO loading, Mg–Al composite had $E_{a,release}$ = 145.9 (vs. 162.6 kJ/mol for pristine Mg–Al alloy). $E_{a,uptake}$ = 54.3 kJ/mol for Y_2O_3 catalyzed G.	[89]
MgH_2	FeNi	MgH_2@FeNi/rGO	5 wt.% FeNi/rGO modified MgH_2 released 6.5 wt.% H_2 at 300 °C (onset 230 °C); uptake: 5.4 wt.%, 20 min, 125 °C, 32 bar H_2	[90]
MgH_2	Ni–CeO_x	MgH_2–Ni–CeO_x/GNS	MgH_2 catalyzed by Ni–CeOx/GNS; graphene nanosheets supported nanoscale Ni&CeOx (x = 1.69)	[91]
MgH_2	TiB_2	MgH_2–TiB_2/GNS	MgH_2 catalyzed by TiB_2/GNS	[92]
MgH_2	NiCu	MgH_2–NiCu/rGO	MgH_2 catalyzed by NiCu/rGO from double layer hydroxide	[93]
MgH_2	Fe–Ni	MgH_2 + 10 wt.% Fe–LiCo 3D G	MgH_2@Fe–Ni/G, namely MgH_2 + 10 wt.% Fe–LiCo 3D G (3D-graphene)	[94]
MgH_2	NiFe–LDH	MgH_2–Ni_3Fe/rGO	NiFe–LDH (layered double hydroxide precursor)/GO yield (Ni_3Fe/rGO) active catalyst	[95]
MgH_2	Al, Cu	$Mg_{90}Al_{10}$–Cu@G	MgH_2→Mg–Al alloys with Cu introduced in situ: $Mg_{90}Al_{10}$–Cu@G nanoplates	[96]

Table 1. *Cont.*

Hydrogen Storing Species	Graphene Dopant	Nanocomposite	Observations	Ref.
MgH_2	–	MgH_2@CA microspheres	abs: 6.2 wt.% H_2 within 5 min at 275 °C; des: 4.9 wt.% H_2 within 100 min at 350 °C; $E_{a,des}$ = 114.8 kJ/mol.	[97]
MgH_2	TiC	MgH_2–TiC@G	desorption at 180 °C by the plasma carbon–modified MgH_2/TiC containing FLGS (few layer graphene sheets)	[98]
MgH_2	SrF_2 and SrF_2	MgH_2–SrF_2(SrF_2)@Gr	MgH_2 cat. by SrF_2 and SrF_2@Gr additives	[99]
MgH_2	FeCoNi	MgH_2@FeCoNi NPs/G	MgH_2@FeCoNi NPs/G as new catalyst	[4]
MgH_2	FeOOH	MgH_2@FeOOH nanodots/G	MgH_2@FeOOH nanodots NDs @G, release H_2 at 229.8 °C (ΔT = 106.8 °C lower than pristine MgH_2), showing good cycling stability (over 20 cycles, 98.5% of initial capacity maintained, while also reducing the activation energy Ea)	[100]
Mg	Ni	Mg/MgH_2 + Ni/GLM	Mg@Ni/G nanocomposites with 5–60 wt.% Ni loading: Mg/MgH_2 + Ni/GLM (graphene-like material)	[101]
MgS	–	MH_2@G, MgS@G	MgS@G prepared by reaction: MH_2@G + S → MgS@G + H_2; for advanced Li-storage	[102]

For instance, Zhang et al. used 2D ZrCo nanosheets to catalyze MgH_2 decomposition [84]. When 10 wt.% ZrCo was added, the composite MgH_2@10 wt.% ZrCo/G released 6.3 wt.% H_2 at 300 °C in 5 min. (Figure 6). When bimetallic catalysts such as ZrCo are homogeneously dispersed on G, they play a role similar to that of Mg_2Ni/Mg_2NiH_4 (as a "hydrogen pump"), easing the dissociation of H_2 as well as its recombination, along with weakening the Mg–H bonds, further improving the hydrogen storage properties of MgH_2 [84].

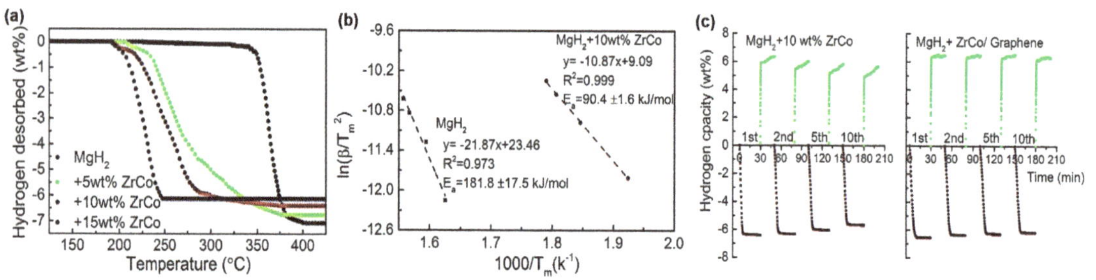

Figure 6. (**a**) DSC curves of MgH_2 + x (x = 5, 10, 15) wt.% ZrCo composites; (**b**) Kissinger's plots (**b**) of the MgH_2 with and without 10 wt.% ZrCo composites; (**c**) Isothermal dehydrogenation/hydrogenation curves of the MgH_2 + 10 wt.% ZrCo and MgH_2 + 10 wt.% ZrCo/5 wt.% Graphene composites as a function of cycling. Reproduced with permission from ref. [84].

4.2. Complex Hydrides Embedded in Graphene Supports: $M(BH_4)_n$, $M(AlH_4)_n$

Nanoconfined complex hydrides have emerged as prime candidates in energy storage applications. They possess the highest H_2 gravimetric storage capacity, and careful tuning of their composition, additives, and morphology could yield highly reversible systems with real prospects in vehicular applications [2]. Graphene (non-catalyzed or catalyzed) has served as the support for manufacturing such stable, high-performance nanocomposites of complex hydride@G. A summary of more representative examples of such composites is given in Table 2.

Table 2. Doped-graphene supports for complex hydrides $M(BH_4)_n$, $M(AlH_4)_n$ and related hydrides.

Hydrogen Storing Species	Graphene Dopant	Nanocomposite	Observations	Ref.
$LiBH_4$, $LiAlH_4$, $NaAlH_4$	(C_3N_4)	(Li, Na)XH_4 (X = B/Al)/C_3N_4	$LiBH_4$, $LiAlH_4$, $NaAlH_4$/C_3N_4 studied by DFT; support offers suitable adsorption site for AlH_x/BH_x (x = 3, 2, 1)	[103]
$LiBH_4$, $LiAlH_4$, $NaAlH_4$	–	(Li, Na)XH_4 (X = B/Al)/G	$LiBH_4$, $LiAlH_4$, $NaAlH_4$/Graphene and Graphdiyne; DFT study shows strong support interaction due to well-defined pore structure	[104]
$LiBH_4$	N	$LiBH_4$–N-doped G/MC	$LiBH_4$ in 10 at%N-doped Graphene/resorcinol formaldehyde; impregnation degree: 30, 50 and 70 vol%; XRD diffraction 2θ = 12.6° that can be attributed to Li–B–N(G)–H interaction	[105]
$LiBH_4$	Ni/Co	$LiBH_4$–Ni/Co NPs–N-doped G aerogels	$LiBH_4$ catalyzed by NiCo NPs–N-doped G aerogels; Co-decorated, des.: 8 wt.% H_2 at 325 °C (1st cycle; with 1 wt.% at 226 °C). Ni-decorated, des: 8 wt.% H_2,	[106]
$LiBH_4$	–	$LiBH_4$–G	$LiBH_4$ wrapped in G	[107]
$LiBH_4$	Fe_3O_4	$LiBH_4$–Fe_3O_4/porous G	$LiBH_4$ catalyzed by (Fe_3O_4 dispersed on porous G)	[108]
$LiBH_4$	–	$LiBH_4$–(mesoporous resorcinol-formaldehyde/G)	$LiBH_4$ nanoconfined in (mesoporous resorcinol-formaldehyde/G) entangled supports; des: 13 wt.% at 400 °C; recharging at 400 °C, 5 h, 60 bar H_2; 6 wt.% reversible storage capacity;	[109]
$LiBH_4$	Ni nanocrystals (2–4 nm)	$LiBH_4$/G	$LiBH_4$ (5–10 nm)–Ni(2–4 nm)/G, affording 9.2 wt.% H_2 due to by-passing B_2H_6 and $B_{12}H_{12}^{2-}$ anion formation	[110]
$LiAlH_4$	$NiCo_2O_4$	$LiAlH_4$–$NiCo_2O_4$ @ rGO	$LiAlH_4$ dehydrogenation by $NiCo_2O_4$ nanorods@rGO nanocomposites by ball milling; $LiAlH_4$ + 7 wt.% $NiCo_2O_4$@rGO des. onset at 62.7 °C (in total 6.28 wt.% H_2); 4.0 wt.% hydrogen within 20 min at 150 °C (isothermal)	[111]
$LiBH_4$	$NiFe_2O_4$	$LiBH_4$-graphene-$NiFe_2O_4$ (Ar)	E_A = 127 kJ/mol (vs. 170 kJ/mol for pristine $LiBH_4$); after 5 a/d cycles. H_2 storage was ~6.14 wt.%; des. onset at 349 °C	[112]
$NaBH_4$	–	nano-$NaBH_4$@GNs	Ultrasmall (6–10 nm) nano-$NaBH_4$@GNs by MFSP (mechanical-force-driven self-printing, a technique similar to 3D printing), for scalable fabrication of 0D complex hydrides in 2D supports; ~5 wt.% stable H_2 storage capacity	[113]
$NaAlH_4$	–	$NaAlH_4$@G	$NaAlH_4$ (12 nm)@G–50; G weakens Al–H bonds of $NaAlH_4$; des.: 5.6 wt.% at 300 °C; 3.8 wt.% (120 °C); Ea = 68.23 kJ/mol (vs. 128 kJ/mol for bulk)	[114]
$NaAlH_4$	NP–TiH_2	$NaAlH_4$– 7 wt.% NP–TiH_2@ G	$NaAlH_4$ catalyzed by 7 wt.% NP–TiH_2@G as active catalyst; 5 wt.% H_2 is achieved reversibly, with onset below 80 °C; TiH_2 (~50 nm lateral, ~15 nm thick) produced my metathesis $TiCl_4$ + LiH in THF with G support	[115]

Table 2. *Cont.*

Hydrogen Storing Species	Graphene Dopant	Nanocomposite	Observations	Ref.
$NaAlH_4$	N. CNTs	$NaAlH_4$@N-doped G/CNTs	$NaAlH_4$@N-doped G and CNTs (produced by NH_3 treatment of G, 600 °C, 30 min); 1.8 wt.% reversible H_2 storage	[116]
$NaAlH_4$	–	$NaAlH_4$/GNs	$NaAlH_4$/GNs graphene nanosheets, C60 fullerenes and MC mesoporous carbon; $NaAlH_4$–support interactions revealed by FE–SEM and ^{27}Al solid state NMR.	[117]
$NaAlH_4$	–	$NaAlH_4$@rGO	$NaAlH_4$ @rGO framework, GOF with $NaAlH_4$@GOF (1 M), desorbing 1.01 wt.% (20.0 wt.% $NaAlH_4$), 16.6% of bulk $NaAlH_4$ due to oxygen functional groups in the GOF reduced by BH_4^-.,	[118]
$NaAlH_4$	TiO_2	TiO_2–$NaAlH_4$@G	Layer-by-layer TiO_2–$NaAlH_4$@G composites with 90% high loading (des. peak at 191.6 °C vs. 286.5 °C for bulk $NaAlH_4$)	[119]
$NaAlH_4$	–	$NaAlH_4$@G	$NaAlH_4$@G nanofibers	[120]
$NaAlH_4$	Al, $TiCl_3$	Al, Ti-doped $NaAlH_4$@G	Ti-doped $NaAlH_4$@G: $NaAlH_4$ co-doped 2 mol% $TiCl_3$, 10 mol% G, 5 mol% Al (ball milling)	[121]
$NaAlH_4$	$CeH_{2.51}$	$NaAlH_4$@FLG/Ce ($CeH_{2.51}$)	$NaAlH_4$@FLG/Ce ($CeH_{2.51}$); onset at 85 °C, 5.06 wt.% H_2 at 200 °C; 4.91 wt.% reversible after 8 a/d cycles. Ce activated G surface.	[122]
$NaAlH_4$	Heteroatoms	$NaAlH_4$@dopant/G	$NaAlH_4$@dopant/G (dopant: S-, N-, vacancies, B, N, O-, P-, F-, HO-), investigated by means of DFT	[123]
$Mg(BH_4)_2$	–	$Mg(BH_4)_2$@G	$Mg(BH_4)_2$@G synthesized by [MgH_2@G + B_2H_6]; it was shown that G weakens the Mg–H and B–H bonds, affording a desorption of H_2 starting at 154 °C (onset) and up to 225 °C (end)	[124]
$Mg(BH_4)_2$	–	$Mg(BH_4)_2$@rGO	$Mg(BH_4)_2$@rGO (1–4 wt.%), investigation of mechanism and phase evolution	[125]
$Mg(BH_4)_2$	–	$Mg(BH_4)_2$@G	$Mg(BH_4)_2$@G; the 3 polymorphs of Mg were produced from $MgBu_2$ by tuning reaction conditions (11.2, 10.3, and 9.9 wt % H for the γ, β, and α phases)	[126]
$NaMgH_3$	–	$NaMgH_3$ @GO	$NaMgH_3$ @GO and $NaMgH_3$ @ MWCNTs	[127]
K_2NaAlH_6	–	K_2NaAlH_6–GS	GS graphene sheets as catalysts for sodium potassium alanate K_2NaAlH_6	[128]
$Mn(BH_4)_2$	Ni, $LiNH_2$	$Mn(BH_4)_2$@(Ni, $LiNH_2$)/G	$Mn(BH_4)_2$@G (Ni, $LiNH_2$ additives) by solvent infiltration/extraction	[129]
Mg_2Ni	–	Mg_2Ni–rGO/MWCNTs	Mg_2Ni–rGO/MWCNTs for battery—supercapacitor hybrid device (high discharge capacity 644 mAh/g)	[130]
Mg_2Ni	–	Mg_2Ni/rGO	rGO/Mg_2Ni—enhanced cycling stability	[131]
Mg_2NiH_4	–	Mg_2NiH_4@GS	Mg_2NiH_4@GS (surface graphene nanosheets)	[132]

Huang et al. have described an interesting method to prepare $NaAlH_4$-based composites with an average hydride size of 12.4 nm, while highlighting the advantage of supporting the complex hydride on graphene (0.31 eV vs. 1.34 eV) (Figure 7) [114]. With high theoretical storage capacities (7.5 wt.% gravimetric, 94 g H_2/L volumetric), $NaAlH_4$ could desorb H_2 in three steps (via Na_3AlH_6—3.7 wt.%, further decomposition to NaH—1.9 wt.% and the often overlooked third step to Na metal); however, due to thermodynamic considerations, only the first two steps are achievable within reasonable temperature ranges.

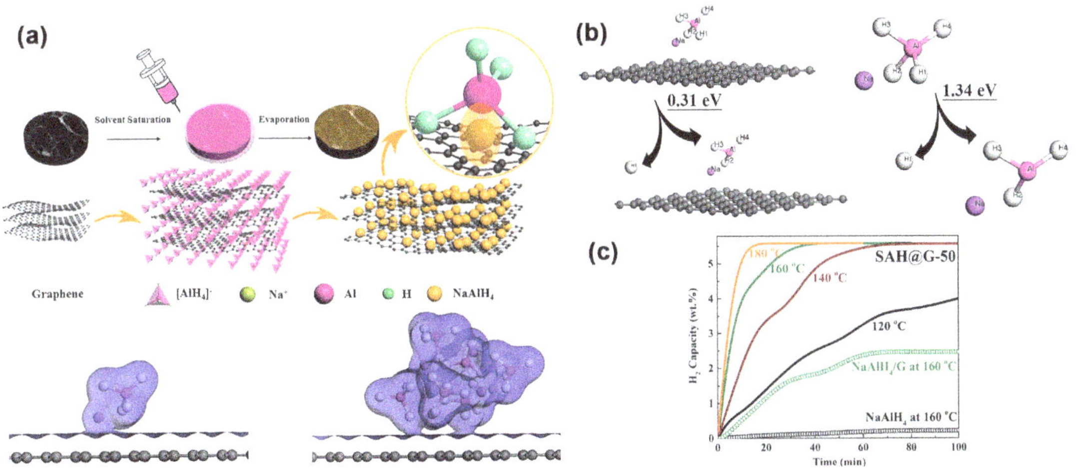

Figure 7. (**a**) Schematic illustration of the synthesis of SAH@G-50. Isosurface of electron density of the $NaAlH_4$ on graphene and $(NaAlH_4)_6$ on graphene; (**b**) Energies required for the removal of hydrogen from $NaAlH_4$ with (0.31 eV) and without the presence of graphene (1.34 eV); (**c**) Isothermal dehydrogenation of SAH@G-50 in comparison with $NaAlH_4$ and the ball-milled $NaAlH_4$/G composite at various temperatures. Reprinted with permission from Ref. [114].

While mechanochemistry or solvent infiltration are typically utilized to confine metal hydrides in nanoporous supports, it is widely accepted that the in situ generation of hydride species can provide more intimate contact between the active hydrogen storage species and the support, resulting in enhanced hydrogenation parameters. Zhang et al., for instance, synthesized the $Mg(BH_4)_2$@G (MBH–Gr) composite, which has the ability to start releasing H_2 at 154 °C vs. mechanochemically milled BM–MBH–Gr (257 °C) or pristine $Mg(BH_4)_2$ (296 °C), showing a tangible ~6 wt.% storage around 200 °C (Figure 8). The synthesis proceeded via the hydrogenation of $MgBu_2$, which shows favorable adsorption of graphene, and β–MgH_2 NPs of ~8 nm are formed highly dispersed on the substrate ("MH-Gr") as a result of this interaction. The large spacing between graphene layers could be responsible for the lack of agglomeration during the solid–gas reaction with B_2H_6 in order to generate $Mg(BH_4)_2$ NPs (reaction (1)) [124].

$$MgBu_2 \overset{50\ barH_2,170°C}{\rightarrow} MgH_2 \overset{+B_2H_6,G}{\rightarrow} Mg(BH_4)_2 \quad (1)$$

However, the authors also noted simultaneous formation of $MgB_{12}H_{12}$ (2478 cm^{-1} in FTIR spectra), and DFT data showed that the synthesis of $MgB_{12}H_{12}$ ($\Delta H = -320$ kJ/mol) and $Mg(BH_4)_2$ (($\Delta H = -308$ kJ/mol) is fairly similar. These values suggest that the presence of graphene favors the formation of $Mg(BH_4)_2$. The optimized geometries (Figure 8d) also reveal that graphene weakens Mg–H and B–B bonds (they are longer in the presence of graphene vs. pristine), congruent with the easier formation of $Mg(BH_4)_2$ [124].

4.3. Reactive Hydride Composites (RHCs) in G Hosts

There are only a few examples of MgH_2-based nancomposites; for instance, the MgH_2–$LiBH_4$ RHC [133] and 2 $LiBH_4$–MgH_2 @ G (80 wt.% loading capacity) [134]. In the case of the latter composite, the synthesis method started from MgH_2 @ G [134]. Not surprisingly, the MgH_2 was compounded with $LiBH_4$, another key component and widely-explored hydrogen storage complex hydride (18.5 wt.%, 121 kg H_2/m^3), with the positive results making use of $LiBH_4$ destabilization by binary metal hydrides such as MgH_2, in addition to the advantageous nanoconfinement effect in the porous graphene support [133]. The mutual destabilization of the system $2LIBH_4$–MgH_2 is depicted in reaction (2), where the final decomposition products are LiH and MgB_2. Even without graphene supports, the RHC system $2LIBH_4$–MgH_2 showed reversibility under moderate conditions (50 bar H_2, 300 °C), considerably lower than the harsh rehydrogenation requirements from spent $LiBH_4$ (200 bar H_2, 700 °C).

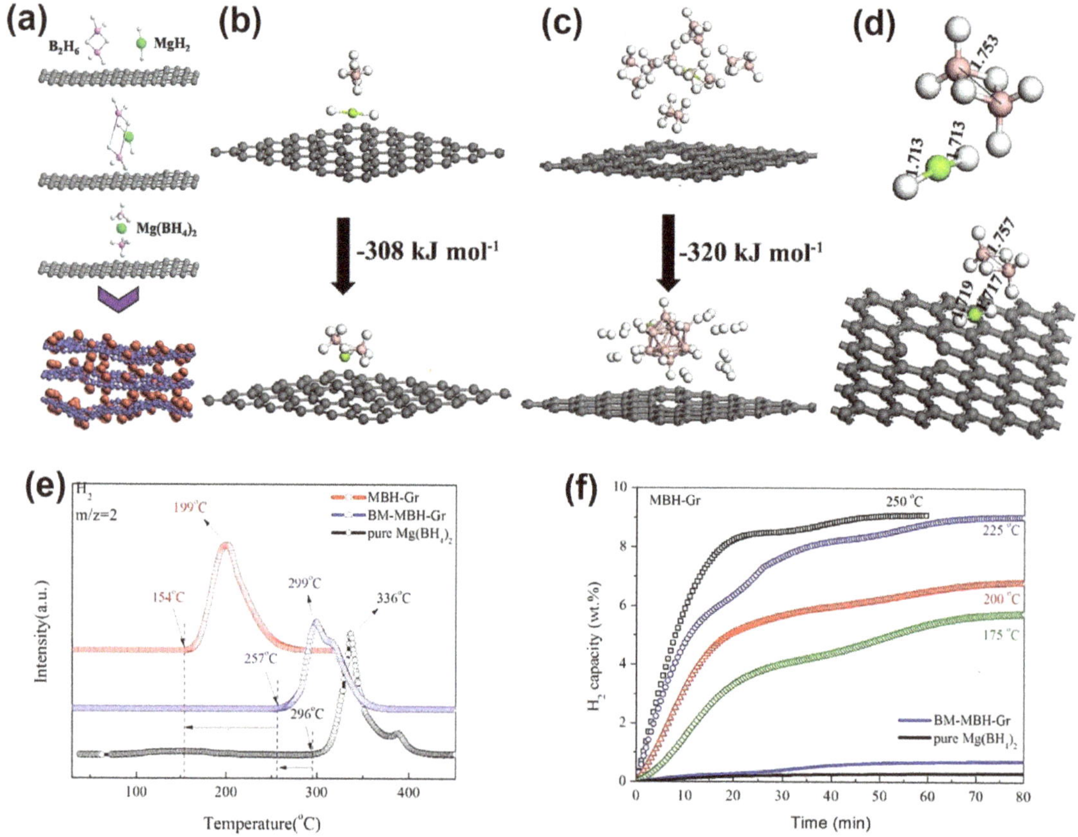

Figure 8. (**a**) Schematic illustration of the fabrication of MBH-Gr via a space-confined solid–gas reaction: the graphene-tailored solid–gas reaction between MH-Gr and B_2H_6 to synthesize $Mg(BH_4)_2$ NPs that are uniformly distributed on graphene (MBH-Gr) under the protection of hydrogen; (**b**–**d**): Relaxed atomic configurations of MgH_2 and B_2H_6, and calculated reaction enthalpies for the formation of (**b**) $Mg(BH_4)_2$ and (**c**) $MgB_{12}H_{12}$ from the reaction between MgH_2 and B_2H_6 with the support of graphene. (**d**) Relaxed atomic configurations of MgH_2 and B_2H_6 without and with the support of graphene. Gray, white, green, and pink spheres are C, H, Mg, and B atoms, respectively; (**e**) Mass spectra of the as-synthesized MBH-Gr compared with BM-MBH-Gr and pure $Mg(BH_4)_2$; (**f**) Isothermal dehydrogenation of the as-synthesized MBH-Gr at various temperatures, with BM-MBH-Gr and pure $Mg(BH_4)_2$ at 175 °C included for comparison. Reprinted with permission from Ref. [124].

$$2\,LiBH_4 + MgH_2 \leftrightarrow 2\,LiH + MgB_2 + 4\,H_2 \tag{2}$$

Xia et al. have demonstrated the solvothermal procedure to impregnate graphene with RHC precursors that finally yielded the binary composite $2LiBH_4$–MgH_2, with homogeneous particle size (10.5 nm) uniformly dispersed on graphene support even at very high loadings, which afforded a reversible 9.1 wt.% hydrogen capacity that did not degrade below 8.9 wt.% after 25 a/d cycles (Figure 9) [134].

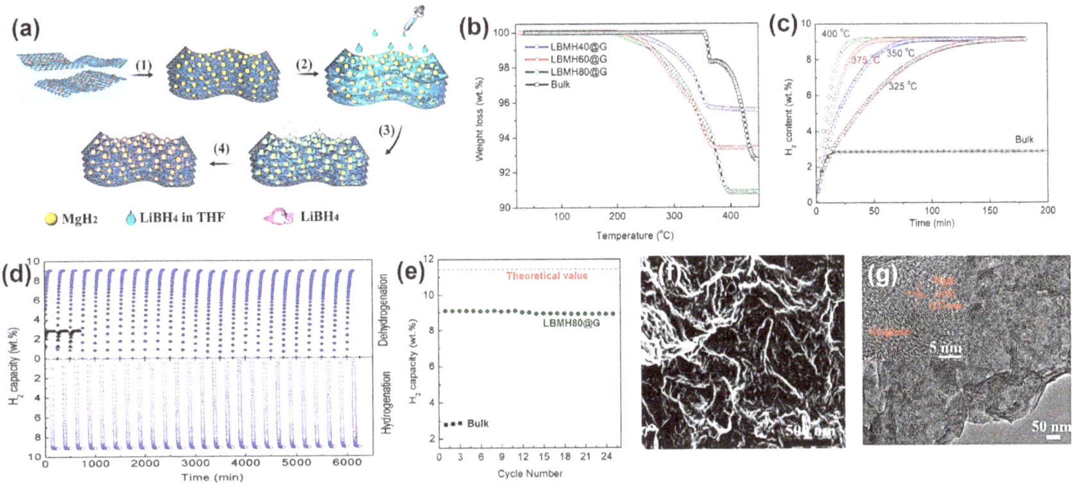

Figure 9. (**a**) Schematic illustration of the fabrication of graphene-supported $2LiBH_4$-MgH_2 nanocomposite: (1) self-assembly of uniform MgH_2 NPs on graphene via solvothermal treatment; (2) infiltration of $LiBH_4$ solution; (3, 4) removal of solvent and heterogeneous nucleation of $LiBH_4$ on MgH_2 NPs.; (**b**) Thermogravimetric analysis curves of the as-prepared $2LiBH_4$-MgH_2 nanocomposites anchored on graphene with various loadings compared with the ball-milled composite (bulk); (**c**) isothermal dehydrogenation of LBMH80@G at various temperatures under a back pressure of 0.3 MPa.; (**d**) Long-term cycling performance of the dehydrogenation (under a back pressure of 0.3 MPa) and hydrogenation for LBMH80@G and bulk $2LiBH_4$-MgH_2 composite at 350 °C; (**e**) normalized H_2 capacity as a function of cycle number, where the hydrogen capacities are normalized to the theoretical value of $2LiBH_4$-MgH_2 composite; (**f**) SEM and (**g**) TEM images of LBMH80@G after 15 cycles of dehydrogenation. Reprinted with permission from Ref. [134].

While other C-based materials supporting RHCs are rather common, the same cannot be said about G-supported RHCs. The improvements observed in lowered dehydrogenation temperature and parameters can be also traced back to the thermal conductivity of the graphene support, which enables the effective dissipation of heat and uniform heating throughout the composite [133]. Interestingly, nanoconfinement might not be solely responsible for the observed effects, as XRD data showed characteristic hydride peaks; hence, these were not fully confined, and the main role might actually be reserved to the highly specific surface area of graphene as well as to the nanosized metal hydrides.

4.4. Other Hydride Systems Confined in 2D-Graphene

Among hydrogen storing materials, ammonia borane (AB) plays a special role. While it has been investigated thoroughly, the mechanism of dehydrogenation is still not very clear, presumably due to polymerization processes occurring during thermal decomposition, and, perhaps more importantly, the system is not reversible. However, there have been attempts to incorporate AB into the porosity of catalyzed graphene supports, which yielded composites such as AB–Ni-Co/rGO catalyst [135], AB/Cu@CuCoOx supported on G [136], or AB/rGO (or GO) produced by ice-templating sheets of GO, serving as supports for the one-step integration of AB. The AB@rGO nanocomposite released H_2 as low as 70 °C [137]. A closely-related lithiated species of AB, namely lithium amidoborane $LiNH_2BH_3$, has been utilized to produce $LiNH_2BH_3/G$ nanocomposites that released 10.9 wt.% H_2 without $(BN)_x$ byproducts such as borazine [138].

Other 2D materials such as MXenes have also been incorporated in hydrogen storing materials such as hydrides, providing excellent improvements, presumably due to the presence of catalytic Ti sites and superior morphological features such as its highly-specific surface area [139].

Liquid organic hydrides can also release H_2 under G–CNT composite catalysts [140]. Substituted alkyl (R = Me) and aryl (R = Ph) silanes R_xSiH_{4-x} (R = Ph, Me) have also been employed in H_2 generation from MeOH. In this instance, however, the Ru-based catalyst was immobilized in the rGO support, producing the highly-active catalyst [Ru]–rGO (0.05 mol%), where [Ru] = [Ru(p-cym)Cl_2(NHCl)] complex [141]. To sum up, there are various approaches spanning metal salt catalysts [142], heterostructured supports [143], Ni-based catalysts supported by DFT computations [144], trimetallic catalysts FeCoNi NPs [145], or bimetallic PdRu [146] that have all played an essential role in the development of efficient catalysts and supports for metal hydrides and magnesium hydride in particular.

5. Conclusions

Featuring high hydrogen storage capacity and electrochemical properties, metal hydrides embedded in graphene supports have emerged as a new class of nanocomposite systems with promising features for energy storage systems. However, their development and integration in practical systems still face challenges to be overcome for their widespread adoption and long-term viability. The cost-effectiveness of these composite systems is an essential factor in their widespread adoption and increased lifespan viability. The cost of raw materials as well as production and energy costs are still too high, prompting the development of cost-effective and scalable fabrication processes.

Regarding the environmental impacts and ways to limit their availability, the use of CRM (critical raw materials) such as Mg, Ni, and Ti and their continuing depletion worldwide are of concern. The development of alternative materials that do not rely on CRMs is crucial for the sustainability of this technology. The long-term stability and durability of metal hydrides embedded in graphene supports are crucial for their sustainability, as frequent replacements can result in the generation of more waste and have a significant environmental impact. The optimization of the synthesis process and the development of durable and stable materials can improve the lifespan and performance of metal hydrides embedded in graphene supports, making them more sustainable. The disposal of metal hydrides embedded in graphene supports at the end of their lifecycle is also a concern. The development of recycling processes for metal hydrides and graphene supports can reduce the environmental impact of their disposal.

6. Current Challenges and Future Outlook

Energy storage systems are today in need of technological breakthroughs to revolutionize the way energy is produced, distributed, and stored. Various reviews have presented the critical shortcomings of current technology [3]. Nanocomposites based on metal hydrides and graphene currently feature hydrogen storage capacities lower than the targets set by the Department of Energy (DOE) for practical hydrogen storage (6.5 wt.% as the ultimate goal). Moreover, the cycling stability of metal hydrides needs to be improved to ensure the long-term performance of the composite material, while the potential side reactions occurring at the nanoscale must be further tuned and reduced. The scalability and cost-effectiveness of the synthesis methods used to prepare graphene-metal hydride composites need further refinement. Current methods such as ball milling (most commonly used) and in situ growth are time-consuming and require expensive equipment with high associated energy bills, which limits their practical application for large-scale production, which is compulsory for industrial adoption of this technology.

Future research efforts addressing these challenges ought to focus on developing new metal hydrides with higher hydrogen storage capacities and improved cycling stability.

In addition, new synthesis methods should be explored to improve the scalability and cost-effectiveness of the process. Advances in computational modeling and simulation can also be used to predict the performance of new metal hydride candidates and optimize their synthesis routes. In this regard, thermodynamic considerations are an essential starting point for the successful selection of future hydride-based materials.

Additionally, while graphene does feature many desirable properties, other carbon-based materials can be explored for the development of new hybrid materials with enhanced properties for energy storage applications, such as carbon nanotubes (MWCNTs) or other forms of porous carbon (OMC, CMK-3, etc.). Carbon materials can be further activated to enhance hydrogen-support interactions, which in turn will lead to a more effective H_2 storage material. Hence, current strategies need to focus on enhancing H_2-support interactions [4] by any of the following approaches: heteroatom-doping of porous G/other nanoporous materials; coordinatively unsaturated metal sites in MOFs; nanoconfinement of H_2 into nanopores (<1 nm); and enhanced orbital interactions H_2-adsorbent [4].

As the global demand for clean energy solutions continues to grow, the development of efficient energy storage systems such as graphene-metal hydride composites can play a crucial role in accelerating the transition towards a sustainable energy future.

Funding: This work was supported by the Romanian Ministry of Research and Innovation through the Project No. PN-III-P1-1.1-TE-2021-1657 (TE 84/2022) and by the Core Program of the National Institute of Materials Physics, granted by the Romanian Ministry of Research, Innovation and Digitization through the Project PC3-PN23080303.

Data Availability Statement: Not applicable.

Acknowledgments: This work was supported by the Romanian Ministry of Research and Innovation through the Project No. PN-III-P1-1.1-TE-2021-1657 (TE 84/2022) and Project PC3-PN23080303.

Conflicts of Interest: The authors declare no conflict of interest. The funders had no role in the design of the study; in the collection, analyses, or interpretation of data; in the writing of the manuscript; or in the decision to publish the results.

References

1. Zhang, J.; Yan, S.; Qu, H. Recent progress in magnesium hydride modified through catalysis and nanoconfinement. *Int. J. Hydrogen Energy* **2018**, *43*, 1545–1565. [CrossRef]
2. Comanescu, C. Recent Development in Nanoconfined Hydrides for Energy Storage. *Int. J. Mol. Sci.* **2022**, *23*, 7111. [CrossRef] [PubMed]
3. Sreedhar, I.; Kamani, K.M.; Kamani, B.M.; Reddy, B.M.; Venugopal, A. A Bird's Eye view on process and engineering aspects of hydrogen storage. *Renew. Sustain. Energy Rev.* **2018**, *91*, 838–860. [CrossRef]
4. So, S.H.; Sung, S.J.; Yang, S.J.; Park, C.R. Where to go for the Development of High-Performance H-2 Storage Materials at Ambient Conditions? *Electron. Mater. Lett.* **2023**, *19*, 1–18. [CrossRef]
5. Comanescu, C. Paving the Way to the Fuel of the Future—Nanostructured Complex Hydrides. *Int. J. Mol. Sci.* **2023**, *24*, 143. [CrossRef]
6. Comanescu, C. Complex Metal Borohydrides: From Laboratory Oddities to Prime Candidates in Energy Storage Applications. *Materials* **2022**, *15*, 2286. [CrossRef]
7. Ding, F.; Yakobson, B.I. Challenges in hydrogen adsorptions: From physisorption to chemisorption. *Front. Phys.* **2011**, *6*, 142–150. [CrossRef]
8. Feng, D.C.; Zhou, D.S.; Zhao, Z.Y.; Zhai, T.T.; Yuan, Z.M.; Sun, H.; Ren, H.P.; Zhang, Y.H. Progress of graphene and loaded transition metals on Mg-based hydrogen storage alloys. *Int. J. Hydrogen Energy* **2021**, *46*, 33468–33485. [CrossRef]
9. Free, Z.; Hernandez, M.; Mashal, M.; Mondal, K. A Review on Advanced Manufacturing for Hydrogen Storage Applications. *Energies* **2021**, *14*, 8513. [CrossRef]
10. Boateng, E.; Chen, A.C. Recent advances in nanomaterial-based solid-state hydrogen storage. *Mater. Today Adv.* **2020**, *6*, 100022. [CrossRef]
11. Jawhari, A.H. Novel Nanomaterials for Hydrogen Production and Storage: Evaluating the Futurity of Graphene/Graphene Composites in Hydrogen Energy. *Energies* **2022**, *15*, 9085. [CrossRef]
12. Arbuzov, A.A.; Mozhzhukhin, S.A.; Volodin, A.A.; Fursikov, P.V.; Tarasov, B.P. Graphene-like nanostructures: Synthesis and use for preparation of catalysts and hydrogen storage composites. *Russ. Chem. B+* **2016**, *65*, 1893–1901. [CrossRef]

13. Xia, G.; Zhang, B.; Chen, X.; Sun, D.; Guo, Z.; Liang, F.; Yu, X. Graphene for Energy Storage and Conversion: Synthesis and Interdisciplinary Applications. *Electrochem. Energy Rev.* **2020**, *3*, 395–430. [CrossRef]
14. Liu, W.; Setijadi, E.; Crema, L.; Bartali, R.; Laidani, N.; Aguey-Zinsou, K.F.; Speranza, G. Carbon nanostructures/Mg hybrid materials for hydrogen storage. *Diam. Relat. Mater.* **2018**, *82*, 19–24. [CrossRef]
15. Yao, Q.L.; Ding, Y.Y.; Lu, Z.H. Noble-metal-free nanocatalysts for hydrogen generation from boron- and nitrogen-based hydrides. *Inorg. Chem. Front.* **2020**, *7*, 3837–3874. [CrossRef]
16. Tarasov, B.P.; Arbuzov, A.A.; Volodin, A.A.; Fursikov, P.V.; Mozhzhuhin, S.A.; Lototskyy, M.V.; Yartys, V.A. Metal hydride–Graphene composites for hydrogen based Energy storage. *J. Alloys Compd.* **2022**, *896*, 162881. [CrossRef]
17. Aditya, M.V.V.S.; Panda, S.; Tatiparti, S.S.V. Boron from net charge acceptor to donor and its effect on hydrogen uptake by novel Mg-B-electrochemically synthesized reduced graphene oxide. *Sci. Rep.* **2021**, *11*, 10995. [CrossRef]
18. Tan, X.; Tahini, H.A.; Smith, S.C. Conductive Boron-Doped Graphene as an Ideal Material for Electrocatalytically Switchable and High-Capacity Hydrogen Storage. *ACS Appl. Mater. Interfaces* **2016**, *8*, 32815–32822. [CrossRef]
19. Wang, Y.; Wang, K.; Guan, C.; He, Z.M.; Lu, Z.S.; Chen, T.; Liu, J.; Tan, X.; Tan, T.T.Y.; Li, C.M. Surface functionalization-enhanced spillover effect on hydrogen storage of Ni-B nanoalloy-doped activated carbon. *Int. J. Hydrogen Energy* **2011**, *36*, 13663–13668. [CrossRef]
20. Wan, L.F.; Cho, E.S.; Marangoni, T.; Shea, P.; Kang, S.Y.; Rogers, C.; Zaia, E.; Cloke, R.R.; Wood, B.C.; Fischer, F.R.; et al. EdgeFunctionalized Graphene Nanoribbon Encapsulation to Enhance Stability and Control Kinetics of Hydrogen Storage Materials. *Chem. Mater.* **2019**, *31*, 2960–2970. [CrossRef]
21. Asefa, T.; Koh, K.; Yoon, C.W. CO_2-Mediated H-2 Storage-Release with Nanostructured Catalysts: Recent Progresses, Challenges, and Perspectives. *Adv. Energy Mater.* **2019**, *9*, 1158. [CrossRef]
22. Shevlin, S.A.; Guo, Z.X. Density functional theory simulations of complex hydride and carbon-based hydrogen storage materials. *Chem. Soc. Rev.* **2009**, *38*, 211–225. [CrossRef] [PubMed]
23. Naqvi, S.R.; Hussain, T.; Luo, W.; Ahuja, R. Metallized siligraphene nanosheets (SiC7) as high capacity hydrogen storage materials. *Nano Res.* **2018**, *11*, 3802–3813. [CrossRef]
24. Aydin, S.; Simsek, M. The enhancement of hydrogen storage capacity in Li, Na and Mg-decorated BC3 graphene by CLICH and RICH algorithms. *Int. J. Hydrogen Energy* **2019**, *44*, 7354–7370. [CrossRef]
25. Tokarev, A.; Avdeenkov, A.V.; Langmi, H.; Bessarabov, D.G. Modeling hydrogen storage in boron-substituted graphene decorated with potassium metal atoms. *Int. J. Energy Res.* **2015**, *39*, 524–528. [CrossRef]
26. Ao, Z.M.; Dou, S.X.; Xu, Z.M.; Jiang, Q.G.; Wang, G.X. Hydrogen storage in porous graphene with Al decoration. *Int. J. Hydrogen Energy* **2014**, *39*, 16244–16251. [CrossRef]
27. Costanzo, F.; van Hemert, M.C.; Kroes, G.J. Promoting Effect of Carbon Surfaces on H-2 Dissociation on Al-n Clusters by First Principles Calculations. *J. Phys. Chem. C* **2014**, *118*, 513–522. [CrossRef]
28. Zhang, Y.; Cui, H.; Tian, W.Z.; Liu, T.; Wang, Y.Z. Effect of hydrogen adsorption Energy on the electronic and optical properties of Si-modified single-layer graphene with an Al decoration. *AIP Adv.* **2020**, *10*, 045012. [CrossRef]
29. Bakhshi, F.; Farhadian, N. Improvement of hydrogen storage capacity on the palladium-decorated N-doped graphene sheets as a novel adsorbent: A hybrid MD-GCMC simulation study. *Int. J. Hydrogen Energy* **2019**, *44*, 13655–13665. [CrossRef]
30. Ramos-Castillo, C.M.; Reveles, J.U.; Zope, R.R.; de Coss, R. Palladium Clusters Supported on Graphene Monovacancies for Hydrogen Storage. *J. Phys. Chem. C* **2015**, *119*, 8402–8409. [CrossRef]
31. Choudhary, A.; Malakkal, L.; Siripurapu, R.K.; Szpunar, B.; Szpunar, J. First principles calculations of hydrogen storage on Cu and Pd-decorated graphene. *Int. J. Hydrogen Energy* **2016**, *41*, 17652–17656. [CrossRef]
32. Bora, P.L.; Ahmad, R.; Singh, A.K. Remarkable enhancement in hydrogen storage on free-standing Ti3B and BC3 supported Ti-3 clusters. *Int. J. Hydrogen Energy* **2015**, *40*, 1054–1061. [CrossRef]
33. Intayot, R.; Rungnim, C.; Namuangruk, S.; Yodsin, N.; Jungsuttiwong, S. Ti-4-Decorated B/N-doped graphene as a high-capacity hydrogen storage material: A DFT study. *Dalton Trans.* **2021**, *50*, 11398–11411. [CrossRef] [PubMed]
34. Ramos-Castillo, C.M.; Reveles, J.U.; Cifuentes-Quintal, M.E.; Zope, R.R.; de Coss, R. Ti-4- and Ni-4-Doped Defective Graphene Nanoplatelets as Efficient Materials for Hydrogen Storage. *J. Phys. Chem. C* **2016**, *120*, 5001–5009. [CrossRef]
35. Ramos-Castillo, C.M.; Reveles, J.U.; Cifuentes-Quintal, M.E.; Zope, R.R.; de Coss, R. Hydrogen storage in bimetallic Ti-Al sub-nanoclusters supported on graphene. *Phys. Chem. Chem. Phys.* **2017**, *19*, 21174–21184. [CrossRef]
36. Bartali, R.; Speranza, G.; Aguey-Zinsou, K.F.; Testi, M.; Micheli, V.; Canteri, R.; Fedrizzi, M.; Gottardi, G.; Coser, G.; Crema, L.; et al. Efficient hydrogen generation from water using nanocomposite flakes based on graphene and magnesium. *Sustain. Energy Fuels* **2018**, *2*, 2516–2525. [CrossRef]
37. Bouaricha, S.; Dodelet, J.P.; Guay, D.; Huot, J.; Schulz, R. Study of the activation process of Mg-based hydrogen storage materials modified by graphite and other carbonaceous compounds. *J. Mater. Res.* **2001**, *16*, 2893–2905. [CrossRef]
38. Cho, E.S.; Ruminski, A.M.; Liu, Y.S.; Shea, P.T.; Kang, S.Y.; Zaia, E.W.; Park, J.Y.; Chuang, Y.D.; Yuk, J.M.; Zhou, X.W.; et al. Hierarchically Controlled Inside-Out Doping of Mg Nanocomposites for Moderate Temperature Hydrogen Storage. *Adv. Funct. Mater.* **2017**, *27*, 1704316. [CrossRef]
39. Dun, C.; Jeong, S.; Kwon, D.H.; Kang, S.; Stavila, V.; Zhang, Z.L.; Lee, J.W.; Mattox, T.M.; Heo, T.W.; Wood, B.C.; et al. Hydrogen Storage Performance of Preferentially Oriented Mg/rGO Hybrids. *Chem. Mater.* **2022**, *34*, 2963–2971. [CrossRef]

40. Cho, Y.; Kang, S.; Wood, B.C.; Cho, E.S. Heteroatom-Doped Graphenes as Actively Interacting 2D Encapsulation Media for Mg-Based Hydrogen Storage. *ACS Appl. Mater. Interfaces* **2022**, *14*, 20823–20834. [CrossRef]
41. Dong, S.; Lv, E.F.; Wang, J.H.; Li, C.Q.; Ma, K.; Gao, Z.Y.; Yang, W.J.; Ding, Z.; Wu, C.C.; Gates, I.D. Construction of transition metal-decorated boron doped twin-graphene for hydrogen storage: A theoretical prediction. *Fuel* **2021**, *304*, 121351. [CrossRef]
42. Han, D.J.; Kim, S.; Cho, E.S. Revealing the role of defects in graphene oxide in the evolution of magnesium nanocrystals and the resulting effects on hydrogen storage. *J. Mater. Chem. A* **2021**, *9*, 9875–9881. [CrossRef]
43. Du, J.Q.; Lan, Z.Q.; Zhang, H.; Lu, S.X.; Liu, H.Z.; Guo, J. Catalytic enhanced hydrogen storage properties of Mg-based alloy by the addition of reduced graphene oxide supported V_2O_3 nanocomposite. *J. Alloys Compd.* **2019**, *802*, 660–667. [CrossRef]
44. Bhatnagar, A.; Gupta, B.K.; Tripathi, P.; Veziroglu, A.; Hudson, M.S.L.; Abu Shaz, M.; Srivastava, O.N. Development and Demonstration of Air Stable rGO-EC@AB(5) Type Hydrogenated Intermetallic Hybrid for Hydrogen Fuelled Devices. *Adv. Sustain. Syst.* **2017**, *1*, 1700087. [CrossRef]
45. Cui, R.C.; Yang, C.C.; Li, M.M.; Jin, B.; Ding, X.D.; Jiang, Q. Enhanced high-rate performance of ball-milled MmNi(3.55)Co(0.75)Mn(0.4)Al(0.3) hydrogen storage alloys with graphene nanoplatelets. *J. Alloys Compd.* **2017**, *693*, 126–131. [CrossRef]
46. Morse, J.R.; Zugell, D.A.; Patterson, E.; Baldwin, J.W.; Willauer, H.D. Hydrogenated graphene: Important material properties regarding its application for hydrogen storage. *J. Power Sources* **2021**, *494*, 229734. [CrossRef]
47. Wu, X.J.; Fei, Z.J.; Liu, W.G.; Tan, J.; Wang, G.H.; Xia, D.Q.; Deng, K.; Chen, X.K.; Xiao, D.T.; Wu, S.W.; et al. Adsorption and desorption of hydrogen on/from single-vacancy and double-vacancy graphenes. *Nucl. Sci. Tech.* **2019**, *30*, 69. [CrossRef]
48. Akilan, R.; Vinnarasi, S.; Mohanapriya, S.; Shankar, R. Adsorption of H(2)molecules on B/N-doped defected graphene sheets—A DFT study. *Struct. Chem.* **2020**, *31*, 2413–2434. [CrossRef]
49. Lotfi, R.; Saboohi, Y. A comparative study on hydrogen interaction with defective graphene structures doped by transition metals. *Phys. E* **2014**, *60*, 104–111. [CrossRef]
50. Lototskyy, M.; Sibanyoni, J.M.; Denys, R.V.; Williams, M.; Pollet, B.G.; Yartys, V.A. Magnesium-carbon hydrogen storage hybrid materials produced by reactive ball milling in hydrogen. *Carbon* **2013**, *57*, 146–160. [CrossRef]
51. Ruse, E.; Buzaglo, M.; Pri-Bar, I.; Shunak, L.; Nadiv, R.; Pevzner, S.; Siton-Mendelson, O.; Skripnyuk, V.M.; Rabkin, E.; Regev, O. Hydrogen storage kinetics: The graphene nanoplatelet size effect. *Carbon* **2018**, *130*, 369–376. [CrossRef]
52. Bahuguna, A.; Sasson, Y. Formate-Bicarbonate Cycle as a Vehicle for Hydrogen and Energy Storage. *Chemsuschem* **2021**, *14*, 1258–1283. [CrossRef] [PubMed]
53. Gupta, S.; Heintzman, E.; Price, C. Electrostatic Layer-By-Layer Self-Assembled Graphene/Multi-Walled Carbon Nanotubes Hybrid Multilayers as Efficient 'All Carbon' Supercapacitors. *J. Nanosci. Nanotechnol.* **2016**, *16*, 4771–4782. [CrossRef] [PubMed]
54. Ghalami, Z.; Ghoulipour, V.; Khanchi, A. Hydrogen and deuterium adsorption on uranium decorated graphene nanosheets: A combined molecular dynamics and density functional theory study. *Curr. Appl. Phys.* **2019**, *19*, 536–541. [CrossRef]
55. Lang, C.G.; Ouyang, L.Z.; Yang, L.L.; Dai, L.Y.; Wu, D.F.; Shao, H.Y.; Zhu, M. Enhanced hydrogen storage kinetics in Mg@FLG composite synthesized by plasma assisted milling. *Int. J. Hydrogen Energy* **2018**, *43*, 17346–17352. [CrossRef]
56. Lee, J.; Sung, D.; Chung, Y.K.; Bin Song, S.; Huh, J. Unveiling two-dimensional magnesium hydride as a hydrogen storage material via a generative adversarial network. *Nanoscale Adv.* **2022**, *4*, 2332–2338. [CrossRef]
57. Ates, M.; Chebil, A.; Yoruk, O.; Dridi, C.; Turkyilmaz, M. Reliability of electrode materials for supercapacitors and batteries in Energy storage applications: A review. *Ionics* **2022**, *28*, 27–52. [CrossRef]
58. Chaichi, A.; Wang, Y.; Gartia, M.R. Substrate Engineered Interconnected Graphene Electrodes with Ultrahigh Energy and Power Densities for Energy Storage Applications. *ACS Appl. Mater. Interfaces* **2018**, *10*, 21235–21245. [CrossRef]
59. Xia, G.; Zhang, B.; Chen, X.; Sun, D.; Guo, Z.; Liang, F.; Yu, X. Molecular-Scale Functionality on Graphene To Unlock the Energy Capabilities of Metal Hydrides for High-Capacity Lithium-Ion Batteries. *ACS Nano* **2018**, *12*, 8177–8186. [CrossRef]
60. Li, M.M.; Wang, Y.; Yang, C.C.; Jiang, Q. In situ grown Co_3O_4 nanocubes on N-doped graphene as a synergistic hybrid for applications in nickel metal hydride batteries. *Int. J. Hydrogen Energy* **2018**, *43*, 18421–18435. [CrossRef]
61. Sirisinudomkit, P.; Iamprasertkun, P.; Krittayavathananon, A.; Pettong, T.; Dittanet, P.; Sawangphruk, M. Hybrid Energy Storage of Ni(OH)(2)-coated N-doped Graphene Aerogel//N-doped Graphene Aerogel for the Replacement of NiCd and NiMH Batteries. *Sci. Rep.* **2017**, *7*, 1124. [CrossRef] [PubMed]
62. Zhang, B.; Xia, G.; Sun, D.; Fang, F.; Yu, X. Magnesium Hydride Nanoparticles Self-Assembled on Graphene as Anode Material for High-Performance Lithium-Ion Batteries. *ACS Nano* **2018**, *12*, 3816–3824. [CrossRef] [PubMed]
63. Zhang, S.J.; Li, C.; Zhang, X.; Sun, X.Z.; Wang, K.; Ma, Y.W. High Performance Lithium-Ion Hybrid Capacitors Employing Fe_3O_4-Graphene Composite Anode and Activated Carbon Cathode. *ACS Appl. Mater. Interfaces* **2017**, *9*, 17137–17145. [CrossRef] [PubMed]
64. He, H.; Xu, X.D.; Liu, D.; Li, J.; Wei, Y.H.; Tang, H.L.; Li, J.S.; Li, X.; Xie, Z.Z.; Qu, D.Y. The impacts of nitrogen doping on the electrochemical hydrogen storage in a carbon. *Int. J. Energy Res.* **2021**, *45*, 9326–9339. [CrossRef]
65. Lan, Z.Q.; Zeng, K.; Wei, B.; Li, G.X.; Ning, H.; Guo, J. Nickel-graphene nanocomposite with improved electrochemical performance for La0.7Mg0.3(Ni0.85Co0.15)(3.5) electrode. *Int. J. Hydrogen Energy* **2017**, *42*, 12458–12466. [CrossRef]
66. Xia, G.; Tan, Y.; Chen, X.; Fang, F.; Sun, D.; Li, X.; Guo, Z.; Yu, X. Oxygen-free Layer-by-Layer Assembly of Lithiated Composites on Graphene for Advanced Hydrogen Storage. *Adv. Sci.* **2016**, *4*, 1600257. [CrossRef]

67. Zhang, Q.; Huang, Y.; Xu, L.; Zang, L.; Guo, H.; Jiao, L.; Yuan, H.; Wang, Y. Highly Dispersed MgH_2 Nanoparticle–Graphene Nanosheet Composites for Hydrogen Storage. *ACS Appl. Nano Mater.* **2019**, *2*, 3828–3835. [CrossRef]
68. Huang, Y.Q.; Xia, G.L.; Chen, J.; Zhang, B.P.; Li, Q.; Yu, X.B. One-step uniform growth of magnesium hydride nanoparticles on graphene. *Prog. Nat. Sci.-Mater.* **2017**, *27*, 87–93. [CrossRef]
69. Jang, M.H.; Park, S.H.; Hong, T.W. Hydrogenation Behaviors of MgHx-Graphene Composites by Reactive Mechanical Grinding. *Korean J. Met. Mater.* **2016**, *54*, 288–294. [CrossRef]
70. Cho, E.S.; Ruminski, A.M.; Aloni, S.; Liu, Y.-S.; Guo, J.; Urban, J.J. Graphene oxide/metal nanocrystal multilaminates as the atomic limit for safe and selective hydrogen storage. *Nat. Commun.* **2016**, *7*, 10804. [CrossRef]
71. Pukazhselvan, D.; Shaula, A.L.; Mikhalev, S.M.; Bdikin, I.; Fagg, D.P. Elucidating Evidence for the In Situ Reduction of Graphene Oxide by Magnesium Hydride and the Consequence of Reduction on Hydrogen Storage. *Catalysts* **2022**, *12*, 735. [CrossRef]
72. Han, D.J.; Bang, K.R.; Cho, H.; Cho, E.S. Effect of carbon nanoscaffolds on hydrogen storage performance of magnesium hydride. *Korean J. Chem. Eng.* **2020**, *37*, 1306–1316. [CrossRef]
73. Singh, M.K.; Bhatnagar, A.; Pandey, S.K.; Mishra, P.C.; Srivastava, O.N. Experimental and first principle studies on hydrogen desorption behavior of graphene nanofibre catalyzed MgH_2. *Int. J. Hydrogen Energy* **2017**, *42*, 960–968. [CrossRef]
74. Zhang, J.; Yu, X.F.; Mao, C.; Long, C.G.; Chen, J.; Zhou, D.W. Influences and mechanisms of graphene-doping on dehydrogenation properties of MgH_2: Experimental and first-principles studies. *Energy* **2015**, *89*, 957–964. [CrossRef]
75. Zhang, L.T.; Chen, L.X.; Xiao, X.Z.; Fan, X.L.; Shao, J.; Li, S.Q.; Ge, H.W.; Wang, Q.D. Fluorographene nanosheets enhanced hydrogen absorption and desorption performances of magnesium hydride. *Int. J. Hydrogen Energy* **2014**, *39*, 12715–12726. [CrossRef]
76. Verma, S.K.; Bhatnagar, A.; Shukla, V.; Soni, P.K.; Pandey, A.P.; Yadav, T.P.; Srivastava, O.N. Multiple improvements of hydrogen sorption and their mechanism for MgH_2 catalyzed through TiH_2@Gr. *Int. J. Hydrogen Energy* **2020**, *45*, 19516–19530. [CrossRef]
77. Wang, K.K.; Wu, G.L.; Cao, H.J.; Li, H.L.; Zhao, X.S. Improved reversible dehydrogenation properties of MgH_2 by the synergetic effects of graphene oxide-based porous carbon and TiCl3. *Int. J. Hydrogen Energy* **2018**, *43*, 7440–7446. [CrossRef]
78. Wang, Y.; Ding, Z.M.; Li, X.J.; Ren, S.Q.; Zhou, S.H.; Zhang, H.M.; Li, Y.; Han, S.M. Improved hydrogen storage properties of MgH_2 by nickel@nitrogen-doped carbon spheres. *Dalton Trans.* **2020**, *49*, 3495–3502. [CrossRef]
79. Wang, Y.; Liu, G.; An, C.H.; Li, L.; Qiu, F.Y.; Wang, Y.J.; Jiao, L.F.; Yuan, H.T. Bimetallic NiCo Functional Graphene: An Efficient Catalyst for Hydrogen-Storage Properties of MgH_2. *Chem.-Asian J.* **2014**, *9*, 2576–2583. [CrossRef]
80. Zhang, J.; Qu, H.; Yan, S.; Wu, G.; Yu, X.F.; Zhou, D.W. Catalytic effect of nickel phthalocyanine on hydrogen storage properties of magnesium hydride: Experimental and first-principles studies. *Int. J. Hydrogen Energy* **2017**, *42*, 28485–28497. [CrossRef]
81. Xie, X.B.; Chen, M.; Liu, P.; Shang, J.X.; Liu, T. High hydrogen desorption properties of Mg-based nanocomposite at moderate temperatures: The effects of multiple catalysts in situ formed by adding nickel sulfides/graphene. *J. Power Sources* **2017**, *371*, 112–118. [CrossRef]
82. Zhang, J.; Qu, H.; Wu, G.; Song, L.B.; Yu, X.F.; Zhou, D.W. Remarkably enhanced dehydrogenation properties and mechanisms of MgH_2 by sequential-doping of nickel and graphene. *Int. J. Hydrogen Energy* **2016**, *41*, 17433–17441. [CrossRef]
83. Zhang, J.G.; Zhu, Y.F.; Zang, X.X.; Huan, Q.Q.; Su, W.; Zhu, D.L.; Li, L.Q. Nickel-decorated graphene nanoplates for enhanced H-2 sorption properties of magnesium hydride at moderate temperatures. *J. Mater. Chem. A* **2016**, *4*, 2560–2570. [CrossRef]
84. Zhang, L.T.; Cai, Z.L.; Zhu, X.Q.; Yao, Z.D.; Sun, Z.; Ji, L.; Yan, N.H.; Xiao, B.B.; Chen, L.X. Two-dimensional ZrCo nanosheets as highly effective catalyst for hydrogen storage in MgH_2. *J. Alloys Compd.* **2019**, *805*, 295–302. [CrossRef]
85. Leng, H.Y.; Miao, N.; Li, Q. Improved hydrogen storage properties of MgH_2 by the addition of KOH and graphene. *Int. J. Hydrogen Energy* **2020**, *45*, 28183–28189. [CrossRef]
86. Li, M.M.; Yang, C.C.; Chen, L.X.; Jiang, Q. Hydrogen storage alloys/reduced graphite oxide: An efficient hybrid electrode with enhanced high-rate dischargeability. *Electrochim. Acta* **2016**, *200*, 59–65. [CrossRef]
87. Huang, H.X.; Yuan, J.G.; Zhang, B.; Zhang, J.G.; Zhu, Y.F.; Li, L.Q.; Wu, Y.; Zhou, S.X. Improvement in the hydrogenation-dehydrogenation performance of a Mg-Al alloy by graphene supported Ni. *Int. J. Hydrogen Energy* **2020**, *45*, 798–808. [CrossRef]
88. Huang, X.T.; Tao, A.P.; Guo, J.; Wei, W.L.; Guo, J.; Lan, Z.Q. Synergistic effect of TiF3@ graphene on the hydrogen storage properties of Mg-Al alloy. *Int. J. Hydrogen Energy* **2018**, *43*, 1651–1657. [CrossRef]
89. Lan, Z.Q.; Sun, Z.Z.; Ding, Y.C.; Ning, H.; Wei, W.L.; Guo, J. Catalytic action of Y2O3@graphene nanocomposites on the hydrogen-storage properties of Mg-Al alloys. *J. Mater. Chem. A* **2017**, *5*, 15200–15207. [CrossRef]
90. Ji, L.; Zhang, L.T.; Yang, X.L.; Zhu, X.Q.; Chen, L.X. The remarkably improved hydrogen storage performance of MgH_2 by the synergetic effect of an FeNi/rGO nanocomposite. *Dalton Trans.* **2020**, *49*, 4146–4154. [CrossRef]
91. Liu, G.; Wang, K.F.; Li, J.P.; Wang, Y.J.; Yuan, H.T. Enhancement of hydrogen desorption in magnesium hydride catalyzed by graphene nanosheets supported Ni-CeOx hybrid nanocatalyst. *Int. J. Hydrogen Energy* **2016**, *41*, 10786–10794. [CrossRef]
92. Liu, G.; Wang, Y.J.; Jiao, L.F.; Yuan, H.T. Solid-state synthesis of amorphous TiB2 nanoparticles on graphene nanosheets with enhanced catalytic dehydrogenation of MgH_2. *Int. J. Hydrogen Energy* **2014**, *39*, 3822–3829. [CrossRef]
93. Liu, J.C.; Liu, Y.N.; Liu, Z.B.; Ma, Z.L.; Ding, Y.J.; Zhu, Y.F.; Zhang, Y.; Zhang, J.G.; Li, L.Q. Effect of rGO supported NiCu derived from layered double hydroxide on hydrogen sorption kinetics of MgH_2. *J. Alloys Compd.* **2019**, *789*, 768–776. [CrossRef]

94. Zhou, D.M.; Cui, K.X.; Zhou, Z.W.; Liu, C.R.; Zhao, W.; Li, P.; Qu, X.H. Enhanced hydrogen-storage properties of MgH_2 by Fe-Ni catalyst modified three-dimensional graphene. *Int. J. Hydrogen Energy* **2021**, *46*, 34369–34380. [CrossRef]
95. Liu, J.C.; Ma, Z.L.; Liu, Z.B.; Tang, Q.K.; Zhu, Y.F.; Lin, H.J.; Zhang, Y.; Zhang, J.G.; Liu, Y.; Li, L.Q. Synergistic effect of rGO supported Ni_3Fe on hydrogen storage performance of MgH_2. *Int. J. Hydrogen Energy* **2020**, *45*, 16622–16633. [CrossRef]
96. Luo, S.C.; Li, S.J.; Liu, Y.N.; Zhang, J.G.; Zhu, Y.F.; Zhang, Y.; Lin, H.J.; Li, L.Q. Synergistically tuned hydrogen storage thermodynamics and kinetics of Mg-Al alloys by Cu formed in situ mechanochemically. *J. Alloys Compd.* **2019**, *806*, 370–377. [CrossRef]
97. Peng, D.D.; Ding, Z.M.; Zhang, L.; Fu, Y.K.; Wang, J.S.; Li, Y.; Han, S.M. Remarkable hydrogen storage properties and mechanisms of the shell-core MgH_2@carbon aerogel microspheres. *Int. J. Hydrogen Energy* **2018**, *43*, 3731–3740. [CrossRef]
98. Tian, M.; Shang, C.X. Mg-based composites for enhanced hydrogen storage performance. *Int. J. Hydrogen Energy* **2019**, *44*, 338–344. [CrossRef]
99. Shukla, V.; Bhatnagar, A.; Verma, S.K.; Pandey, A.P.; Vishwakarma, A.K.; Srivastava, P.; Yadav, T.P.; Srivastava, O.N. Simultaneous improvement of kinetics and thermodynamics based on SrF2 and SrF2@Gr additives on hydrogen sorption in MgH_2. *Mater. Adv.* **2021**, *2*, 4277–4290. [CrossRef]
100. Song, M.C.; Zhang, L.T.; Zheng, J.G.; Yu, Z.D.; Wang, S.N. Constructing graphene nanosheet-supported FeOOH nanodots for hydrogen storage of MgH_2. *Int. J. Miner. Met. Mater.* **2022**, *29*, 1464–1473. [CrossRef]
101. Tarasov, B.P.; Arbuzov, A.A.; Mozhzhuhin, S.A.; Volodin, A.A.; Fursikov, P.V.; Lototskyy, M.V.; Yartys, V.A. Hydrogen storage behavior of magnesium catalyzed by nickel-graphene nanocomposites. *Int. J. Hydrogen Energy* **2019**, *44*, 29212–29223. [CrossRef]
102. Zhang, B.P.; Xia, G.L.; Chen, W.; Gu, Q.F.; Sun, D.L.; Yu, X.B. Controlled-Size Hollow Magnesium Sulfide Nanocrystals Anchored on Graphene for Advanced Lithium Storage. *ACS Nano* **2018**, *12*, 12741–12750. [CrossRef] [PubMed]
103. Zhao, X.D.; Ma, Z.N.; Wu, D.H.; Zhang, X.; Jing, Y.; Zhou, Z. Computational study of catalytic effect of C3N4 on H-2 release from complex hydrides. *Int. J. Hydrogen Energy* **2015**, *40*, 8897–8902. [CrossRef]
104. Yu, H.Z.; Du, A.J.; Song, Y.; Searles, D.J. Graphyne and Graphdiyne: Versatile Catalysts for Dehydrogenation of Light Metal Complex Hydrides. *J. Phys. Chem. C* **2013**, *117*, 21643–21650. [CrossRef]
105. Gasnier, A.; Luguet, M.; Pereira, A.G.; Troiani, H.; Zampieri, G.; Gennari, F.C. Entanglement of N-doped graphene in resorcinol-formaldehyde: Effect over nanoconfined LiBH4 for hydrogen storage. *Carbon* **2019**, *147*, 284–294. [CrossRef]
106. Gasnier, A.; Amica, G.; Juan, J.; Troiani, H.; Gennari, F.C. N-Doped Graphene-Rich Aerogels Decorated with Nickel and Cobalt Nanoparticles: Effect on Hydrogen Storage Properties of Nanoconfined LiBH4. *J. Phys. Chem. C* **2020**, *124*, 115–125. [CrossRef]
107. Ye, J.K.; Xia, G.L.; Yu, X.B. In-situ constructed destabilization reaction of LiBH4 wrapped with graphene toward stable hydrogen storage reversibility. *Mater. Today Energy* **2021**, *22*, 100885. [CrossRef]
108. Xu, G.; Zhang, W.; Zhang, Y.; Zhao, X.; Wen, P.; Ma, D. Fe3O4 Nanoclusters Highly Dispersed on a Porous Graphene Support as an Additive for Improving the Hydrogen Storage Properties of LiBH4. *RSC Adv.* **2018**, *8*, 19353–19361. [CrossRef]
109. Gasnier, A.; Gennari, F.C. Graphene Entanglement in a Mesoporous Resorcinol-Formaldehyde Matrix Applied to the Nanoconfinement of LiBH4 for Hydrogen Storage. *RSC Adv.* **2017**, *7*, 27905–27912. [CrossRef]
110. Zhang, X.; Zhang, L.; Zhang, W.; Ren, Z.; Huang, Z.; Hu, J.; Gao, M.; Pan, H.; Liu, Y. Nano-synergy enables highly reversible storage of 9.2 wt% hydrogen at mild conditions with lithium borohydride. *Nano Energy* **2021**, *83*, 105839. [CrossRef]
111. Xia, Y.P.; Wei, S.; Huang, Q.; Li, J.Q.; Cen, X.H.; Zhang, H.Z.; Chu, H.L.; Sun, L.X.; Xu, F.; Huang, P.R. Facile synthesis of NiCo2O4-anchored reduced graphene oxide nanocomposites as efficient additives for improving the dehydrogenation behavior of lithium alanate. *Inorg. Chem. Front.* **2020**, *7*, 1257–1272. [CrossRef]
112. Palade, P.; Comanescu, C.; Radu, C. Synthesis of Nickel and Cobalt Ferrite-Doped Graphene as Efficient Catalysts for Improving the Hydrogen Storage Kinetics of Lithium Borohydride. *Materials* **2023**, *16*, 427. [CrossRef] [PubMed]
113. Li, Y.T.; Ding, X.L.; Zhang, Q.G. Self-Printing on Graphitic Nanosheets with Metal Borohydride Nanodots for Hydrogen Storage. *Sci. Rep.* **2016**, *6*, 1–10. [CrossRef] [PubMed]
114. Huang, Y.Q.; Xia, G.L.; Zhang, J.; Guo, Z.P.; Yu, X.B. Graphene-tailored molecular bonds for advanced hydrogen and lithium storage performance. *Energy Storage Mater.* **2019**, *17*, 178–185. [CrossRef]
115. Ren, Z.; Zhang, X.; Li, H.-W.; Huang, Z.; Hu, J.; Gao, M.; Pan, H.; Liu, Y. Titanium Hydride Nanoplates Enable 5 Wt% of Reversible Hydrogen Storage by Sodium Alanate below 80 °C. *Research* **2021**, *2021*, 9819176. [CrossRef]
116. Kumar, L.H.; Rao, C.V.; Viswanathan, B. Catalytic effects of nitrogen-doped graphene and carbon nanotube additives on hydrogen storage properties of sodium alanate. *J. Mater. Chem. A* **2013**, *1*, 3355–3361. [CrossRef]
117. Li, Y.T.; Fang, F.; Fu, H.L.; Qiu, J.M.; Song, Y.; Li, Y.S.; Sun, D.L.; Zhang, Q.G.; Ouyang, L.Z.; Zhu, M. Carbon nanomaterial-assisted morphological tuning for thermodynamic and kinetic destabilization in sodium alanates. *J. Mater. Chem. A* **2013**, *1*, 5238–5246. [CrossRef]
118. Do, H.W.; Kim, H.; Cho, E.S. Enhanced hydrogen storage kinetics and air stability of nanoconfined $NaAlH_4$ in graphene oxide framework. *RSC Adv.* **2021**, *11*, 32533–32540. [CrossRef]
119. Huang, Y.K.; Shao, H.X.; Zhang, Q.Y.; Zang, L.; Guo, H.N.; Liu, Y.F.; Jiao, L.F.; Yuan, H.T.; Wang, Y.J. Layer-by-layer uniformly confined Graphene-$NaAlH_4$ composites and hydrogen storage performance. *Int. J. Hydrogen Energy* **2020**, *45*, 28116–28122. [CrossRef]

120. Qian, Z.; Hudson, M.S.L.; Raghubanshi, H.; Scheicher, R.H.; Pathak, B.; Araujo, C.M.; Blomqvist, A.; Johansson, B.; Srivastava, O.N.; Ahuja, R. Excellent Catalytic Effects of Graphene Nanofibers on Hydrogen Release of Sodium alanate. *J. Phys. Chem. C* **2012**, *116*, 10861–10866. [CrossRef]
121. Wang, J.; Ebner, A.D.; Prozorov, T.; Zidan, R.; Ritter, J.A. Effect of graphite as a co-dopant on the dehydrogenation and hydrogenation kinetics of Ti-doped sodium aluminum hydride. *J. Alloys Compd.* **2005**, *395*, 252–262. [CrossRef]
122. Zheng, H.Y.; Ding, Z.Q.; Xie, Y.J.; Li, J.F.; Huang, C.K.; Cai, W.T.; Liu, H.Z.; Guo, J. Cerium hydride generated during ball milling and enhanced by graphene for tailoring hydrogen sorption properties of sodium alanate. *Int. J. Hydrogen Energy* **2021**, *46*, 4168–4180. [CrossRef]
123. Xu, L.Y.; Ge, Q.F. Effect of defects and dopants in graphene on hydrogen interaction in graphene-supported NaAlH4. *Int. J. Hydrogen Energy* **2013**, *38*, 3670–3680. [CrossRef]
124. Zhang, H.Y.; Xia, G.L.; Zhang, J.; Sun, D.L.; Guo, Z.P.; Yu, X.B. Graphene-Tailored Thermodynamics and Kinetics to Fabricate Metal Borohydride Nanoparticles with High Purity and Enhanced Reversibility. *Adv. Energy Mater.* **2018**, *8*, 1702975. [CrossRef]
125. Jeong, S.; Heo, T.W.; Oktawiec, J.; Shi, R.; Kang, S.Y.; White, J.L.; Schneemann, A.; Zaia, E.W.; Wan, L.F.; Ray, K.G.; et al. A Mechanistic Analysis of Phase Evolution and Hydrogen Storage Behavior in Nanocrystalline $Mg(BH_4)_2$ within Reduced Graphene Oxide. *ACS Nano* **2020**, *14*, 1745–1756. [CrossRef] [PubMed]
126. White, J.L.; Strange, N.A.; Sugar, J.D.; Snider, J.L.; Schneemann, A.; Lipton, A.S.; Toney, M.F.; Allendorf, M.D.; Stavila, V. Melting of Magnesium Borohydride under High Hydrogen Pressure: Thermodynamic Stability and Effects of Nanoconfinement. *Chem. Mater.* **2020**, *32*, 5604–5615. [CrossRef]
127. Wang, Z.M.; Tao, S.; Li, J.J.; Deng, J.Q.; Zhou, H.Y.; Yao, Q.R. The Improvement of Dehydriding the Kinetics of NaMgH3 Hydride via Doping with Carbon Nanomaterials. *Metal* **2017**, *7*, 9. [CrossRef]
128. Bhatnagar, A.; Pandey, S.K.; Shahi, R.R.; Hudson, M.S.L.; Shaz, M.A.; Srivastava, O.N. Synthesis, characterization and hydrogen sorption studies of mixed sodium-potassium alanate. *Cryst. Res. Technol.* **2013**, *48*, 520–531. [CrossRef]
129. Varin, R.A.; Bidabadi, A.S.; Polanski, M.; Biglari, M.; Stobinski, L. The effects of filamentary Ni, graphene and lithium amide (LiNH2) additives on the dehydrogenation behavior of mechano-chemically synthesized crystalline manganese borohydride (Mn(BH4)(2)) and its solvent filtration/extraction. *Mater. Res. Bull.* **2018**, *100*, 394–406. [CrossRef]
130. Li, N.; Du, Y.; Feng, Q.P.; Huang, G.W.; Xiao, H.M.; Fu, S.Y. A Novel Type of Battery-Supercapacitor Hybrid Device with Highly Switchable Dual Performances Based on a Carbon Skeleton/Mg2Ni Free-Standing Hydrogen Storage Electrode. *ACS Appl. Mater. Interfaces* **2017**, *9*, 44828–44838. [CrossRef]
131. Du, Y.; Li, N.; Zhang, T.L.; Feng, Q.P.; Du, Q.; Wu, X.H.; Huang, G.W. Reduced Graphene Oxide Coating with Anticorrosion and Electrochemical Property-Enhancing Effects Applied in Hydrogen Storage System. *ACS Appl. Mater. Interfaces* **2017**, *9*, 28980–28989. [CrossRef] [PubMed]
132. Zhang, J.G.; Zhu, Y.F.; Lin, H.J.; Liu, Y.N.; Zhang, Y.; Li, S.Y.; Ma, Z.L.; Li, L.Q. Metal Hydride Nanoparticles with Ultrahigh Structural Stability and Hydrogen Storage Activity Derived from Microencapsulated Nanoconfinement. *Adv. Mater.* **2017**, *29*, 1700760. [CrossRef] [PubMed]
133. Puszkiel, J.; Gasnier, A.; Amica, G.; Gennari, F. Tuning LiBH4 for Hydrogen Storage: Destabilization, Additive, and Nanoconfinement Approaches. *Molecules* **2020**, *25*, 163. [CrossRef] [PubMed]
134. Xia, G.L.; Tan, Y.B.; Wu, F.L.; Fang, F.; Sun, D.L.; Guo, Z.P.; Huang, Z.G.; Yu, X.B. Graphene-wrapped reversible reaction for advanced hydrogen storage. *Nano Energy* **2016**, *26*, 488–495. [CrossRef]
135. Chou, C.C.; Chen, B.H.; Lee, D.J. Hydrogen storage in a chemical hydride fuel system containing ammonia borane and Ni-Co/r-GO catalyst. *Enregy Procedia* **2014**, *61*, 142–145. [CrossRef]
136. Li, J.L.; Ren, X.Y.; Lv, H.; Wang, Y.Y.; Li, Y.F.; Liu, B. Highly efficient hydrogen production from hydrolysis of ammonia borane over nanostructured Cu@CuCoOx supported on graphene oxide. *J. Hazard. Mater.* **2020**, *391*, 122199. [CrossRef]
137. Champet, S.; van den Berg, J.; Szczesny, R.; Godula-Jopek, A.; Gregory, D.H. Nano-inclusion in one step: Spontaneous icetemplating of porous hierarchical nanocomposites for selective hydrogen release. *Sustain. Energy Fuels* **2019**, *3*, 396–400. [CrossRef]
138. Liu, X.R.; Wu, Y.F.; Wang, S.M.; Li, Z.N.; Guo, X.M.; Ye, J.H.; Jiang, L.J. Current progress and research trends on lithium amidoborane for hydrogen storage. *J. Mater. Sci.* **2020**, *55*, 2645–2660. [CrossRef]
139. Kumar, P.; Singh, S.; Hashmi, S.A.R.; Kim, K.H. MXenes: Emerging 2D materials for hydrogen storage. *Nano Energy* **2021**, *85*, 105989. [CrossRef]
140. Shi, L.J.; Liu, X.J.; Tuo, Y.X.; Xu, J.; Li, P.; Han, Y.F. Graphene CNT composite as catalyst support for microwave-assisted hydrogen releasing from liquid organic hydride. *Int. J. Hydrogen Energy* **2017**, *42*, 17403–17413. [CrossRef]
141. Ventura-Espinosa, D.; Carretero-Cerdan, A.; Baya, M.; Garcia, H.; Mata, J.A. Catalytic Dehydrogenative Coupling of Hydrosilanes with Alcohols for the Production of Hydrogen On-demand: Application of a Silane/Alcohol Pair as a Liquid Organic Hydrogen Carrier. *Chem.-Eur. J.* **2017**, *23*, 10815–10821. [CrossRef] [PubMed]
142. He, T.; Wang, X.H.; Liu, H.Z.; Gao, S.C.; Wang, Y.Y.; Li, S.Q.; Yan, M. Enhanced hydrogen desorption/absorption properties of magnesium hydride with CeF3@Gn. *Int. J. Hydrogen Energy* **2020**, *45*, 4754–4764. [CrossRef]
143. Li, Q.Y.; Qiu, S.Y.; Wu, C.Z.; Lau, T.; Sun, C.H.; Jia, B.H. Computational Investigation of MgH_2/Graphene Heterojunctions for Hydrogen Storage. *J. Phys. Chem. C* **2021**, *125*, 2357–2363. [CrossRef]
144. Sigal, A.; Rojas, M.I.; Leiva, E.P.M. Interferents for hydrogen storage on a graphene sheet decorated with nickel: A DFT study. *Int. J. Hydrogen Energy* **2011**, *36*, 3537–3546. [CrossRef]

145. Singh, S.; Bhatnagar, A.; Shukla, V.; Vishwakarma, A.K.; Soni, P.K.; Verma, S.K.; Shaz, M.A.; Sinha, A.S.K.; Srivastava, O.N. Ternary transition metal alloy FeCoNi nanoparticles on graphene as new catalyst for hydrogen sorption in MgH_2. *Int. J. Hydrogen Energy* **2020**, *45*, 774–786. [CrossRef]
146. Al-Msrhad, T.M.H.; Devrim, Y.; Uzundurukan, A.; Budak, Y. Investigation of hydrogen production from sodium borohydride by carbon nano tube-graphene supported PdRu bimetallic catalyst for PEM fuel cell application. *Int. J. Energy Res.* **2021**, *46*, 4156–4173. [CrossRef]

crystals

MDPI

Article

Influence of Graphene Oxide Concentration and Ultrasonication Energy on Fracture Behavior of Nano-Reinforced Cement Pastes

José D. Ríos [1], Carlos Leiva [2], Antonio Martinez de la Concha [1], María Pilar Ariza [1] and Héctor Cifuentes [1,*]

[1] Department of Continuum Mechanics and Structural Analysis, ETS de Ingeniería, Universidad de Sevilla, Camino de los Descubrimientos, s/n, 41092 Seville, Spain; jdrios@us.es (J.D.R.); delaconcha@us.es (A.M.d.l.C.); mpariza@us.es (M.P.A.)

[2] Department of Chemical and Environmental Engineering, ETS de Ingeniería, Universidad de Sevilla, Camino de los Descubrimientos, s/n, 41092 Seville, Spain; cleiva@us.es

* Correspondence: bulte@us.es

Abstract: The aim of this study is twofold. First, to assess the effect of the sonication process on the optimal dispersion of GO sheets for nanostructural reinforcement of cement pastes, as there is currently no clear criterion on this effect in the literature. For this purpose, in the first stage, the GO content in distilled water was fixed at 0.03% by weight, and the sheets were dispersed using different levels of ultrasonic energy, ranging from 0 J/mL to 2582 J/mL. In the second stage, to analyze the modification of pore structure due to the addition of GO sheets in different ratios (0–0.06% by weight) and its relationship with the mechanical and fracture properties of reinforced cement pastes. According to the results, it has been determined that the incorporation of GO sheets into the matrix alters the mechanical and fracture behavior, varying depending on matrix pore size and GO particle size. The addition of GO leads to a reduction in the average size of macropores (greater than 8 μm) of 13% for a dosage of 0.45% in weight and micropores (between 8 and 0.5 μm) in a 64% for the same composition with non-sonicated GO, although the total volume of pores in these ranges only decreased slightly. This reduction is more pronounced when the GO has not been sonicated and has larger particle size. Sonicated GO primarily modifies the range of capillary pores (<0.5 μm). The addition of GO with the highest degree of dispersion (465 nm) did not show significant improvements in compressive strength or Young's modulus, as the cement used contains a significant volume of macropores that are not substantially reduced in any composition. Adding 0.030% ultrasonicated GO achieved a 7.8% increase in fracture energy, while an addition of 0.045% resulted in a 13.3% decrease in characteristic length, primarily due to the effect of capillary and micropores.

Keywords: graphene oxide; sonication; cement paste; fracture energy; porosity; microstructure; compressive strength

Citation: Ríos, J.D.; Leiva, C.; Martinez de la Concha, A.; Ariza, M.P.; Cifuentes, H. Influence of Graphene Oxide Concentration and Ultrasonication Energy on Fracture Behavior of Nano-Reinforced Cement Pastes. *Crystals* **2024**, *14*, 707. https://doi.org/10.3390/cryst14080707

Academic Editor: Justina Gaidukevic

Received: 1 July 2024
Revised: 29 July 2024
Accepted: 29 July 2024
Published: 5 August 2024

1. Introduction

The fracture behavior of cement-based materials is fundamentally governed by the process of generation, coalescence, and propagation of cracks in the microstructure [1,2]. The amount, dimension, geometry, and distribution of defects in the material matrix have a remarkable influence on the cracking propagation and the subsequent failure mechanism at the macroscopic level [3–5]. In fact, the purpose of steel fiber-reinforced concretes, commonly reinforced with fiber lengths around 6 and 60 mm, is to act as barriers that hinder the crack propagation in the matrix and transmit stresses through a bridging effect linking the cracked zones, improving the fracture behavior. However, the effectiveness of steel fibers is limited to crack sizes fundamentally between microns and a few millimeters, depending on the type of fiber used [6,7]. Since cementitious materials are multi-scale flawed materials, which range in flaw size from a few nanometers to tens of millimeters, the steel fibers are not sufficiently effective for nanocracks, and so the reinforcement of the matrix by adding nanoparticles should be conducted for the enhancement of their mechanical properties and fracture behavior.

Improvements in the science of materials have made possible the use of nanoscale reinforcements, such as carbon nano-fibers [8,9], carbon nanotubes (CNTs) [10,11], graphene sheets [12–16], carbon dots [17], graphene oxide sheets (GO) [18–20], reduced graphene oxide sheets (rGO) [21–24], and chemically functionalized graphene oxide [25], with the aim of improving the overall mechanical properties of materials. Nanoparticles are highly effective additives for the modification of cement products, even at low concentrations ($\leq$1%). Nevertheless, there is a great controversy between results of GO reinforced cement products reported by different researchers [16,26–28]. Shamsaei et al. [29] and Lu et al. [30] also noticed these high variations in increase of performance. In fact, there are some results that show great improvement and tangible changes with addition of GO [27,28] and others that show indifference or even disadvantages [18,26].

Graphene oxide is an interesting material to improve the mechanical and fracture properties of cement-based materials due to its high tensile strength of 130 GPa and large specific area of 2630 (m^2/g) [31]. Additionally, GO can be well dispersed in water solutions due to the presence of oxygen functional groups [19,32]. GO has a high reactivity with cement products (portlandite and C-S-H), influencing chemical reactions and the subsequent macroscopic behavior of cement materials [33]. Thus, the addition of GO modifies the properties in a fresh and hardened state. On the other hand, GO sheets can fill nanopores and provide a higher density cement matrix. GO forms strong covalent bonds to improve structural interfaces and the performance of hydration products [28,34,35]. The addition of graphene oxide can strengthen the material matrix at a nanoscale level where conventional steel fibers are ineffective.

Studies of graphene oxide have been going on for 100 years since GO was described for the first time [36,37]. In the last twenty years, there has been an increase in studies in cementitious materials, pastes [36,38,39], mortars [13,31], and concretes [39–41]. Many studies have focused on rheological properties [37,42–44], hydration [19,21,35,37,45,46], and strength development [19,21,28,37], whereas other works have focused on the influence of the process of sonicating of nanoparticles [34,46,47]. Others have focused on mechanical properties [14,25,48], but few have focused on detailed studies of the evolution of fracture properties with the influence of sonicate content and energy on the microstructure at the nano- and micro-level.

Most usually studied matters are variations in mechanical strength, porosity, and microstructure when GO is added like reinforcement. It could be thought that discrepancies between results are caused by the existence of significant differences between experimental procedures [46–49]. Additionally, since GO sheets are micro or nanoscale, it is possible that differences in the manufacturing process apparently irrelevant for researchers might be determinant [50]. All production processes are different from one another, which has created substantial differences that can make the results of the studies incomparable. Some points of discordance or disagreement between different studies are the use of surfactants in the mixture, the revolution level of mixers, the ultrasonication energy transfer to the mixture to optimize the GO sheets dispersion, the control of temperature generated in the mixture during the mixing process, the use of tape or distilled water, and the water to cement ratio [24,35,44,46,49].

The objective of this paper is twofold. First, the assessment of the effect of the sonication process on the suitable dispersion of GO sheets. For this, the content of GO in distilled water was fixed to 0.03 wt% and the sheets were dispersed by ultrasonication. The solutions were submitted to different ultrasonication energy levels from 0 J/mL to 2582 J/mL following the same procedure in all cases. The influence of ultrasonication energy on the GO sheet sizes achieved was directly measured by a particle size analyzer and the subsequent suitable sonication energy was determined. Second, the effect of different contents of GO sheets on the capillarity, micro, and macro pore structure of the cement paste matrices were analyzed by mercury intrusion porosimetry (MIP). An assessment of the porosity ranges that are affected by the addition of GO sheets was discussed. Finally, the mechanical and fracture behavior of the various pastes were experimentally assessed. In this way, the mechanisms by which changes in the matrix's microstructure caused by the presence of GO sheets affects the mechanical and fracture properties of concrete have been thoroughly investigated.

2. Materials and Methods

2.1. Materials

Ordinary Portland cement, CEM I 52.5 R/SR, was used in this study. The particle size distribution of cement measured by laser diffraction is shown in Figure 1. As observed, the cement has a particle size distribution between 0.04 and 60 μm [51]. The mean and median particle sizes are 8.09 μm and 5.1 μm, respectively. The range of suitable particle sizes for the hydration process of cement is between 3 and 32 μm (red lines in Figure 1) [52,53]. This cement presents a percentage of 90.61% of particle sizes in the suitable range. Thus, the particle size is significantly suitable to achieve very efficient hydration reactions.

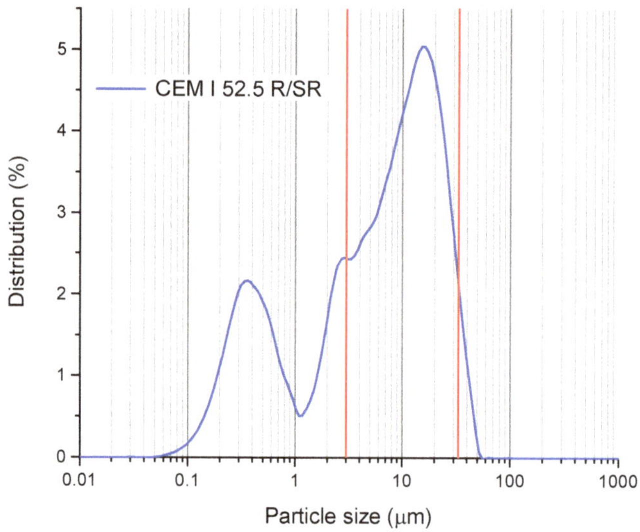

Figure 1. Particle size distribution of cement as measured by laser diffraction.

Additionally, the XRF chemical composition of cement is presented in Table 1. The predominant compound is CaO from the cement clinker.

Table 1. XRF chemical composition of cement [wt%].

CaO	Al_2O_3	SiO_2	SO_3	Fe_2O_3	K_2O	MgO	Na_2O	TiO_2
45.61	6.59	18.29	4.02	2.85	1.09	1	0.29	0.41
P_2O_5	Cl_2O_3	BaO	MnO_2	SrO	CuO	ZnO	LOI	Total
0.13	0.07	0.06	0.05	0.05	0.04	0.02	19.43	100

A dispersed GO in water with a concentration of 0.03 wt% was employed in this study from Graphenea. The elemental analysis in weight of GO presented is C (49–56%), H (0–1%), N (0–1%), S (2–4%) and O (41–50%). In Figure 2 the incremental volume percentage of GO sheet sizes in water suspension when it has not been yet sonicated or turned over is shown. As shown, the high variability in the results of the particle size distribution measurements indicate that the GO sheets in solution are randomly attached to each other, showing particle sizes between 0.42 and 10.1 μm. Three particle size measurements were performed on different mixtures with the same concentration to ensure the reliability of the results. It is worth noting the significant deviation of the particle size results when the dilution is not previously sonicated. The random adherence of the particles to each other makes the mixture totally inhomogeneous. This demonstrates how essential it is to properly sonicate the GO and to control the process. In this work, an exhaustive analysis of the ultrasonication process and its influence on the results was performed in Section 2.2.

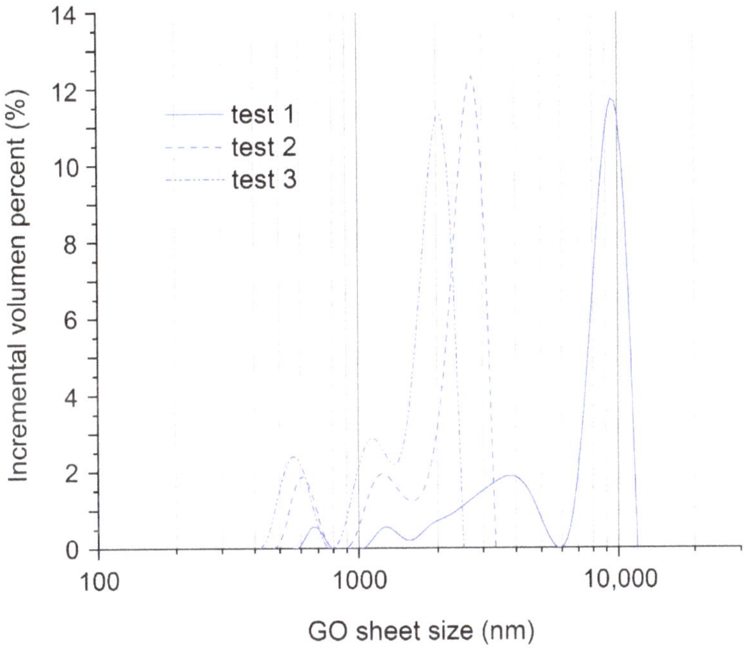

Figure 2. Particle size distribution of GO as measured by laser diffraction.

2.2. GO Dispersion Analyses

The GO dispersion analyses were performed for the same aqueous GO dispersion, 0.03 wt%. Although the effectiveness may be slightly affected for other concentrations, the study was carried out for the most common concentration according to other authors [19,54]. An exhaustive analysis of the influence of ultrasonication energy was performed to ensure the optimum dispersion of GO sheets. It should be clarified that the efficiency of the sonication process can vary depending on numerous factors such as the concentration of GO water solution, the sonication energy, the type and depth of the probe in the solution, the total volume of solution to sonicate, and the arrangement of that volume in the vessel [19,22,47]. Having the same volume in vessels of different width–height ratios modifies the conditions of sonication and its outcome. The heating generated during the sonication process must be controlled to avoid or reduce water evaporation that changes the initial GO–water concentration [19,55]. All this implies that the sonication time alone is not relevant data to be able to reproduce a sonication process as observed in some studies. It is much more relevant to talk about the sonication energy transferred to the fluid.

Various samples were prepared in order to determine the optimum ultrasonication energy to be transferred to the sample. The sonication process was carried out using a sonicator model Sonics Vibra-Cell vcx-750 (Llinars del Vallés, Catalonia, Spain) of 20 kHz with a 19 mm diameter probe. The amplitude set was 70% and pulse of intervals of 20″ to avoid overheating. The GO suspension was sonicated in a glass container 8 cm in diameter and 15 cm in height. Cold water at 6 °C was poured around the vessel to mitigate the heating of the sample and subsequent evaporation. The ultrasonication energy transferred to fluid can be seen in Table 2. Before ultrasonication, the solution was mixed by a tumbler for 24 h.

Table 2. Ultrasonication analyses parameters and results of dispersion.

Time (min)	Ultrasonication Energy (J/mL)	Min. GO Sheet Size (μm)	Max. GO Sheet Size (μm)	Mean GO Sheet Size (μm)
0	0	0.483 ± 14.7%	5.90 ± 87.7%	3.10 ± 74.4%
10	224	0.255 ± 3.4%	0.99 ± 0.3%	0.572 ± 0.1%
15	299	0.250 ± 0.0%	0.99 ± 0.3%	0.544 ± 0.5%
30	663	0.211 ± 0.0%	0.89 ± 0.0%	0.465 ± 0.1%
45	1204	0.199 ± 0.0%	0.99 ± 0.3%	0.492 ± 0.1%
60	1518	0.208 ± 12.1%	0.99 ± 0.3%	0.491 ± 1.1%
90	1831	0.203 ± 6.5%	1.03 ± 3.4%	0.536 ± 4.9%
120	2582	0.223 ± 0.0%	1.17 ± 6.8%	0.639 ± 8.4%

To determine the best GO dispersion according to the different ultrasonication energy levels, a direct measurement of the GO sheet size was measured by a particle size analyzer (Mastersizer 3000, Malvern, UK). Three samples under the same test conditions were carried out to determine the specific GO sheet size distribution. The results of the average GO sheet size distribution can be seen in Figure 2 and the values for mean, minimum and maximum GO sheet sizes and their standard deviations are presented in Table 2.

For the ultrasonication energy of 224 J/mL, the range of GO sheet sizes is 0.255–0.99 μm (Figure 3 and Table 2). This range is much smaller than the interval obtained for the reference sample without sonicating, 0.42–10 μm (Figure 2), so sonication is indispensable. The GO sheet size decreases from a mean value of 3.10 μm to 0.572 μm, a decrease of 81.5% (Table 2). What is even more relevant is that the standard deviation goes from 74.4%, without sonication, to 0.1% with 224 J/mL of ultrasonication energy. This factor is extremely important for the repeatability and reliability of characterization results. For the 299 J/mL ultrasonication energy, a slight shift of the maximum peak of the GO sheet size distribution curve towards the lower particle size zone is observed. This leads to a lower mean particle size, as can be seen in Table 2, from 0.572 to 0.544 μm. It can be seen how the decrease in relative terms of the mean GO sheet size is much smaller (4.9%) and how the deviation of the results, although larger than for 224 J/mL, is still very small (0.5%).

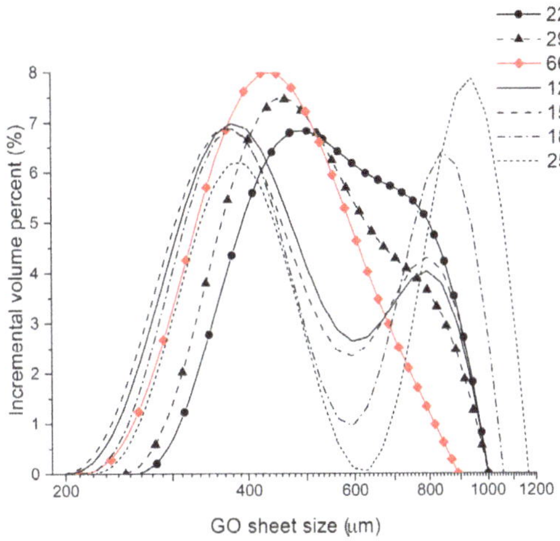

Figure 3. GO sheet size distribution in the 0.03 wt% suspension at different ultrasonication energy levels.

Regarding the ultrasonication energy results of 663 J/mL, this is the case for which the maximum GO sheet size is significantly smaller, 0.890 μm, and where the lower limit is one of the smallest, 0.211 μm. This is reflected in the lowest mean GO sheet size being reached, at 0.465 μm. All this indicates that this amount of energy transferred to solution is, between the values analyzed, the best GO dispersion. In addition, the deviation of results is the most reduced, so the pastes manufactured with this sonication energy should provide the most reliable results.

For sonication energies between 1204 and 2582 J/mL, the trend observed is increasing in mean GO sheet size from 0.492 to 0.639 μm, and the deviation of the results from 0.1 to 8.4%. It is worth noting that providing ultrasonication energy higher than 663 J/mL does not improve GO dispersion. Too-large power sonication results in a scissoring effect that breaks up the nanoparticles. Consequently, the power should be decreased as much as possible to guarantee a more homogeneous dispersion [56]. Based on the GO dispersion analysis performed, the most suitable ultrasonication energy has been set at 663 J/mL.

In Figure 4, the minimum, mean, and maximum values of GO sheet sizes and their deviation versus the ultrasonication energy transferred to suspension of 0.03 wt% (Table 2) is represented. As seen, infra-sonication and supra-sonication generate results of GO sheet sizes with higher deviation; this effect is more significant for the minimum and mean curves of particle size. On the other hand, a peak is clearly observed in the curves for the 663 J/mL energy case, which implies the smallest GO sheet size and the results with the lowest deviation.

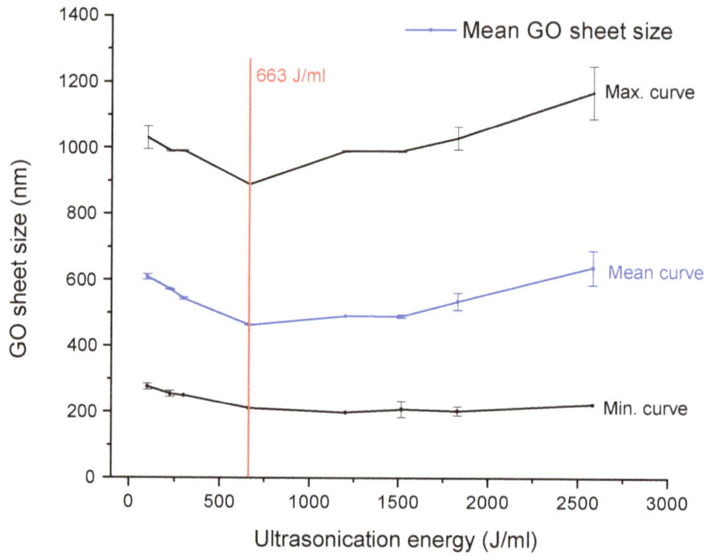

Figure 4. Evolution of GO sheet sizes with increasing ultrasonication energy for 0.030 wt%.

2.3. *Mix Proportions and Specimen Preparation*

In this work, five different types of cement pastes reinforced with GO have been examined. The first paste, without GO addition, was used as the reference mix. The remaining four pastes had GO contents by weight of cement of 0.015%, 0.030%, 0.045%, and 0.060%. For the GO-reinforced pastes, two batches of specimens were produced for each type of mix, with one batch to analyze the fracture behavior without sonication, 0 J/mL, and sonicated with 663 J/mL (suitable value determined in Section 2.2). A total of nine batches of specimens were manufactured. The mix proportions, nomenclature and ultrasonication energy transferred to the sample are shown in Table 3.

Table 3. Mix proportions, nomenclature, and sonication energy. Note: The generic nomenclature used in this work is P–X1–X2, in which P denotes paste, X1 denotes the GO concentration in percentage by cement weight, and X2 denotes the ultrasonication energy transferred to the solution used for manufacturing each paste.

	Cement (kg/m^3)	w/c Ratio	GO (%)	Ultrasonication Energy (J/mL)
P–0–0	1400	0.35	0	0
P–0.015–0	1400	0.35	0.015	0
P–0.015–663	1400	0.35	0.015	663
P–0.03–0	1400	0.35	0.03	0
P–0.03–663	1400	0.35	0.03	663
P–0.045–0	1400	0.35	0.045	0
P–0.045–663	1400	0.35	0.045	663
P–0.06–0	1400	0.35	0.06	0
P–0.06–663	1400	0.35	0.06	663

Each batch of specimens consists of three beams of 40 mm × 40 mm × 100 mm and three cubes of 40 mm per side. In total, 27 prismatic and 27 cubic specimens were manufactured. The GO–water solutions were properly sonicated just before the manufacturing process. Subsequently, the pastes were manufactured with a 5-liter mixer. First, the cement was added and mixed for 3 min, then the dispersed GO water solution was added and mixed for 5 min more. During this time, a fresh fluid paste was obtained and then poured into the molds. Next, the specimens were left for 24 h air curing, covered with plastic to reduce the loss of hydration. Afterwards, the specimens were demolded and cured in water at 20 °C for 27 days more. After this time, they were removed from the water and left to dry in the open air for 24 h. Finally, the experimental tests were carried out.

As it can be seen in Figure 5, The probe used was 19 mm (3/4″), following the manufacturer's recommendation based on the volume of mixture to be sonicated. The sonicator was placed in a soundproof box within a climate-controlled room, maintaining a temperature of 24 °C ± 2 °C. Additionally, the sonication container was placed inside another container with water to mitigate any abrupt temperature changes during sonication. Sonication was performed prior to starting the paste fabrication process. Once the aqueous solution with graphene oxide was sonicated, fabrication proceeded.

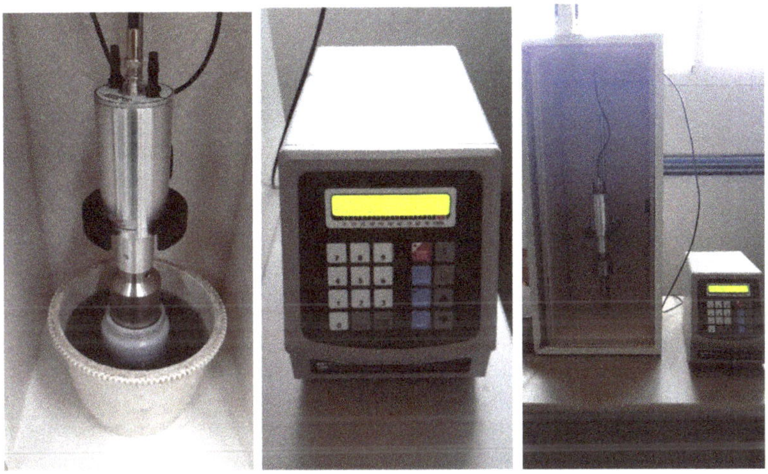

Figure 5. Probe of ultrasonic device.

3. Experimental Program

3.1. Fracture Behavior

Three notched prismatic specimens of 40 × 40 × 100 mm^3 were subjected to three-point bending tests to determine fracture energy using the RILEM work-of-fracture method [57]. Each sample was sawn with a notch to a depth ratio of one-sixth. The samples were fitted with a 10 mm-long LVDT to determine mid-span deflection and a clip gauge to determine crack mouth opening displacement (CMOD).

In addition, an inverse analysis using the non-linear hinge model [58–60] was carried out to obtain the values of the bilinear tension softening graph (Figure 6). The bilinear graph is defined by the failure stress (f_t), the displacement at failure (w_u), the slope of the first branch (a_1), and the slope of the second branch (a_2) [61].

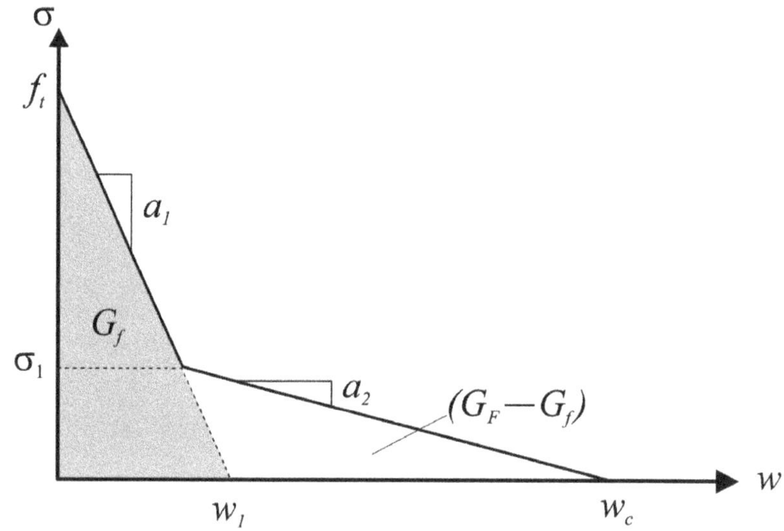

Figure 6. Bilinear tension softening graph.

3.2. *Compressive Strength and Young's Modulus*

The compressive strength was determined from experimental tests on 40 mm cubic specimens following the guidelines of the standard UNE-EN 12390-3 [62]. Three specimens for each type of paste and ultrasonication level were tested. Nine different paste batches were tested (Table 3); thereby, 27 tests were carried out in total. To determine Young's modulus, it was obtained indirectly from the load-CMOD results reported from experimental three-point bending tests on $40 \times 40 \times 100$ mm^3 prismatic specimens notched at one sixth of the height according to the standard RILEM TCM-85 [57]. The three-point bending tests were carried out to also determine the fracture properties of the pastes and will be described in more detail in the following subsection. The method for determining Young's modulus was that described by Swartz et al. [63], which relates the initial slope of the experimental load-CMOD curve to the elasticity modulus of the material. Three tests for each type of paste and ultrasonication level were assessed.

3.3. *Mercury Instrusion Porosimetry (MIP)*

High-pressure mercury intrusion porosimetry (MIP) was used to determine the porosity and pore size distribution in the range between 7 nm and 100 μm. The samples were obtained from the three-point bending specimens. Three samples per composition were carried out.

3.4. *Stereomicroscope*

A SMZ25 stereomicroscope (Nikon Instruments Inc, Melville, New York, USA) was used. The instrument is connected to a computer running NIS Elements BR software, which was used to capture the images. Small specimens measuring 20 mm × 40 mm × 40 mm were fabricated for this test.

3.5. *SEM*

A TENEO equipment (FEI, (Thermo Fisher Scientific, Waltham, MA, USA)) with a field emission scanning electron microscope operated at 5 kV and equipped with an energy-dispersive X-ray spectroscopy system was used.

4. Results

4.1. Porosity of Cement Matrix

The MIP analysis provides useful information on pores between 7 nm and 100 μm. Figure 7 shows the pore size distribution obtained for all pastes with different GO concentrations and ultrasonication levels. Figure 6 is divided into four subfigures corresponding to the four different GO concentrations (0.015%, 0.030%, 0.045%, and 0.060%). In each subfigure, the pore distribution curves of the GO pastes, both sonicated and non-sonicated, as well as that of the reference paste (without GO content) are shown. In this way, the effect of GO concentration and ultrasonication can be compared in each subfigure. The most significant data obtained from the curves are shown in Table 4.

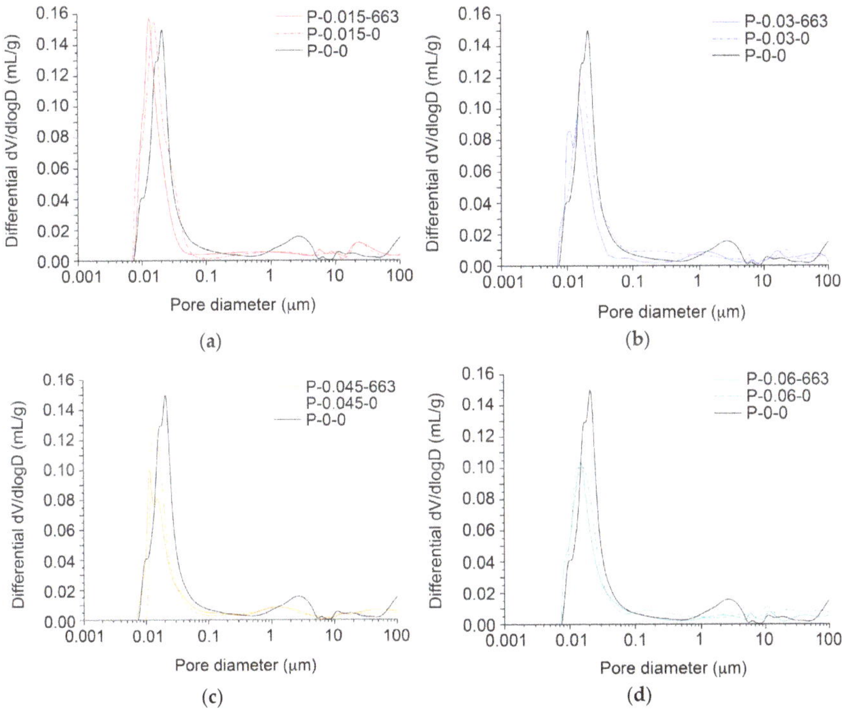

Figure 7. Differential porosity curves for (**a**) P–0.015, (**b**) P–0.030, (**c**) P–0.045, and (**d**) P–0.060.

Table 4. Pore volume for each range of each paste and ultrasonication level.

	Pore Volume (mL/g)		
	<500 nm	500–8000 nm	>8000 nm
P–0–0	1.70 ± 0.05	0.076 ± 0.004	0.062 ± 0.004
P–0.015–0	1.59 ± 0.04	0.049 ± 0.003	0.060 ± 0.005
P–0.015–30	1.26 ± 0.03	0.044 ± 0.003	0.069 ± 0.004
P–0.03–0	1.19 ± 0.04	0.049 ± 0.006	0.061 ± 0.005
P–0.03–30	1.15 ± 0.04	0.041 ± 0.003	0.066 ± 0.005
P–0.045–0	0.85 ± 0.01	0.028 ± 0.002	0.056 ± 0.004
P–0.045–30	0.94 ± 0.01	0.022 ± 0.002	0.061 ± 0.004
P–0.06–0	1.21 ± 0.02	0.047 ± 0.003	0.099 ± 0.006
P–0.06–30	1.11 ± 0.03	0.045 ± 0.005	0.065 ± 0.005

As shown in Figure 7, there are three different pore ranges: capillary pores (less than 0.5 μm in diameter), micropores (0.5–8 μm in diameter), and macropores (greater than 8 μm in diameter). The three main parameters in the pore distribution are as follows: (1) the GO pore size, (2) GO percentage, and (3) pore volume of P–0–0. Figure 5 shows that most of the pores are capillary pores in all dosages, especially in the 0.007–0.1 μm range. P–0–0 has a higher amount of macropores compared to other CEM I 52.5 R/SR cements used in previous studies with GO addition [64].

Regarding macropores, as can be seen in Table 4, the addition of GO can divide large pores into several smaller ones [65], but the total pore volume of macropores (greater than 8 μm) is very similar for additions of 0, 0.15%, and 0.03% of non-sonicated GO. The use of sonicated GO results in a smaller decrease in the total volume of macropores compared with non-sonicated GO (Figure 8a,b). The minimum macropores are reached in P–0.045–0, with 87% of P–0.045–0. The two compositions with 0.06% (with the highest proportion of GO) produce a decrease in the workability of the cement paste, leading to an increase in the macropores present [66,67], as can be seen in Figure 8c.

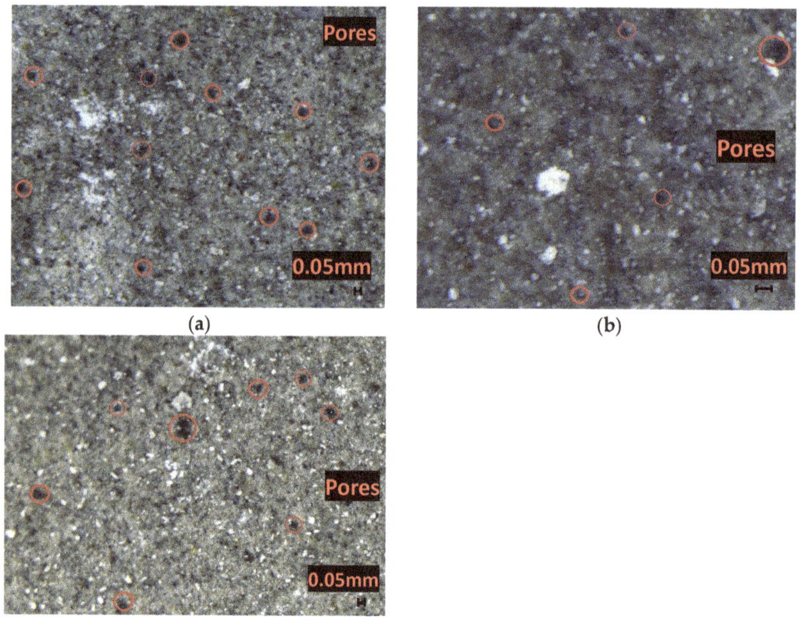

Figure 8. Images from stereomicroscope for (**a**) P–0–0, (**b**) P–0.045–0, (**c**) P–0.06–0.

Regarding micropores, the addition of GO produces a reduction in micropores, and this reduction is greater with higher GO addition because the micropores present in P–0–0 are divided into smaller ones P–0.045–0 (Figure 9a,b), many of which fall into the higher range of capillary pores (0.1–0.5 μm). The addition of sonicated GO results in an even greater reduction in this range of pores than non-sonicated pastes. P–0.045–0 and P–0.045–30 show 36% and 28% of the microporosity of P–0–0, respectively. For the two compositions with 0.06% (with the highest proportion of GO), there is an increase in micropores in the paste, due to the reduced workability of the cement paste.

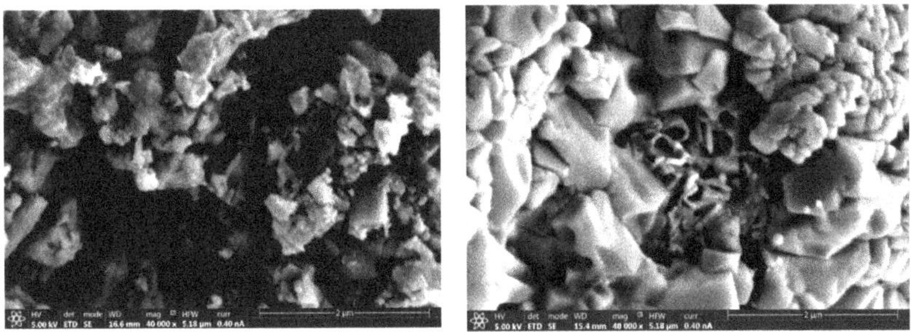

(a) (b)

Figure 9. Image from SEM for (**a**) P–0–0 and (**b**) P–0.045–0.

Regarding capillary pores, the addition of GO, whether sonicated or non-sonicated, reduces the volume of pores present in this range by 0.15%, 0.03%, and 0.045%. Although in the range (0.1–0.5 μm), the pore volume is increased by the division of micropores into smaller pores; the reduction in pore volume smaller than 0.1 μm is especially significant, so much so that even the peak porosity shifts to smaller sizes in all compositions. This is due to the filling effect of small pores by GO, probably producing nanopores smaller than 7 nm. P–0.045–0 and P–0.045–30 present 50% and 55% of the microporosity of P–0–0, respectively. For both 0.06% compositions, the capillary pores are increased due to the reduced workability of the paste. This behavior in the variation of capillary pores is similar to that found in previous works [62].

Considering the results of the MIP test, it is determined that effective sonication involves tailoring the dimensions of the GO sheets to match the pore sizes of the matrix, rather than striving for the smallest possible GO sheet size. When sonication results in a particle size that is excessively small for the pore size of the matrix being reinforced, the impact of GO on mechanical and fracture properties becomes insignificant, as will be discussed in subsequent sections. It explains why numerous studies have demonstrated significant enhancements in mechanical properties following GO addition [14,26,48,65], while others have reported contradictory findings [46,50,66].

4.2. Compressive Strength and Young's Modulus

Figure 10 shows the average values of compressive strength, f_c, and Young's modulus, E_c, for each GO content and ultrasonication level assessed. As shown in Figure 10a, the compressive strength of pastes reinforced with non-sonicated GO solution (0 J/mL) achieves mean values similar to the reference paste (no GO) for all evaluated GO contents (0.015%, 0.030%, 0.045%, and 0.060%). For the sonicated pastes, no significant improvement in compressive strength is observed for any analyzed GO concentration. The explanation lies in the fact that the volume of macropores does not change significantly for any of the GO concentrations (see Table 4). During the compressive strength test, macropores determine the preferential fracture path of the matrix, as cracks propagate between macropores. Since the addition of GO, with the sonication performed in this study, affects the size of capillary pores and these do not influence compressive strength [67], no improvements in this property are obtained.

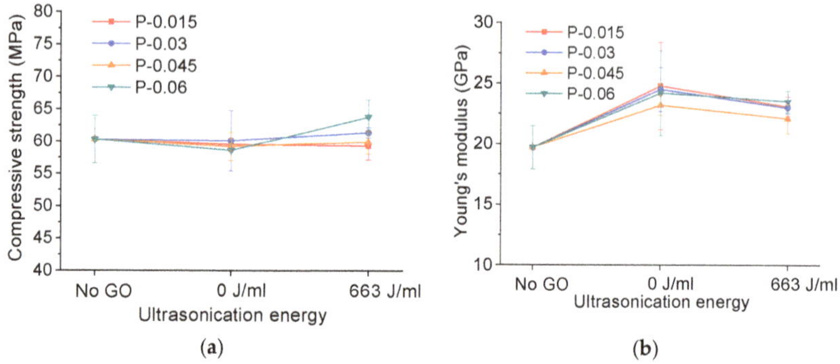

Figure 10. (**a**) Compressive strength and (**b**) Young's modulus for each GO content and ultrasonication.

With respect to the Young's modulus, all the pastes showed an increase in Young's modulus with an increase between 7.6–25.8% compared to the reference paste. According to Kendall and coauthors, this property is primarily influenced by the existence of a large number of pores with sizes in the nanometer range [68,69], although more research is needed in this area. Table 4 shows an analysis of the pore size distribution below the maximum measured particle size for all samples.

For the reference paste, it is observed that having a significant pore size distribution smaller than 500 nm and between 500–8000 nm (6.3%), in comparison to GO-reinforced pastes, results in lower Young's modulus values. Focusing on the sonicated and non-sonicated GO-reinforced pastes, no clear trend is observed. We believe that the way pores are distributed and connected in the material plays a more relevant role than a specific size range, which is beyond the scope of this study.

4.3. Fracture Behavior

The size-independent fracture energy was calculated using the RILEM work-of-fracture method and experimental three-point bending tests. Additionally, bilinear tension softening diagrams ($\sigma_t - w$) were obtained using the non-linear hinge model described in [69]. The average values of size-independent fracture energy obtained from three experimental tests by mix are shown in Table 5.

Table 5. Size independent fracture energy of each GO content and ultrasonication level.

Mix	Fracture Energy, G_F (N/m)
P–0–0	20.5 ± 3.7
P–0.015–0	24.9 ± 1.5
P–0.015–663	21.8 ± 1.5
P–0.030–0	22.3 ± 2.4
P–0.030–663	21.9 ± 1.6
P–0.045–0	21.6 ± 2.3
P–0.045–663	19.6 ± 1.9
P–0.060–0	21.9 ± 4.4
P–0.060–663	22.2 ± 2.7

In cement pastes, the crack fracture mechanism is a critical aspect that influences the material's structural integrity and durability. The process of crack formation and propagation in cementitious materials can be complex, involving several microstructural factors and environmental conditions [70]. Cracks in cement paste typically initiate at points of stress concentration, such as pores, microcracks, or inclusions within the matrix. Once a crack initiates, its propagation is governed by the fracture mechanics of the material.

Based on the fracture energy results, it is observed that the addition of GO sheets does not generate a clear trend towards improving the overall fracture energy. Nonetheless, a trend is observed with the pore distribution of the pastes, which is partially related to the addition of graphene sheets. The lowest fracture energy value is obtained for the P–0.045–0 paste, which had the lowest value for capillary pores and macropores, and the second lowest in the micropore range (Table 4). The lower overall fracture energy is due to the fact that a denser cement paste matrix exhibits more brittle behavior.

On the other hand, the highest total fracture energy value was achieved for P–0.015–0 (21.4%). This paste exhibits the highest number of capillary pores and micropores, except for the reference paste (Table 4). A greater number of capillary pores and micropores can lead to a less brittle matrix [66], as observed in this case. We will proceed with a more detailed analysis of the remaining properties that influence fracture behavior.

On the other hand, a two-dimensional FE model of the three-point bending test was developed and fitted to the experimental results, as described in Section 3.1, to identify the parameters that define the cohesive laws of each paste. Figure 11 shows the experimental and FE mean load-displacement curves for P–0.045–663. The minimum acceptable requirements for numerical fitting of the cohesive law parameters are an R-squared value greater than 0.95 of the FE results in relation to the experimental load-displacement curves. The cohesive laws and their parameters obtained with this procedure are shown in Figure 12 and Table 6, respectively.

Table 6. Bilinear cohesive law parameters of each GO content and ultrasonication level.

	f_t (MPa)	σ_1 (MPa)	a_1 (mm^{-1})	a_2 (mm^{-1})	w_1 (mm)	w_u (mm)	G_f (N/m)	G_F–G_f (N/m)	E_c (GPa)	l_{ch} (m)
P–0–0	5.3	0.16	771.6	6.7	0.0068	0.030	17.85	1.8	19.7	0.015
P–0.01–0	4.3	0.13	595.9	4.6	0.0072	0.035	15.48	1.8	24.8	0.033
P–0.015–663	4.5	0.18	568.4	5.6	0.0075	0.040	16.87	3.8	23.1	0.025
P–0.030–0	4.0	0.36	527.5	15.6	0.0076	0.030	15.20	4.0	24.5	0.034
P–0.030–663	5.5	0.18	771.6	6.7	0.0070	0.030	19.25	2.2	23.0	0.017
P–0.045–0	4.5	0.27	660.9	10.5	0.0068	0.032	15.30	3.4	22.1	0.024
P–0.045–663	5.7	0.17	921.5	7.1	0.0062	0.030	17.67	2.0	21.2	0.013
P–0.060–0	4.2	0.25	660.9	10.5	0.0067	0.032	14.07	3.2	24.2	0.030
P–0.060–663	4.5	0.27	660.9	10.5	0.0068	0.032	15.30	3.4	23.5	0.026

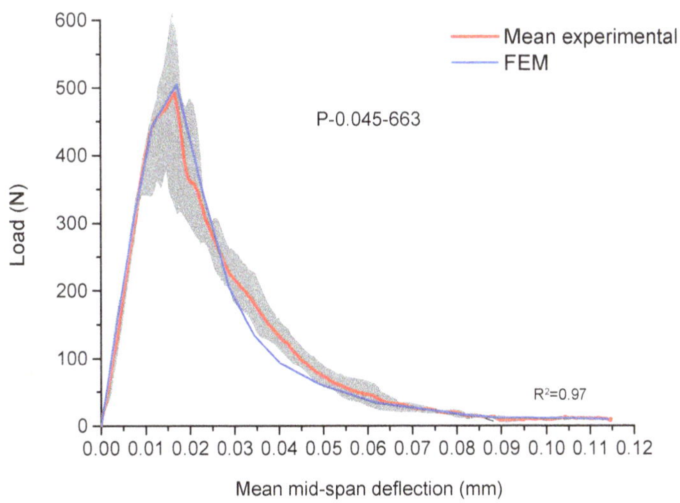

Figure 11. Experimental region, mean experimental, and FEM load vs. mid-span deflection curves of P–0.045–663 with squared error of comparison.

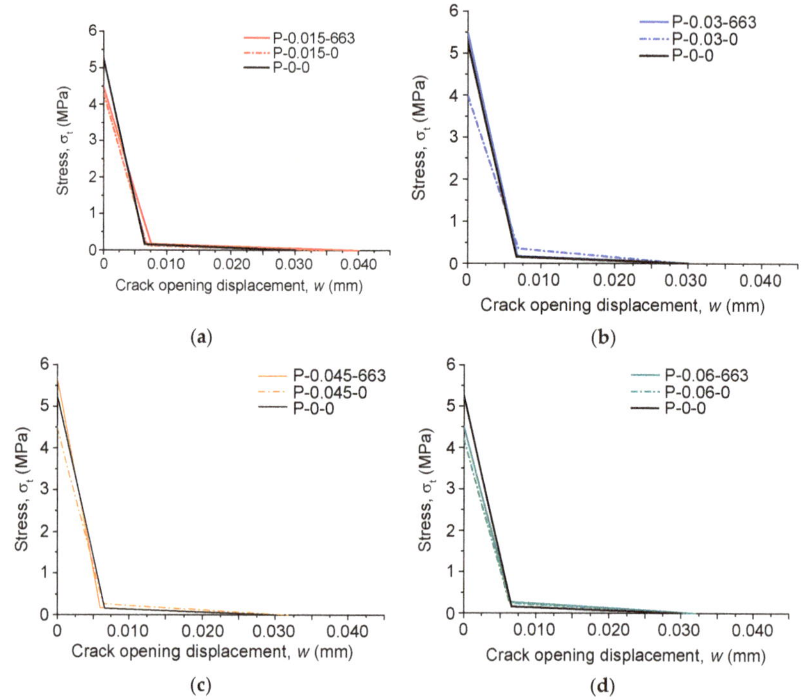

Figure 12. Bilinear stress-crack opening displacement law: (**a**) 0.015%, (**b**) 0.030%, (**c**) 0.045%, and (**d**) 0.060% GO content.

Concrete fracture necessitates the formation of the well-known Fracture Process Zone (FPZ) [56], involving micro-cracking, coalescence, crack branching, and frictional interlocking as indicated by the weak interfacial transition zone (ITZ) between aggregates and cement paste [71]. Based on Hillerborg's fictitious crack model [72], cohesive normal stresses develop within the FPZ prior to a pre-existing crack, with a value equal to the material's tensile strength at the crack tip, followed by a descending part (softening) until the stresses reach zero at the end of the FPZ. The distribution of these cohesive stresses is typically represented by a bilinear softening diagram [73] in which the normal stress, σ, is a function of the crack opening width, w, in the FPZ (Figure 6). The first linear branch of the softening diagram is primarily related to microcracking, whereas the second linear branch is a result of the frictional aggregate interlock [74,75]. It is worth noting that the fracture energy of the second branch, GF-Gf (Table 6), will not have a significant impact on the total fracture energy, as the cement pastes do not contain aggregates nor fibers [76]. Cohesive models can also be used to investigate the fracture behavior of this type of quasi-brittle material when a crack (or notch) already exists [77].

In the case of tensile strength, the maximum value was obtained for the sample P–0.045–663, with an increase of 7.5% compared to the reference paste (Figure 12). This increase is due to the fact that this sample has the lowest amount of micropores and the second lowest amount of capillary and macropores (Table 4). On the other hand, the sample P–0.030–663 showed the second highest tensile strength, with an increase of 3.7% compared to the reference paste. In this case, the lowest presence of pores is found in the micropores. Therefore, it can be confirmed that micropores have a greater impact on the tensile strength of the paste. Additionally, if this paste has a low amount of capillary and macropores, this property is further enhanced.

Regarding the fracture energy (Gf), it is observed that the highest value was obtained for the sample P–0.030–663. This matrix presents a balanced combination of low capillary pore and micropore values, along with high tensile strength. This combination ensures that once the tensile strength of the material is reached, the fracture is not as abrupt, and a greater amount of energy is required to fracture the paste.

As all specimens were the same size, it is feasible to carry out a ductility analysis primarily centered on the results of the characteristic length [76,77], so that the higher the l_{ch}, the higher the ductility of the paste. The characteristic length may be written in terms of three parameters:

$$l_{ch} = \frac{E_c G_F}{f_t^2} \quad (1)$$

where E_c is Young's modulus, G_F is the fracture energy, and f_t is the tensile strength.

The lowest value was obtained for the P–0.045–663 paste, indicating that it has a more brittle fracture process zone. This is because it was the paste with the lowest amount of capillary pores and micropores, combined with the highest tensile strength. It can be observed that, in general, non-sonicated pastes presented a greater characteristic length (P–0.015, P–0.030, and P–0.060), coinciding with lower tensile strength and a higher amount of capillary pores and micropores. Therefore, it can be established that there is a direct relationship between the amount of capillary pores and micropores and the characteristic length of the paste.

5. Conclusions

In this work, the effects of the addition of GO solutions with different concentrations on the microstructure of cement paste and how it affects its mechanical and fracture behavior have been assessed. From the results, the following conclusions can be drawn:

- Proper ultrasonication should adjust the size of the GO to match the pore sizes of the original paste, rather than making the GO as small as possible. If sonication produces a particle size that is too small for the pore size of the matrix intended for reinforcement, the effect of the GO on porosity will be negligible.
- The P–0.03–663 exhibits the highest fracture energy at 7.8% compared to non-reinforced paste. This is primarily attributed to the optimal combination of capillary and micropores, resulting in elevated tensile strength and consequently higher fracture energy.
- The lowest characteristic length was observed for the P–0.045–663 fundamentally influenced by the reduced combination of capillary pores and micropores, which leads to a more brittle matrix.
- The tensile strength of the material is more influenced by the micropores than by the capillary pores; it is higher when the GO is sonicated, and the maximum value was obtained for the sample P–0.045–663.

Author Contributions: Conceptualization, M.P.A. and C.L.; methodology, J.D.R. and H.C.; validation, A.M.d.l.C. and J.D.R.; formal analysis, J.D.R. and C.L.; investigation, J.D.R. and A.M.d.l.C.; resources, M.P.A. and H.C.; data curation, J.D.R. and C.L.; writing—original draft preparation, C.L. and J.D.R.; writing—review and editing, H.C. and M.P.A.; project administration, M.P.A. and H.C. All authors have read and agreed to the published version of the manuscript.

Funding: This research has been funded by the "Consejería de Economía y Conocimiento" of the Andalusian Government under grant number US-1266248 and by the "Ministerio de Ciencia e Innovación" of the Spanish Government under grant number PID2019-110928RB-C33.

Data Availability Statement: The original contributions presented in the study are included in the article; further inquiries can be directed to the corresponding author.

Conflicts of Interest: The authors declare no conflicts of interest.

References

1. Jiang, C.; Fan, K.; Wu, F.; Chen, D. Experimental study on the mechanical properties and microstructure of chopped basalt fibre reinforced concrete. *Mater. Des.* **2014**, *58*, 187–193. [CrossRef]

2. Van Mier, J.G.M.; Van Vliet, M.R.A. Influence of microstructure of concrete on size/scale effects in tensile fracture. *Eng. Fract. Mech.* **2003**, *70*, 2281–2306. [CrossRef]
3. Ren, J.; Lai, Y. Study on the durability and failure mechanism of concrete modified with nanoparticles and polypropylene fiber under freeze-thaw cycles and sulfate attack. *Cold Reg. Sci. Technol.* **2021**, *188*, 103301. [CrossRef]
4. Ríos, J.D.; Leiva, C.; Ariza, M.P.; Seitl, S.; Cifuentes, H. Analysis of the tensile fracture properties of ultra-high-strength fiber-reinforced concrete with different types of steel fibers by X-ray tomography. *Mater. Des.* **2019**, *165*, 107582. [CrossRef]
5. Ríos, J.D.; Cifuentes, H.; Leiva, C.; Seitl, S. Analysis of the mechanical and fracture behavior of heated ultra-high-performance fiber-reinforced concrete by X-ray computed tomography. *Cem. Concr. Res.* **2019**, *119*, 77–88. [CrossRef]
6. Cong, X.; Qiu, T.; Xu, J.; Liu, X.; Wang, L.; Wang, Y.; Chen, C.; Zhao, L.; Xing, C.; Tan, Y. Study on the effectiveness of fibre reinforcement on the engineering performance of foamed concrete. *Case Stud. Constr. Mater.* **2022**, *16*, e01015. [CrossRef]
7. Zhang, B.; Feng, Y.; Xie, J.; He, J.; Zhang, Y.; Cai, C.; Huang, D.; Li, L. Effects of fibres on ultra-lightweight high strength concrete: Dynamic behaviour and microstructures. *Cem. Concr. Compos.* **2022**, *128*, 104417. [CrossRef]
8. Puentes, J.; Barluenga, G.; Palomar, I. Effects of nano-components on early age cracking of self-compacting concretes. *Constr. Build. Mater.* **2014**, *73*, 89–96. [CrossRef]
9. Chen, J.; Zhao, D.; Ge, H.; Wang, J. Graphene oxide-deposited carbon fiber/cement composites for electromagnetic interference shielding application. *Constr. Build. Mater.* **2015**, *84*, 66–72. [CrossRef]
10. Hawreen, A.; Bogas, J.A. Creep, shrinkage and mechanical properties of concrete reinforced with different types of carbon nanotubes. *Constr. Build. Mater.* **2019**, *198*, 70–81. [CrossRef]
11. Liew, K.M.; Kai, M.F.; Zhang, L.W. Carbon nanotube reinforced cementitious composites: An overview. *Compos. Part A Appl. Sci. Manuf.* **2016**, *91*, 301–323. [CrossRef]
12. Wang, B.; Zhao, R. Effect of graphene nano-sheets on the chloride penetration and microstructure of the cement based composite. *Constr. Build. Mater.* **2018**, *161*, 715–722. [CrossRef]
13. e Silva, R.A.; de Castro Guetti, P.; da Luz, M.S.; Rouxinol, F.; Gelamo, R.V. Enhanced properties of cement mortars with multilayer graphene nanoparticles. *Constr. Build. Mater.* **2017**, *149*, 378–385. [CrossRef]
14. Li, G.; Yuan, J.B.; Zhang, Y.H.; Zhang, N.; Liew, K.M. Microstructure and mechanical performance of graphene reinforced cementitious composites. *Compos. Part A Appl. Sci. Manuf.* **2018**, *114*, 188–195. [CrossRef]
15. Xu, J.; Zhang, D. Pressure-sensitive properties of emulsion modified graphene nanoplatelets/cement composites. *Cem. Concr. Compos.* **2017**, *84*, 74–82. [CrossRef]
16. Han, B.; Zheng, Q.; Sun, S.; Dong, S.; Zhang, L.; Yu, X.; Ou, J. Enhancing mechanisms of multi-layer graphenes to cementitious composites. *Compos. Part A Appl. Sci. Manuf.* **2017**, *101*, 143–150. [CrossRef]
17. He, H.; Shuang, E.; Lu, D.; Hu, Y.; Yan, C.; Shan, H.; He, C. Deciphering size−induced influence of carbon dots on mechanical performance of cement composites. *Constr. Build. Mater.* **2024**, *425*, 136030. [CrossRef]
18. Pan, Z.; He, L.; Qiu, L.; Korayem, A.H.; Li, G.; Zhu, J.W.; Collins, F.; Li, D.; Duan, W.H.; Wang, M.C. Mechanical properties and microstructure of a graphene oxide-cement composite. *Cem. Concr. Compos.* **2015**, *58*, 140–147. [CrossRef]
19. Li, X.; Habibnejad, A.; Li, C.; Liu, Y.; He, H.; Sanjayan, J.G. Incorporation of graphene oxide and silica fume into cement paste: A study of dispersion and compressive strength. *Constr. Build. Mater.* **2016**, *123*, 327–335. [CrossRef]
20. Lv, S.; Ma, Y.; Qiu, C.; Sun, T.; Liu, J.; Zhou, Q. Effect of graphene oxide nanosheets of microstructure and mechanical properties of cement composites. *Constr. Build. Mater.* **2013**, *49*, 121–127. [CrossRef]
21. Jing, G.; Wu, J.; Lei, T.; Wang, S.; Strokova, V.; Nelyubova, V.; Wang, M.; Ye, Z. From graphene oxide to reduced graphene oxide: Enhanced hydration and compressive strength of cement composites. *Constr. Build. Mater.* **2020**, *248*, 118699. [CrossRef]
22. Chintalapudi, K.; Mohan, R.; Pannem, R. An intense review on the performance of Graphene Oxide and reduced Graphene Oxide in an admixed cement system. *Constr. Build. Mater.* **2020**, *259*, 120598. [CrossRef]
23. Yan, S.; He, P.; Jia, D.; Yang, Z.; Duan, X.; Wang, S.; Zhou, Y. Effect of reduced graphene oxide content on the microstructure and mechanical properties of graphene–geopolymer nanocomposites. *Ceram. Int.* **2016**, *42*, 752–758. [CrossRef]
24. Jing, G.; Ye, Z.; Wu, J.; Wang, S.; Cheng, X.; Strokova, V. Introducing reduced graphene oxide to enhance the thermal properties of cement composites. *Cem. Concr. Compos.* **2020**, *109*, 103559. [CrossRef]
25. Wang, M.; Yao, H.; Wang, R.; Zheng, S. Chemically functionalized graphene oxide as the additive for cement–matrix composite with enhanced fluidity and toughness. *Constr. Build. Mater.* **2017**, *150*, 150–156. [CrossRef]
26. Qian, Y.; Abdallah, M.Y.; Kawashima, S. Characterization of Cement-Based Materials Modified with Graphene-Oxide. In *Nanotechnology in Construction*; Springer: Berlin/Heidelberg, Germany, 2015; pp. 259–264. [CrossRef]
27. Zhao, L.; Guo, X.; Liu, Y.; Ge, C.; Chen, Z.; Guo, L.; Shu, X.; Liu, J. Investigation of dispersion behavior of GO modified by different water reducing agents in cement pore solution. *Carbon* **2018**, *127*, 255–269. [CrossRef]
28. Kang, D.; Seo, K.S.; Lee, H.Y.; Chung, W. Experimental study on mechanical strength of GO-cement composites. *Constr. Build. Mater.* **2017**, *131*, 303–308. [CrossRef]
29. Shamsaei, E.; de Souza, F.B.; Yao, X.; Benhelal, E.; Akbari, A.; Duan, W. Graphene-based nanosheets for stronger and more durable concrete: A review. *Constr. Build. Mater.* **2018**, *183*, 642–660. [CrossRef]
30. Lu, Z.; Chen, B.; Leung, C.K.Y.; Li, Z.; Sun, G. Aggregation size effect of graphene oxide on its reinforcing efficiency to cement-based materials. *Cem. Concr. Compos.* **2019**, *100*, 85–91. [CrossRef]

31. Murugan, M.; Santhanam, M.; Sen Gupta, S.; Pradeep, T.; Shah, S.P. Influence of 2D rGO nanosheets on the properties of OPC paste. *Cem. Concr. Compos.* **2016**, *70*, 48–59. [CrossRef]
32. Guan, X.; Yu, L.; Li, H. Experimental study on fracture mechanics of cementitious materials reinforced by graphene oxide—Silica nanocomposites. *Constr. Build. Mater.* **2022**, *325*, 126758. [CrossRef]
33. Zhao, L.; Guo, X.; Song, L.; Song, Y.; Dai, G.; Liu, J. An intensive review on the role of graphene oxide in cement-based materials. *Constr. Build. Mater.* **2020**, *241*, 117939. [CrossRef]
34. Li, X.; Liu, Y.M.; Li, W.G.; Li, C.Y.; Sanjayan, J.G.; Duan, W.H.; Li, Z. Effects of graphene oxide agglomerates on workability, hydration, microstructure and compressive strength of cement paste. *Constr. Build. Mater.* **2017**, *145*, 402–410. [CrossRef]
35. Yang, H.; Cui, H.; Tang, W.; Li, Z.; Han, N.; Xing, F. Composites: Part A A critical review on research progress of graphene/cement based composites. *Compos. Part A Appl. Sci. Manuf.* **2017**, *102*, 273–296. [CrossRef]
36. Stobinski, L.; Lesiak, B.; Malolepszy, A.; Mazurkiewicz, M.; Mierzwa, B.; Zemek, J.; Jiricek, P.; Bieloshapka, I. Graphene oxide and reduced graphene oxide studied by the XRD, TEM and electron spectroscopy methods. *J. Electron. Spectrosc. Relat. Phenom.* **2014**, *195*, 145–154. [CrossRef]
37. Vallurupalli, K.; Meng, W.; Liu, J.; Khayat, K.H. Effect of graphene oxide on rheology, hydration and strength development of cement paste. *Constr. Build. Mater.* **2020**, *265*, 120311. [CrossRef]
38. Du, Y.; Yang, J.; Skariah, B.; Li, L.; Li, H. Influence of hybrid graphene oxide/carbon nanotubes on the mechanical properties and microstructure of magnesium potassium phosphate cement paste. *Constr. Build. Mater.* **2020**, *260*, 120449. [CrossRef]
39. Akarsh, P.K.; Marathe, S.; Bhat, A.K. Influence of graphene oxide on properties of concrete in the presence of silica fumes and M-sand. *Constr. Build. Mater.* **2021**, *268*, 121093. [CrossRef]
40. Akarsh, P.K.; Shrinidhi, D.; Marathe, S.; Bhat, A.K. Graphene oxide as nano-material in developing sustainable concrete—A brief review. *Mater. Today Proc.* **2022**, *60*, 234–246. [CrossRef]
41. Rajesh, V.; Narendra Kumar, B. Influence of nano-structured graphene oxide on strength and performance characteristics of high strength fiber reinforced self compacting concrete. *Mater. Today Proc.* **2022**, *60*, 694–702. [CrossRef]
42. Shang, Y.; Zhang, D.; Yang, C.; Liu, Y.; Liu, Y. Effect of graphene oxide on the rheological properties of cement pastes. *Constr. Build. Mater.* **2015**, *96*, 20–28. [CrossRef]
43. Wang, Q.; Cui, X.; Wang, J.; Li, S.; Lv, C.; Dong, Y. Effect of fly ash on rheological properties of graphene oxide cement paste. *Constr. Build. Mater.* **2017**, *138*, 35–44. [CrossRef]
44. Long, W.J.; Li, H.D.; Fang, C.L.; Xing, F. Uniformly dispersed and re-agglomerated graphene oxide-based cement pastes: A comparison of rheological properties, mechanical properties and microstructure. *Nanomaterials* **2018**, *8*, 31. [CrossRef] [PubMed]
45. Mowlaei, R.; Lin, J.; Basquiroto, F.; Souza, D.; Fouladi, A.; Habibnejad, A.; Shamsaei, E.; Duan, W. The effects of graphene oxide-silica nanohybrids on the workability, hydration, and mechanical properties of Portland cement paste. *Constr. Build. Mater.* **2021**, *266*, 121016. [CrossRef]
46. Jian, S.; Zou, B.; Collins, F.; Ling, X.; Majumber, M.; Hui, W. Predicting the influence of ultrasonication energy on the reinforcing efficiency of carbon nanotubes. *Carbon* **2014**, *77*, 1–10. [CrossRef]
47. Gao, Y.; Jing, H.; Zhao, Z.; Shi, X.; Li, L. Influence of ultrasonication energy on reinforcing-roles of CNTs to strengthen ITZ and corresponding anti-permeability properties of concrete. *Constr. Build. Mater.* **2021**, *303*, 124451. [CrossRef]
48. Peng, H.; Ge, Y.; Cai, C.S.; Zhang, Y.; Liu, Z. Mechanical properties and microstructure of graphene oxide cement-based composites. *Constr. Build. Mater.* **2019**, *194*, 102–109. [CrossRef]
49. Jing, G.; Xu, K.; Feng, H.; Wu, J.; Wang, S.; Li, Q.; Cheng, X.; Ye, Z. The non-uniform spatial dispersion of graphene oxide: A step forward to understand the inconsistent properties of cement composites. *Constr. Build. Mater.* **2020**, *264*, 120729. [CrossRef]
50. Lee, A.Y.; Yang, K.; Anh, N.D.; Park, C.; Lee, S.M.; Lee, T.G.; Jeong, M.S. Raman study of D* band in graphene oxide and its correlation with reduction. *Appl. Surf. Sci.* **2021**, *536*, 147990. [CrossRef]
51. Suescum-Morales, D.; Ríos, J.D.; Martínez-De La Concha, A.; Cifuentes, H.; Jiménez, J.R.; Fernández, J.M. Cement and Concrete Research Effect of moderate temperatures on compressive strength of ultra-high-performance concrete: A microstructural analysis. *Cem. Concr. Res.* **2021**, *140*, 106303. [CrossRef]
52. Esquinas, A.R.; Álvarez, J.I.; Jiménez, J.R.; Fernández, J.M. Durability of selfcompacting concrete made from non-conforming fly ash from coal-fired power plants. *Constr. Build. Mater.* **2018**, *189*, 993–1006. [CrossRef]
53. Bentz, D.P.; Haecker, C.J. An argument for using coarse cements in high-performance concretes. *Cem. Concr. Res.* **1999**, *29*, 615–618. [CrossRef]
54. Qureshi, T.S.; Panesar, D.K. Nano reinforced cement paste composite with functionalized graphene and pristine graphene nanoplatelets. *Compos. Part B* **2020**, *197*, 108063. [CrossRef]
55. Cai, Y.; Hou, P.; Cheng, X.; Du, P.; Ye, Z. The effects of nanoSiO_2 on the properties of fresh and hardened cement-based materials through its dispersion with silica fume. *Constr. Build. Mater.* **2017**, *148*, 770–780. [CrossRef]
56. Gao, Y.; Jing, H.W.; Chen, S.J.; Du, M.R.; Chen, W.Q.; Duan, W.H. Influence of ultrasonication on the dispersion and enhancing effect of graphene oxide–carbon nanotube hybrid nanoreinforcement in cementitious composite. *Compos. Part B Eng.* **2019**, *164*, 45–53. [CrossRef]
57. Recommendation, R.D. Determination of the Fracture Energy of Mortar and Concrete by Means of Three-Point Bend Tests on Notched Beams. *Mater. Struct.* **1985**, *18*, 285–290. [CrossRef]

58. Ramachandra Murthy, A.; Karihaloo, B.L.; Iyer, N.R.; Raghu Prasad, B.K. Bilinear tension softening diagrams of concrete mixes corresponding to their size−independent specific fracture energy. *Constr. Build. Mater.* **2013**, *47*, 1160–1166. [CrossRef]
59. Abdalla, H.M.; Karihaloo, B.L. A method for constructing the bilinear tension softening diagram of concrete corresponding to its true fracture energy. *Mag. Concr. Res.* **2004**, *56*, 597–604. [CrossRef]
60. Cifuentes, H.; Ríos, J.D.; Martínez-De La Concha, A. Analysis of the behavior of very high strength concrete subjected to thermal shock loading. *Hormigón Y Acero* **2022**, *73*, 73–79. [CrossRef]
61. Martínez-de-la-Concha, A.; Ríos, J.D.; Cifuentes, H. Numerical Study of the Shear Behavior of Ultra-High-Performance Concrete Beams. *Hormigón Y Acero* **2023**, *75*, 157–162. [CrossRef]
62. *EN-12390-3: 2009*; Testing Hardened Concrete Part 3: Compressive Strength of Test Specimens, AENOR. European Committee for Standarization (CEN): Brussels, Brussels, 2009.
63. Swartz, S.E.; Shah, S.P.; Ouyang, C. *Fracture Mechanics of Concrete*; John Wiley & Sons: Hoboken, NJ, USA, 1995.
64. Xiong, G.; Ren, Y.; Wang, C.; Zhang, Z.; Zhou, S.; Kuang, C.; Zhao, Y.; Guo, B.; Hong, S. Effect of power ultrasound assisted mixing on graphene oxide in cement paste: Dispersion, microstructure and mechanical properties. *J. Build. Eng.* **2023**, *69*, 106321. [CrossRef]
65. Chacón Bonet, C.; Cifuentes Bulte, H.; Luna-Galiano, Y.; Rios, J.D.; Ariza, P.; Carlos, C.L. Exploring the impact of graphene oxide on mechanical and durability properties of mortars incorporating demolition waste: Micro and nano-pore structure effects. *Mater. Construcción* **2023**, *73*, e327. [CrossRef]
66. Hong, X.; Lee, J.C.; Ng, J.L.; Md Yusof, Z.; He, Q.; Li, Q. Effect of Graphene Oxide on the Mechanical Properties and Durability of High-Strength Lightweight Concrete Containing Shale Ceramsite. *Materials* **2023**, *16*, 2756. [CrossRef] [PubMed]
67. Ruiz Martinez, J.D.; Cifuentes, H.; Rios, J.D.; Ariza, P.; Leiva, C. Development of Mortars That Use Recycled Aggregates from a Sodium Silicate Process and the Influence of Graphene Oxide as a Nano-Addition. *Materials* **2023**, *16*, 7167. [CrossRef] [PubMed]
68. Kendall, K.; Howard, A.J.; Birchall, J.D.; Pratt, P.L.; Proctor, B.A.; Jefferis, S.A.; Hirsch, P.B.; Birchall, J.D.; Double, D.D.; Kelly, A.; et al. The relation between porosity, microstructure and strength, and the approach to advanced cement-based materials. *Philos. Trans. R. Soc. London. Ser. A Math. Phys. Sci.* **1983**, *310*, 139–153. [CrossRef]
69. Chen, X.; Wu, S.; Zhou, J. Influence of porosity on compressive and tensile strength of cement mortar. *Constr. Build. Mater.* **2013**, *40*, 869–874. [CrossRef]
70. Fang, B.; Qian, Z.; Song, Y.; Diao, X.; Shi, T.; Cai, X.; Wang, L. Evaluation of early crack resistance performance of concrete mixed with ternary minerals using temperature stress testing machine (TSTM). *J. Clean. Prod.* **2024**, *465*, 142780. [CrossRef]
71. Scrivener, K.L.; Crumbie, A.K.; Laugesen, P. The Interfacial Transition Zone (ITZ) Between Cement Paste and Aggregate in Concrete. *Interface Sci.* **2004**, *12*, 411–421. [CrossRef]
72. Hillerborg, A.; Modéer, M.; Peterson, P.E. Analysis of crack formation and crack growth by means of fracture mechanics and finite elements. *Cem. Concr. Res.* **1976**, *6*, 773–781. [CrossRef]
73. Bazant, Z.P. Concrete fracture models: Testing and practice. *Eng. Fract. Mech.* **2002**, *69*, 165–205. [CrossRef]
74. Alyhya, W.S.; Dhaheer, M.S.A.; Karihaloo, B.L. Influence of mix composition and strength on the fracture properties of self−compacting concrete. *Constr. Build. Mater.* **2016**, *110*, 312–322. [CrossRef]
75. Qureshi, T.S.; Panesar, D. Impact of graphene oxide and highly reduced graphene oxide on cement based composites. *Constr. Build. Mater.* **2019**, *206*, 71–83. [CrossRef]
76. Xiao, J.; Lv, Z.; Duan, Z.; Zhang, C. Pore structure characteristics, modulation and its effect on concrete properties: A review. *Constr. Build. Mater.* **2023**, *397*, 132430. [CrossRef]
77. Cifuentes, H.; Karihaloo, B.H. Determination of size−independent specific fracture energy of normal− and high−strength self−compacting concrete from wedge splitting tests. *Constr. Build. Mater.* **2013**, *48*, 548–553. [CrossRef]

crystals

MDPI

Article

The Influence of a Commercial Few-Layer Graphene on Electrical Conductivity, Mechanical Reinforcement and Photodegradation Resistance of Polyolefin Blends

S. M. Nourin Sultana [1], Emna Helal [1,2], Giovanna Gutiérrez [2], Eric David [1], Nima Moghimian [2] and Nicole R. Demarquette [1,*]

[1] Mechanical Engineering Department, Ecole de Technologie Superieure, 1100 Notre-Dame Street West, Montreal, QC H3C 1K3, Canada

[2] NanoXplore Inc., 4500 Thimens Blvd, Saint-Laurent, QC H4R 2P2, Canada

* Correspondence: nicoler.demarquette@etsmtl.ca

Abstract: This work demonstrates the potentials of a commercially available few-layer graphene (FLG) in enhancing the electro-dissipative properties, mechanical strength, and UV protection of polyolefin blend composites; interesting features of electronic packaging materials. Polyethylene (PE)/ polypropylene (PP)/ FLG blend composites were prepared following two steps. Firstly, different concentrations of FLG were mixed with either the PE or PP phases. Subsequently, in the second step, this pre-mixture was melt-blended with the other phase of the blend. FLG-filled composites were characterized in terms of electrical conductivity, morphological evolution upon shear-induced deformation, mechanical properties, and UV stability of polyolefin blend composites. Premixing of FLG with the PP phase has been observed to be a better mixing strategy to attain higher electrical conductivity in PE/PP/FLG blend composite. This observation is attributed to the influential effect of FLG migration from a thermodynamically less favourable PP phase to a favourable PE phase via the PE/PP interface. Interestingly, the addition of 4 wt.% (~2 vol.%) and 5 wt.% (~2.5 vol.%) of FLG increased an electrical conductivity of ~10 orders of magnitude in PE/PP—60/40 (1.87×10^{-5} S/cm) and PE/PP—20/80 (1.25×10^{-5} S/cm) blends, respectively. Furthermore, shear-induced deformation did not alter the electrical conductivity of the FLG-filled composite, indicating that the conductive FLG network within the composite is resilient to such deformation. In addition, 1 wt.% FLG was observed to be sufficient to retain the original mechanical properties in UV-exposed polyolefin composites. FLG exhibited pronounced UV stabilizing effects, particularly in PE-rich blends, mitigating surface cracking and preserving ductility.

Keywords: graphene; polyolefin; electrical conductivity; photoprotection; mechanical properties

Citation: Sultana, S.M.N.; Helal, E.; Gutiérrez, G.; David, E.; Moghimian, N.; Demarquette, N.R. The Influence of a Commercial Few-Layer Graphene on Electrical Conductivity, Mechanical Reinforcement and Photodegradation Resistance of Polyolefin Blends. *Crystals* **2024**, *14*, 687. https://doi.org/10.3390/cryst14080687

Academic Editor: Conrad Becker

Received: 12 June 2024
Revised: 20 July 2024
Accepted: 24 July 2024
Published: 27 July 2024

1. Introduction

Currently, the incorporation of carbonaceous filler [1–9] in the polymer matrix is a promising field of research, aiming at the improvement of the properties of polymers. Among these carbon-based fillers, carbon black is zero-dimensional, carbon nanotube is one-dimensional, and graphene is two-dimensional [10]. Due to the two-dimensional geometry of graphene, the extent of dermal effect, inhalation problem, and gene toxicity associated with graphene is much less than that of carbon black or carbon nanotube [11]. This ensures the safer handling of graphene. Moreover, graphene [12] stands out as a highly promising candidate. Theoretically, sp^2-hybridized carbon atoms are arranged in a hexagonal lattice within atomic thickness to form pristine graphene [13]. This unique structure offers remarkable characteristics. A perfect single-layer graphene is not only the lightest (weighing merely 0.77 milligrams per square meter) but also the strongest (300 times stronger than Kevlar) nanofiller discovered to date [14]. Furthermore, graphene demonstrates exceptional electrical and thermal conductivity, along with remarkable flexibility, capable of stretching up to 25% of its original length without fracturing. These remarkable attributes refer to graphene as a wonder material with the capability of enhancing its properties (processability [15], photostability [16], mechanical [17], and electrical [18]) of polymer composites.

In general, graphene has been observed to enhance the tensile strength and stiffness of polymeric systems, often at the expense of ductility and impact strength [17–19]. Nevertheless, some exceptional observations can be found in the literature [20,21]. Due to graphene's intrinsic electrical conductivity, it can significantly improve the electrical conductivity of polymer composites [22,23]. Furthermore, polymer blend composites often exhibit an even higher electrical conductivity than single polymer/conductive filler composites. This is attributed to the double percolation effect [24–29], which is achievable in co-continuous polymer blend composites. Moreover, the migration of electrically conductive filler from one phase to another via the polymer phase interface (achievable via strategic mixing conditions) can also facilitate the improvement of electrical conductivity in polymer blend composites [30].

Numerous studies [31–34] have investigated the influence of laboratory-processed graphene on the mechanical and electrical properties of various polymer composites. However, the large-scale production of very pure and fully exfoliated graphene is challenging [35]. In contrast, mass production of few-layer graphene (FLG) is obtainable following the cost-effective mechanochemical exfoliation of graphite [36], which involves a lower carbon footprint in comparison to mass production of carbon black [37]. Despite exhibiting a wider size distribution and increased defects compared to lab-grade graphene, particles produced through this process still retain many of graphene's notable properties. Accordingly, better electrical conductivity, improved mechanical strength, and improved photostability of commercial-grade few-layer graphene-filled (FLG) polymer composites have been reported in the literature [16,38]. Nevertheless, research aiming to achieve a combination of improved electrical conductivity, mechanical strength, and photoprotection in the same polymer blend composite remains limited in the literature.

In this connection, polyolefin systems were chosen due to their versatility in domestic and industrial applications. Moreover, approximately 50% of global plastic waste is composed of PE/PP mixtures [39]. A significant portion of this plastic waste can be recovered as a PE/PP mixture from post-consumer waste using simple separation steps. However, further separating PE from PP is both complex and costly [40]. This motivates research into enhancing the properties of controlled PE/PP blend systems to gain a deeper understanding. Such knowledge could be valuable in guiding the upcycling of polyolefin mixtures recovered from post-consumer waste. In this study, polyethylene (PE), polypropylene (PP), and PE/PP blend systems were selected to explore the potential of incorporating commercially available FLG. Different concentrations of FLG were melt-mixed with PE, PP, and PE/PP blend systems. For the PE/PP blend system, two different FLG mixing strategies were employed to investigate their influence on the electrical properties of the composite. However, if the filler fails to stabilize the morphology, subsequent morphological evolution during post-processing is a likely phenomenon for an immiscible polymer blend composite. Shear-induced deformation followed by small-amplitude oscillatory shear (SAOS), a rheological approach [41], was conducted to study the morphological stability/evolution of the PE/PP/FLG blend composite. After the deformation step, SAOS was performed to reach an equilibrium state of the deformed morphology. A similar approach has been extensively studied for droplet-dispersed type polymer blend systems [42]. Furthermore, the influence of adding FLG on the mechanical properties of the composites before and after UV exposure was documented.

2. Materials and Methods

2.1. *Materials*

GrapheneBlack 3X, a few-layer (6 to 10) graphene powder, was received from NanoXplore Inc., Montreal, QC, Canada, to prepare the composites for this study.

Two homopolymers, high-density polyethylene (HDPE) and polypropylene (PP), were used as the thermoplastic components of the composites. Table 1 outlines the specifications of the polymers.

Table 1. Identification, MFI, and density of the polymers used in this work.

Polymer	Commercial Name	MFI (g/10 min)	Density (g/cm^3)
HDPE	Alathon H5618	17 (190 °C, 2.16 kg)	0.955
PP	Polypropylene 3720 WZ	20 (230 °C, 2.16 kg)	0.905

2.2. Methods

To prepare the composites for this study, master batch (MB) pellets of either PE/FLG or PP/FLG were used for the preparation of composites. The MB pellets contained 30 wt.% FLG, provided by NanoXplore, Inc. The samples were produced by using a HAAKE twin-screw extruder (Rheomex OS PTW16/40).

PE and PP were separately blended at 200 °C and 150 rpm with their respective MB pellets to create the PE/FLG and PP/FLG composites. Each composite consisted of 1 and 5 wt.% of FLG, respectively.

In addition, PE/PP—20/80 and PE/PP—60/40 blend composites with the FLG concentration ranging from 1 to 5 wt.% were prepared in two steps. First step: PE/FLG or PP/FLG MB pellets were blended with PE or PP at 200 °C and 150 rpm, referred to as the premixing step. Second step: these premixtures were melt-blended with the other respective polymer blend at 190 °C and 115 rpm.

Finally, extruded pellets were injection molded in an Arburg Allrounder 221K-350-100 Injection molding machine to prepare dog-bone-shaped specimens and rectangular specimens for the characterization of tensile and impact properties, respectively. During injection molding, a certain temperature profile (zone 1: 175 °C, zone 2—180 °C, zone 3—180 °C, zone 4—180 °C, zone 5—185 °C) was followed at a 0.3 in^3/sec injection speed and 8500 Psi injection pressure. It is worth mentioning that the injection duration was 30 s, pressing was continued for 5 s, clamping pressure was maintained at 1000 psi, and the injected samples were cooled for 25 s while the mold temperature was maintained at 70 °C.

Moreover, disks of 1 mm thickness and 25 mm diameter were prepared by compressing the composites at 200 °C and 10 MPa for 10 min in a compression mold.

2.3. Photodegradation Process

The injection-molded composites were weathered in a QUV chamber. The samples were treated under UV radiation for a total of 2 weeks and 4 weeks by following the alternating two steps mentioned in Table 2.

Table 2. Weathering conditions of the specimens in QUV chamber.

Step	Function	Irradiance (W/m^2)	Temperature (°C)	Time (h:m)
1	Exposure to UV radiation	0.89	60	8:00
2	No UV radiation	n/a	50	4:00

2.4. Characterizations

Scanning electron microscope (SEM) images were captured by using a Hitachi SEM S3600-N (Model: MEB-3600-N, manufactured by Hitachi Science Systems, Ltd., Tokyo, Japan). Prior to SEM imaging, the surface (cryo-fracture or UV exposed) of the samples was gold coated by employing a Gold Sputter Coater (Model: K550X, manufactured by Quorum Technologies Ltd, East Sussex, Lewes, UK).

The electrical conductivity of the composites was assessed using a broadband dielectric spectrometer (BDS) from Novocontrol Technologies GmbH & Co. KG, Montabaur, Germany.

The investigation was conducted under an excitation voltage of 3 VRMS across a frequency range from 1×10^{-2} to 3×10^{5} Hz. Gold-coated disk-shaped composites of 1 mm thickness were sandwiched between two brass disks (electrodes) of 25 mm diameter in order to create a plane capacitor structure for the analysis of the electrical properties. The real part of the complex electrical conductivity at the lowest frequency (1×10^{-2} Hz) has been reported as the electrical conductivity of the composite in the subsequent section of this study.

Shear-induced coalescence followed by SAOS was conducted on the rheometer, MCR 501. Compression-molded neat and FLG-filled PE/PP—20/80 composites were sheared at 0.05 s^{-1} rate for 500 s, 2000 s, and 5000 s, respectively. In other words, the samples were deformed at 25, 100 and 250 strains (shear rate × duration), respectively. After each extent of deformation, the sample was subjected to SAOS for stabilization of the deformed morphology, and the sample was saved for further investigation. A schematic representation of the rheological approach has been provided in Figure 1. The shear rate (0.05 s^{-1}) was chosen low enough to favor coalescence. This shear rate was selected based on the critical capillary number (a function of the viscosity ratio of the dispersed matrix phase, the viscosity of the matrix phase, the interfacial tension and the average droplet size). Relevant theory [43,44], the calculation of the critical capillary number, and the selection of the shear rate to investigate the morphological evolution are provided in the Supplementary section of this work. It is worth mentioning that the deformed samples were saved and cryo-fractured, followed by a gold coating to investigate the resultant morphology using SEM images.

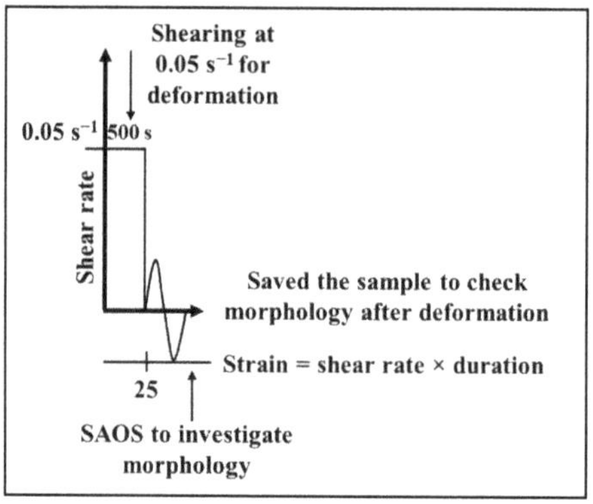

Figure 1. A schematic representation of the shear deformation followed by SAOS, a rheological approach, to study the morphological evolution of neat and FLG-filled PE/PP blends. A similar approach was followed for 2000 s and 5000 s, respectively.

The tensile properties were characterized according to ASTM D638 at room temperature with a 10 kN load cell and a crosshead speed of 50 mm/min. The MTS Alliance RF/200 tensile test apparatus was used for the investigation of tensile property. Dog-bone-shaped samples with 160 mm length, 12 mm width, and 3.5 mm thickness were used to characterize the tensile properties of the samples.

The Izod (notched) impact strength was investigated according to ASTM D256 by using the impact strength tester device (manufactured by International Equipments, Maharashtra, India). Prior to the test, a notch was created on each sample (62 mm × 12 mm) by using a motorized notch cutter, manufactured by International Equipments, India.

At least five replicas were considered for each sample to investigate mechanical properties, and the average value has been reported along with the standard deviation (SD) as an error bar.

A Perkin-Elmer FTIR spectrometer(manufactured by PerkinElmer, Wales, UK), specifically the Spectrum Two™ model equipped with a diamond crystal, was utilized for capturing the Attenuated total reflectance-Fourier transform infrared spectroscopy (ATR-FTIR) spectrum. Thin films collected from the surface of both neat and FLG-filled composites (both pre- and post-UV exposure) were analyzed to obtain their respective FTIR spectra. Each spectrum comprised 10 scans and was collected at a resolution of 4 cm^{-1} across the wave number range of 400–4000 cm^{-1}.

FTIR spectra were utilized to investigate the relative carbonyl content of each sample, employing Equation (1)

$$\text{Relative carbonyl content} = \frac{A_{C=O}\ \text{after UV exposure}}{A_{C=O}\ \text{before exposure}} \quad (1)$$

In this context, $A_{(C=O)}$ denotes the area under the peak (within the wave number range of 1680–1800 cm^{-1}), indicative of the C=O functional group [16]. FTIR spectra of samples before and after UV exposure can be found in the Supplementary Materials (see Figure S1a–e). Evaluating the relative carbonyl content is of interest as it serves as an indicator of polymer degradation over time. To ensure accuracy, a minimum of three spectra from distinct regions of the samples were analyzed to compute and present the average carbonyl content values.

3. Results and Discussion

3.1. Morphology of Neat Blends

Figure 2a–c illustrate the microstructure of neat PE/PP blends with ratios of 20/80, 60/40, and 80/20, respectively. All the blends presented a droplet-dispersed morphology. However, droplet phase and size are governed by the blend composition. For instance, the PE/PP—20/80 blend represents PE droplets in the PP matrix. In contrary, PE/PP—60/40 and PE/PP—80/20 blends show PP droplets dispersed in the PE matrix. However, PP droplet size is bigger in PE/PP—60/40 than that of PE/PP—80/20. This size difference is attributed to the higher proportion of the PP phase (minor phase) in the PE/PP 60/40 blend system. The presence of a comparatively higher portion of the minor phase in an immiscible polymer blend increases the probability of droplet coalescence. This phenomenon may result in bigger droplets in the matrix phase.

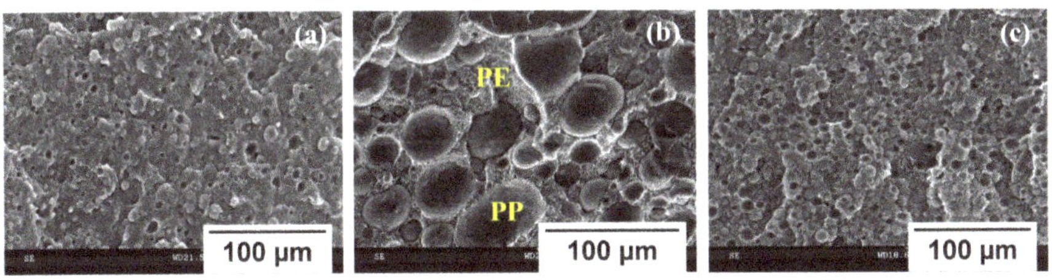

Figure 2. SEM images of PE/PP blends: (**a**) PE/PP—20/80, droplets representing PE phase; (**b**) PE/PP—60/40, droplets representing PP phase; (**c**) PE/PP—80/20, droplets representing PP phase.

3.2. Influence of FLG on Morphology of Blend Composite

Figure 3 illustrates the distribution of PP droplet size (number average droplet radius, R_N) in a neat and FLG-filled PE/PP—20/80 blend composite. The addition of FLG results in a reduced and narrower distribution of PE droplet size. The addition of 2 wt.% and 5 wt.% of FLG results in a notable reduction in droplet size. The formation of finer morphology is attributed to the inhibition effect of rigid filler against the coalescence of droplets to grow bigger. A similar effect of rigid fillers was reported in the literature [45].

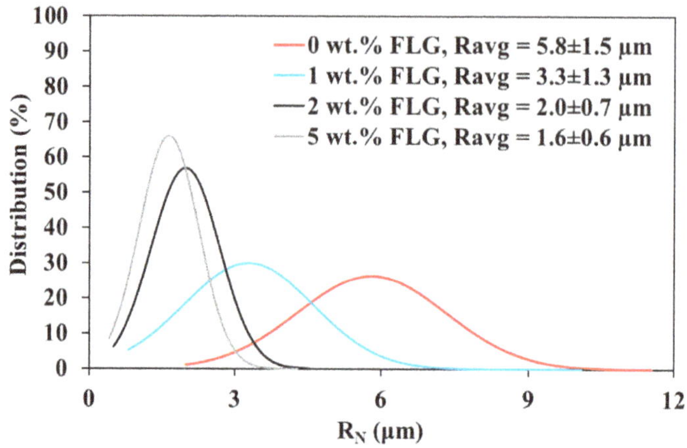

Figure 3. Distribution of PE droplet size in neat and FLG-filled PE/PP—20/80 blend composite with 1, 2 and 5 wt.% of FLG, respectively.

The SEM micrographs in Figure 4a–c depict the influence of the addition of 5 wt.% of FLG on the microstructure of PE/PP blend composites. The introduction of FLG causes deformation of PE droplets in PE/PP—20/80 and PP droplets in PE/PP—60/40 and PE/PP—80/20 blends, respectively. Particularly in the PE/PP—60/40 blend, the deformation of relatively larger PP droplets enhances the likelihood of continuity in the minor PP phase. This phenomenon, in turn, can facilitate the creation of an uninterrupted network of electrically conductive fillers in the polymer blend composite.

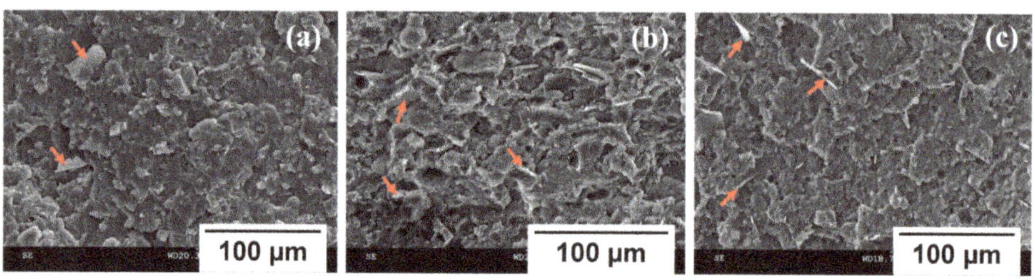

Figure 4. SEM images of the phase morphology of 5 wt.% FLG-filled composites; (**a**) PE/PP—20/80; (**b**) PE/PP—60/40 and (**c**)PE/PP—80/20. The red arrows are used to indicate the presence of FLG in the composites.

3.3. Influence of FLG on Electrical Conductivity of Composites

The impact of different FLG mixing strategies has been studied (see Figure S2 in the Supplementary Materials) on the electrical conductivity of a PE/PP—60/40 blend composite. Notably, the composite exhibits significantly higher electrical conductivity when FLG is pre-mixed with the PP phase compared to when it is pre-mixed with the PE phase. This observation motivated us to carry on the rest of the study with composites, where FLG was pre-mixed with the PP phase. Figure 5 reports the change in electrical conductivity in FLG-filled composites with different concentrations of FLG in the composites. The graph displays that the electrical conductivity of PE/PP—60/40 blend composite is higher than that of PE/PP—20/80 blend composite, specifically up to 4 wt.% of FLG. However, when the FLG concentration reaches 5 wt.%, both blend composites demonstrate similar electrical conductivity. This suggests that the electrical properties are significantly influenced by the morphology, governed by blend compositions. In contrast, FLG-filled PE, PP, and PE/PP—80/20 remain insulative, even at a dosage of 5 wt.% FLG. Notably, the PP/FLG composite exhibits higher electrical conductivity than that of PE/FLG at 9 wt.% concentration of FLG. This observation implies that a relatively lower concentration of FLG is required in the PP matrix compared to the PE matrix to establish a continuous FLG network.

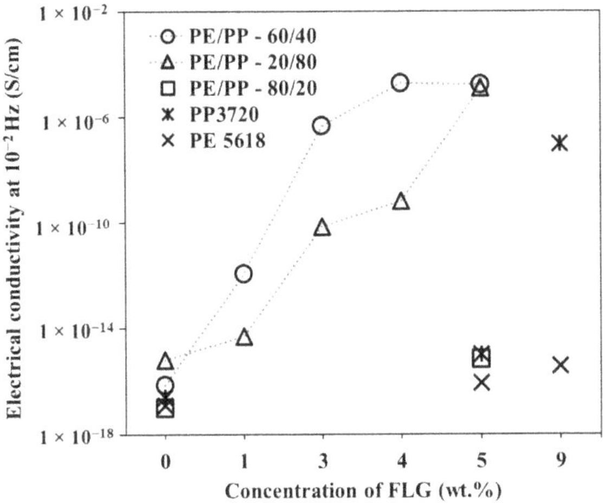

Figure 5. Real part of the conductivity of the compounds at 10^{-2} Hz (the lowest frequency) for different concentrations of FLG. Dotted lines are used to guide the reader's eye.

3.4. Influence of Deformation on Electrical Conductivity of Blend Composite

Figure 6 presents the electrical conductivity of the deformed PE/PP—20/80 blend composite, containing 5 wt.% FLG, as a function of the frequency of the applied electric field. After deforming the composite in molten condition in the Rheometer, the samples were preserved to investigate the electrical conductivity. Intriguingly, despite the deformation in morphology (discussed using Figure S3 in the Supplementary Materials) and a change in filler localization (see Figure S4 in the Supplementary Materials), the electrical conductivity of the composite remained unchanged. This underscores the robustness of the FLG network in withstanding mechanical deformation in molten conditions. This observation is in agreement with the observation reported in the literature [46].

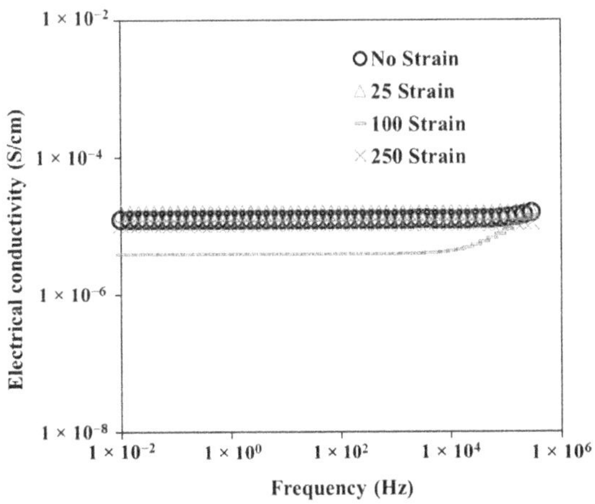

Figure 6. Electrical conductivity of PE/PP -20/80 blend composite (containing 5 wt.% FLG) after deformation by different strains, as a function of frequency.

3.5. *Influence of FLG on Mechanical Properties Polyolefin Composites*

In Figure 7a–c, the mechanical properties of the samples are plotted as a function of FLG concentration. Since the ductility of PE is around 10 times higher than that of neat PP, the ductility of PE has been plotted on the secondary axis of Figure 7c with different concentration of FLG. Figure 7a illustrates that the tensile modulus of neat PP is higher than that of neat PE. In addition, the tensile moduli of PE, PP, and PE/PP blends are increased by the additions of FLG. It is worth mentioning that the tensile strength of the samples as a function of the concentration of FLG has been provided in Figure S5 in the supplementary materials. The stiffness of polymeric materials can be increased due to the reinforcement effect of rigid fillers. Analogous findings can be observed in the literature [47,48]. Figure 7b shows that the impact strength of PE, PP, and PE/PP blend composites was not significantly affected by the presence of FLG. A similar effect of rigid filler on the impact strength of polymer material was reported earlier [49]. However, Figure 7c shows that the addition of FLG decreased the ductility of the composites. The reduction in elongation at break of PE and PE/PP—60/40 (PE- rich blend) composites is statistically more significant than that of PP and PE/PP—20/80 (PP- rich blend) composites. The presence of rigid filler hinders the polymer chain mobility, which ultimately affects the ductility of the composite. Several works in the literature have reported similar findings [2,17,19].

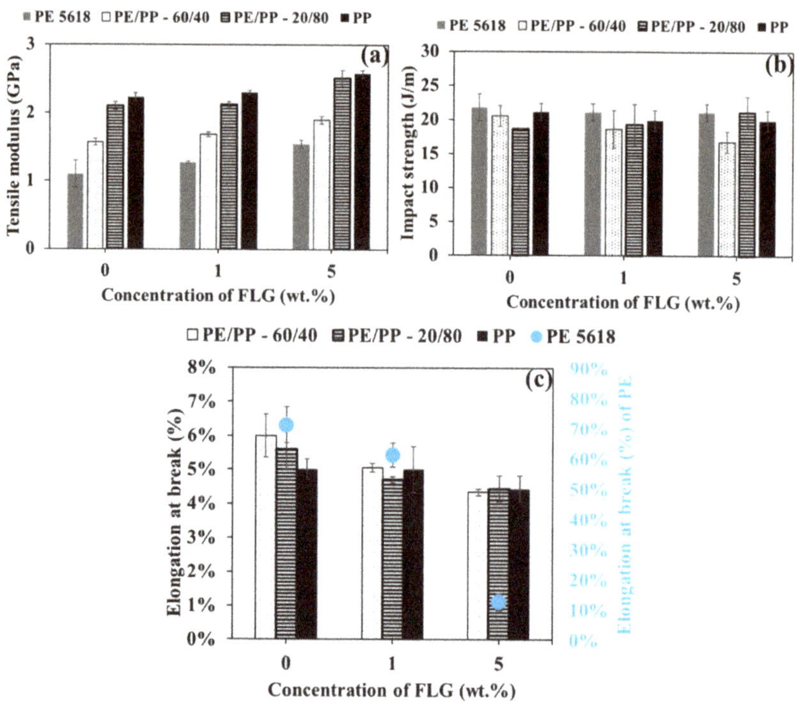

Figure 7. Mechanical properties of: (**a**) tensile modulus, (**b**) impact strength, and (**c**) elongation at break of PE, PE/PP blends, and PP composites with different concentration of FLG.

3.6. *Effect of Adding FLG on UV-Exposed Composites*

3.6.1. Property Retention of UV-Degraded Composites

In Figure 8, the retention of elongation at break (a sensitive property to UV degradation) after 4 weeks of UV exposure for neat and FLG-filled PE, PE/PP—60/40, PE/PP—20/80, and PP is plotted as the FLG concentration was varied in the composites. The figure illustrates that UV-exposed unfilled PE, PP, and PE/PP blends become too brittle to capture the correct elongation at break under the testing condition. This indicates a substantial UV degradation of the pure polyolefins. Conversely, FLG-filled polyolefin composites exhibit enhanced retention of ductility under UV radiation, owing to the photoprotective properties of FLG. Nevertheless, this photoprotective effect of FLG is more pronounced in polymeric systems where PE is the predominant phase compared to those where PP is the major component. Further analyses have been carried out in terms of sample surface imaging and FTIR investigation.

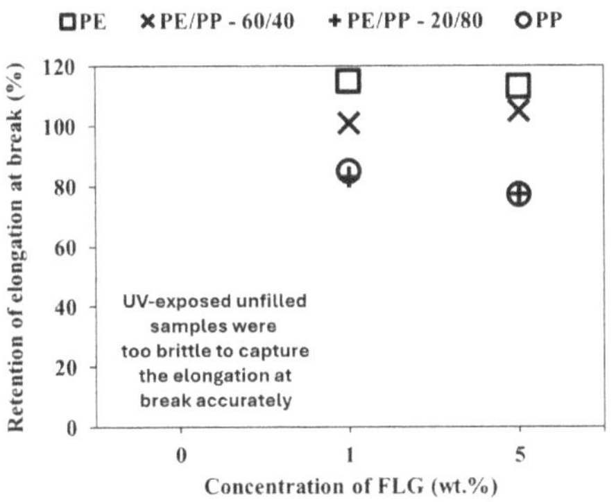

Figure 8. Retention of elongation at break of UV-exposed PE, PE/PP blends, and PP as the FLG concentration increased in the specimens.

3.6.2. Appearance of the Compounds after UV Exposure

The SEM images in Figure 9a–c illustrate the surfaces of neat PE, PE/PP—60/40 blend, and PP samples, which were exposed for 4 weeks under UV radiation. Figure S6 has been provided in the Supplementary section to extend the idea of the appearance of a polyolefin sample before UV exposure. Additionally, Figure 9d–f display the UV-degraded surfaces of PE, PE/PP—60/40 blend, and PP composites, containing 1 wt.% FLG, respectively. Apparently, the presence of FLG lessens the occurrence of UV degradation-driven cracks in the samples' surface compared to their neat counterparts. This suggests that FLG can mitigate the UV degradation process in polyolefins.

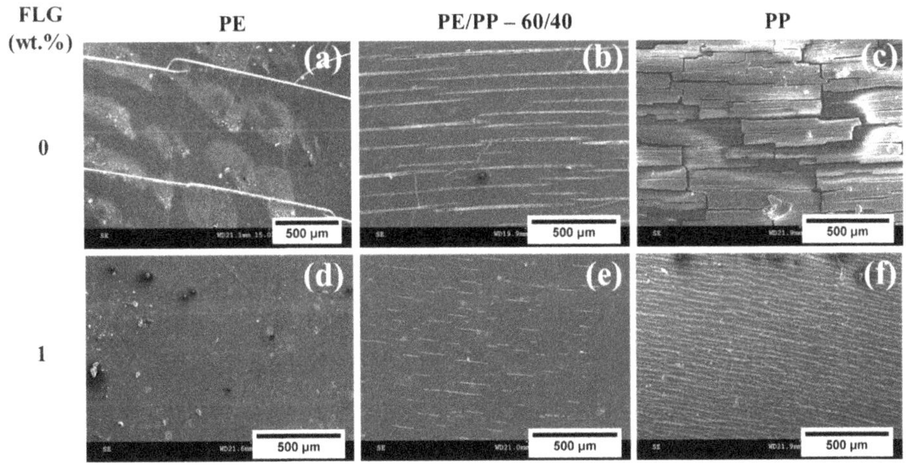

Figure 9. SEM micrographs of the UV-exposed unfilled (**a**) PE, (**b**) PE/PP—60/40, and (**c**) PP along with 1 wt.% FLG-filled (**d**) PE, (**e**) PE/PP—60/40 and (**f**) PP composites.

3.7. Chemical Analysis

In Figure 10, the relative carbonyl content of UV-exposed samples of PE, PE/PP—60/40, PE/PP—20/80 blends, and pure PP, both without and with FLG, is plotted against exposure time. The relative carbonyl content is determined by analyzing the FTIR absorption spectra of each UV-exposed sample and normalizing them against the carbonyl content of the corresponding sample prior to UV exposure. The incorporation of FLG reduces the formation of carbonyl content in UV-exposed composites, indicating the UV-stabilizing effect of FLG in polyolefin systems.

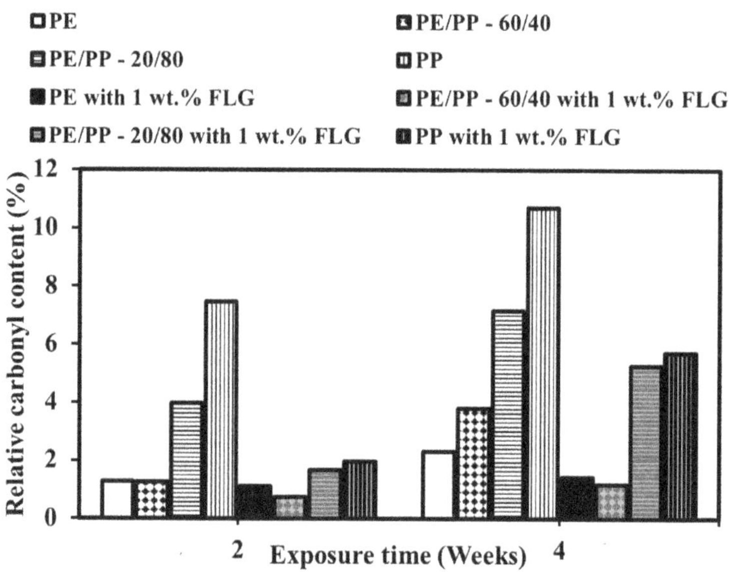

Figure 10. The change in the relative carbonyl content of UV-exposed PE, PE/PP blends, and PP samples with the duration of the UV exposure.

Furthermore, it is evident that the relative carbonyl content in UV-exposed PP is significantly higher compared to that in PE, suggesting that PP is more susceptible to UV degradation than PE. This tendency towards UV degradation is also observed in PE/PP blends; blends with a higher proportion of PP exhibit greater susceptibility to UV degradation compared to those with more PE. An explanation has been extended in the discussion part. Figure 10 illustrates the order of UV degradation tendency in the PE/PP polymer system as follows: PP > PE/PP—20/80 > PE/PP—60/40 > PE. A similar order of UV degradation tendency is noted in FLG-filled PE/PP composites, but the degradation rate is slower, attributed to the UV-stabilizing effect of FLG.

3.8. Discussion

The comprehensive analysis of the morphology, electrical properties, mechanical properties, and UV degradation behavior of FLG-filled PE/PP blend systems provides valuable insights into the multifaceted effects of FLG on polyolefin blend systems.

In a filler-filled, immiscible polymer blend system, double percolation [24–29] is an interesting phenomenon. This phenomenon refers to the percolation of one phase through another phase with the concomitant formation of a filler network through one of the percolating phases. In the case of a conductive filler, the double percolation effect increases the electrical conductivity of the composite by a significant order of magnitude. In this work, Figure 4b shows that the PP phase was not percolated through the PE matrix in the 5 wt.% FLG-filled PE/PP—60/40 blend composite. Despite lacking the double percolation effect, this composite presents higher electrical conductivity. In fact, the predominant presence of FLG in the PP phase of the PE/PP blend composite is likely due to the premixing of FLG with the PP phase. Moreover, FLG migration from the PP to PE phase via PE/PP interface is also realized according to thermodynamic prediction [30]. Therefore, thermodynamic preference-induced FLG migration from the PP to PE phase via the PE/PP interface and volume exclusion effect of the PE phase in PE/PP—60/40 and 20/80 blend are two dominant factors in tailoring the conductive FLG network in PE/PP—60/40 and 20/80 blend composites. It has been reported in the literature [50] that once the conductive filler network is fully developed, further addition of filler does not reflect any significant effect on the conductivity of the composite. This phenomenon can be referred to as the saturated effect of the filler network. Moreover, blends with ratios of 60/40 and 20/80 exhibit superior conductivity compared to PE/FLG or PP/FLG composites, even when FLG concentration is higher (9 wt.%). This difference can be attributed to the poor dispersion of FLG in the PE phase compared to the PP phase under the same processing conditions. Moreover, superior electrical conductivity in the PE/PP—60/40 blend composite is not observed when FLG is pre-mixed with PE (see Figure S2 in the Supplementary section). This could be attributed to the possible wrapping of FLG by insulating PE due to a stronger affinity between FLG and PE. Consequently, an interrupted FLG network results in reduced electrical conductivity in PE, even at higher concentrations of FLG. Additionally, during composite cooling, small PE crystals may form on FLG surfaces, impeding FLG contact and resulting in lower electrical conductivity in FLG-filled PE/PP—60/40 blend composites, prepared by FLG premixing with PE phase. These findings underscore the critical role of FLG dispersion and localization in attaining optimal electrical properties in polyolefin blend composites.

Furthermore, the shear-induced deformation applied in the molten condition of the compound does not affect the electrical conductivity of the composite in its solid state. This indicates that the conductive network established by FLG remains unaffected even under applied shear deformation. This highlights the resilience of FLG-filled composites to mechanical deformation while maintaining electrical performance.

FLG addition enhances the tensile modulus of both PE and PP, indicating improved stiffness attributed to the reinforcement effect of FLG. When mechanically strong, rigid fillers are present in polymer-like materials, the fillers act like stress concentrators, facilitating load transfer. In turn, the filler-filled composite increases. Furthermore, better filler-matrix adhesion results in higher strength in the composite. Moreover, FLG incorporation does not adversely affect the impact resistance of the polyolefin blend composites. In general, rigid fillers have two opposing effects on the impact strength of a filler-filled composite. One effect is its role as a defect center, responsible for reducing the impact strength of the composite. Another role is to resist the crack, which is responsible for increasing the impact strength of the composite. When these two effects counterbalance each other, the impact strength of the filler-filled composite may remain unchanged. However, the impact strength of a filler-filled composite can also be influenced by filler-matrix adhesion, filler loading, and the interfacial properties of the composite. The preservation of mechanical properties coupled with the enhancement of stiffness underscores the potential of FLG as a multifunctional filler in polyolefin blend systems. However, the presence of graphene reduced the ductility of the composite, irrespective of the composition of the sample. In general, the presence of rigid particles like few-layer graphene restrains the mobility of polymer chains, which results in a reduction in the ductility of the filler-filled polymer composite.

Furthermore, FLG demonstrates a UV stabilizing effect on polyolefin composites, reducing the formation of carbonyl content, indicative of UV degradation of polymers. Usually, light-absorbing groups (chromophores), as internal or external components in polymer materials, initiate photodegradation [51]. Initially, chromophores absorb photons from UV radiation and become excited. These excited chromophores then break down into free radicals due to further UV exposure in the presence of oxygen. These free radicals attack the polymer chains and propagate the photodegradation process. The presence of graphene in the polymer matrix can decelerate the photodegradation process by absorbing photons. Additionally, graphene can interact with excited chromophores, absorbing their energy and releasing it as heat, which is known as the quenching effect. Moreover, the π-conjugated electrons in graphene can react with free radicals, neutralizing them and thereby slowing down the photodegradation of the polymer material via the free radical scavenging effect of graphene. Apart from these chemical effects of graphene, it can also act as a physical barrier against oxygen diffusion through the polymer matrix due to its two-dimensional geometry. It has been noticed that the UV stabilizing effect of FLG is more pronounced in PE-rich blends compared to PP-rich blends, highlighting the differential susceptibility of PE and PP to UV degradation. PP exhibits a comparatively higher sensitivity to UV degradation, potentially attributed to the presence of tertiary carbon atoms within the polymer chain [52]. The tertiary carbon atom can be readily attacked by UV radiation, which makes PP more susceptible to UV degradation than PE. The presence of FLG mitigates UV-induced surface cracking and preserves the ductility of the composite materials, further emphasizing its role in enhancing the durability and performance of polyolefin blend composites under UV exposure.

4. Conclusions

This study underscores the notable influence of FLG inclusion on the electrical conductivity, mechanical properties, and UV protection of polyolefin blend composites, as follows:

- The addition of FLG results in electrically conductive, mechanically strong, and more durable PE/PP blend composites.
- The application of shear-induced deformation in molten conditions can change the phase morphology of the blend composite, yet the electrical conductivity remains unaffected, highlighting the resilience of the conductive network within the FLG-filled composites.
- Only 4 wt.% (~2 vol.%) of commercial-grade and low-cost FLG could induce an electrical conductivity of the order of 1.87×10^{-5} S/cm (semi-conductive zone) in PE/PP—60/40.
- As little as 1 wt.% FLG is adequate to retard the UV degradation of polyolefin composite.
- FLG demonstrates a UV stabilizing effect, more pronounced in PE-rich blends, mitigating UV-induced surface cracking and preserving ductility.

In conclusion, FLG offers a versatile solution for designing a reinforced electrostatic dissipative polyolefin composite with improved UV resistance. By understanding the interplay between FLG dispersion, morphology, and filler-polymer interactions, it is possible to tailor the properties of these composites for a variety of applications, from electric applications (semi-conductive separators in power cables, electro-dissipating packaging materials, etc.) to outdoor structural components.

Supplementary Materials: The following supporting information can be downloaded at: https://www.mdpi.com/article/10.3390/cryst14080687/s1. Figure S1: FTIR spectra of neat (a) PE, PE/PP–20/8, PE/PP–60/40 and PP samples, and FTIR spectra of neat (b) PE, (c) PE/PP–20/8, (d) PE/PP–60/40 and (e) PP samples indicating the carbonyl growth after UV exposure within the wave number range of 1680 to 1800 cm^{-1}. Figure S2: Electrical conductivity of PE/PP–60/40 blend composite with 5 wt.% FLG; premixed with PP phase and premixed with PE phase. Figure S3: Distribution of PE droplet size in neat and FLG-filled PE/PP–20/80 blend composite with 1 and 2 wt.% of FLG, respectively, after deformation by 250 strain. Figure S4: SEM image of 2 wt.% FLG-filled PE/PP–20/80 blend composite after deformation, red arrows indicate the deformation-driven migration of FLG from PP phase to PE droplet surface. Figure S5: Tensile strength of PE, PE/PP blends and PP composites as a function of FLG concentration. Figure S6: SEM images of the surface appearance of a polyolefin sample before UV exposure. Table S1: The parameters and corresponding values to calculate the shear rate to ensure droplet coalescence during the coalescence test.

Author Contributions: Conceptualization, S.M.N.S., E.H., G.G., E.D., N.M. and N.R.D.; Methodology, S.M.N.S., E.H., E.D. and N.R.D.; Validation, S.M.N.S., E.H., G.G., E.D., N.M. and N.R.D.; Formal analysis, S.M.N.S., E.H., G.G., E.D., N.M. and N.R.D.; Investigation, S.M.N.S., E.H., E.D. and N.R.D.; Resources, E.H., G.G., E.D., N.M. and N.R.D.; Writing—original draft, S.M.N.S.; Writing—review and editing, E.H., G.G., E.D., N.M. and N.R.D.; Visualization, S.M.N.S., E.H., E.D. and N.R.D.; Supervision, E.H., G.G., E.D., N.M. and N.R.D.; Project administration, E.H., G.G., E.D., N.M. and N.R.D.; Funding acquisition, S.M.N.S., E.D. and N.R.D. All authors have read and agreed to the published version of the manuscript.

Funding: Grants from NanoXplore Inc.; Natural Sciences and Engineering Research Council (grant number CRDPJ 538482—18); PRIMA Quebec (grant number R18-46-001); Fonds de Recherche du Québec-Nature et Technologies (grant number 318813) were received to conduct the research work. The article processing charge was partially waived by the Editorial Office, MDPI and the rest of the amount funded by the IOAP of Ecole de Technologie Superieure (ETS) Montreal.

Data Availability Statement: The author confirms that the data supporting the reported results can be found within the article and its Supporting Materials. Raw data that supports the findings of this work can be inquired to the corresponding authors, upon reasonable request.

Conflicts of Interest: Author Giovanna Gutiérrez and Nima Moghimian were employed by the company NanoXplore Inc.. The remaining authors declare that the research was conducted in the absence of any commercial or financial relationships that could be construed as a potential conflict of interest.

References

1. Zhu, D.; Bin, Y.; Matsuo, M. Electrical Conducting Behaviors in Polymeric Composites with Carbonaceous Fillers. *J. Polym. Sci. Part B Polym. Phys.* **2006**, *45*, 1037–1044. [CrossRef]
2. dos Anjos, E.G.R.; Marini, J.; Gomes, N.A.S.; Rezende, M.C.; Passador, F.R. Synergistic effect of adding graphene nanoplates and carbon nanotubes in polycarbonate/acrylonitrile-styrene-butadiene copolymer blend. *J. Appl. Polym. Sci.* **2022**, *139*, e52873. [CrossRef]
3. dos Anjos, E.G.; Moura, N.K.; Antonelli, E.; Baldan, M.R.; Gomes, N.A.; Braga, N.F.; Santos, A.P.; Rezende, M.C.; A Pessan, L.; Passador, F.R. Role of adding carbon nanotubes in the electric and electromagnetic shielding behaviors of three different types of graphene in hybrid nanocomposites. *J. Thermoplast. Compos. Mater.* **2022**, *36*, 3209–3235. [CrossRef]
4. Chaudhuri, I.; Fruijtier-Pölloth, C.; Ngiewih, Y.; Levy, L. Evaluating the evidence on genotoxicity and reproductive toxicity of carbon black: A critical review. *Crit. Rev. Toxicol.* **2017**, *48*, 143–169. [CrossRef] [PubMed]
5. Chiu, F.-C.; Behera, K.; Cai, H.-J.; Chang, Y.-H. Polycarbonate/Poly(vinylidene fluoride)-blend-based nanocomposites—Effect of adding different carbon nanofillers/organoclay. *Polymers* **2021**, *13*, 2626. [CrossRef]
6. Nunes, M.A.B.S.; Matos, B.R.; Silva, G.G.; Ito, E.N.; Melo, T.J.A.; Fechine, G.J.M. Hybrids nanocomposites based on a polymer blend (linear low-density polyethylene/poly(ethylene-co-methyl acrylate) and carbonaceous fillers (graphene and carbon nanotube). *Polym. Compos.* **2021**, *42*, 661–677. [CrossRef]
7. López-Martínez, E.D.; Martínez-Colunga, J.G.; Ramírez-Vargas, E.; Sanchez-Valdes, S.; Valle, L.F.R.; Benavides-Cantu, R.; Rodríguez-Gonzalez, J.A.; Mata-Padilla, J.M.; Cruz-Delgado, V.J.; Borjas-Ramos, J.J.; et al. Influence of carbon structures on the properties and photodegradation of LDPE/LLDPE films. *Polym. Adv. Technol.* **2022**, *33*, 1727–1741. [CrossRef]
8. Lago, E.D.; Cagnin, E.; Boaretti, C.; Roso, M.; Lorenzetti, A.; Modesti, M. Influence of different carbon-based fillers on electrical and mechanical properties of a PC/ABS blend. *Polymers* **2020**, *12*, 29. [CrossRef]
9. Cui, Y.; Kundalwal, S.; Kumar, S. Gas barrier performance of graphene/polymer nanocomposites. *Carbon* **2016**, *98*, 313–333. [CrossRef]
10. Afzal, A.; Kausar, A.; Siddiq, M. Perspectives of Polystyrene Composite with Fullerene, Carbon Black, Graphene, and Carbon Nanotube: A Review. *Polym. Technol. Eng.* **2016**, *55*, 1988–2011. [CrossRef]
11. Moghimian, N.; Nazarpour, S. The future of carbon: An update on graphene's dermal, inhalation, and gene toxicity. *Crystals* **2020**, *10*, 718. [CrossRef]
12. Sun, X.; Huang, C.; Wang, L.; Liang, L.; Cheng, Y.; Fei, W.; Li, Y. Recent Progress in Graphene/Polymer Nanocomposites. *Adv. Mater.* **2021**, *33*, e2001105. [CrossRef]
13. Choi, W.; Lahiri, I.; Seelaboyina, R.; Kang, Y.S. Synthesis of graphene and its applications: A review. *Crit. Rev. Solid State Mater. Sci.* **2010**, *35*, 52–71. [CrossRef]
14. Chakraborty, M.; Hashmi, M.S.J. Wonder material graphene: Properties, synthesis and practical applications. *Adv. Mater. Process. Technol.* **2018**, *4*, 573–602. [CrossRef]
15. Ferreira, E.H.C.; Andrade, R.J.E.; Fechine, G.J.M. The "superlubricity state" of carbonaceous fillers on polyethylene-based composites in a molten state. *Macromolecules* **2019**, *52*, 9620–9631. [CrossRef]
16. Karimi, S.; Helal, E.; Gutierrez, G.; Moghimian, N.; David, E.; Samara, M.; Demarquette, N. Photo-stabilization mechanisms of high-density polyethylene (HDPE) by a commercial few-layer graphene. *Polym. Eng. Sci.* **2023**, *63*, 3879–3890. [CrossRef]

17. Abbasi, F.; Shojaei, D.A.; Bellah, S.M. The compatibilization effect of exfoliated graphene on rheology, morphology, and mechanical and thermal properties of immiscible polypropylene/polystyrene (PP/PS) polymer blends. *J. Thermoplast. Compos. Mater.* **2019**, *32*, 1378–1392. [CrossRef]
18. Haghnegahdar, M.; Naderi, G.; Ghoreishy, M.H.R. Electrical and thermal properties of a thermoplastic elastomer nanocomposite based on polypropylene/ethylene propylene diene monomer/graphene. *Soft Mater.* **2017**, *15*, 82–94. [CrossRef]
19. Parameswaranpillai, J.; Joseph, G.; Shinu, K.P.; Jose, S.; Salim, N.V.; Hameed, N. Development of hybrid composites for automotive applications: Effect of addition of SEBS on the morphology, mechanical, viscoelastic, crystallization and thermal degradation properties of PP/PS–xGnP composites. *RSC Adv.* **2015**, *5*, 25634–25641. [CrossRef]
20. Pour, R.H.; Hassan, A.; Soheilmoghaddam, M.; Bidsorkhi, H.C. Mechanical, thermal, and morphological properties of graphene reinforced polycarbonate/acrylonitrile butadiene styrene nanocomposites. *Polym. Compos.* **2016**, *37*, 1633–1640. [CrossRef]
21. Bijarimi, M.; Amirul, M.; Norazmi, M.; Ramli, A.; Desa, M.S.Z.; Desa, A.; Abu Samah, M.A. Preparation and characterization of poly (lactic acid) (PLA)/polyamide 6 (PA6)/graphene nanoplatelet (GNP) blends bio-based nanocomposites. *Mater. Res. Express* **2019**, *6*, 055044. [CrossRef]
22. Rafeie, O.; Aghjeh, M.K.R.; Tavakoli, A.; Kalajahi, M.S.; Oskooie, A.J. Conductive poly(vinylidene fluoride)/polyethylene/graphene blend-nanocomposites: Relationship between rheology, morphology, and electrical conductivity. *J. Appl. Polym. Sci.* **2018**, *135*, 46333. [CrossRef]
23. Sadeghi, A.; Moeini, R.; Yeganeh, J.K. Highly conductive PP/PET polymer blends with high electromagnetic interference shielding performances in the presence of thermally reduced graphene nanosheets prepared through melt compounding. *Polym. Compos.* **2019**, *40*, E1461–E1469. [CrossRef]
24. Al-Saleh, M.H.; Al-Anid, H.K.; Hussain, Y.A. Electrical double percolation and carbon nanotubes distribution in solution processed immiscible polymer blend. *Synth. Met.* **2013**, *175*, 75–80. [CrossRef]
25. Sumita, M.; Sakata, K.; Hayakawa, Y.; Asai, S.; Miyasaka, K.; Tanemura, M. Double percolation effect on the electrical conductivity of conductive particles filled polymer blends. *Colloid Polym. Sci.* **1992**, *270*, 134–139. [CrossRef]
26. Jia, L.; Yan, D.; Cui, C.; Ji, X.; Li, Z. A Unique double percolated polymer composite for highly efficient electromagnetic interference shielding. *Macromol. Mater. Eng.* **2016**, *301*, 1232–1241. [CrossRef]
27. Otero-Navas, I.; Arjmand, M.; Sundararaj, U. Carbon nanotube induced double percolation in polymer blends: Morphology, rheology and broadband dielectric properties. *Polymer* **2017**, *114*, 122–134. [CrossRef]
28. Pan, Y.; Liu, X.; Hao, X.; Starý, Z.; Schubert, D.W. Enhancing the electrical conductivity of carbon black-filled immiscible polymer blends by tuning the morphology. *Eur. Polym. J.* **2016**, *78*, 106–115. [CrossRef]
29. Strugova, D.; Junior, J.C.F.; David, É.; Demarquette, N.R. Ultra-low percolation threshold induced by thermal treatments in co-continuous blend-based PP/PS/MWCNTs nanocomposites. *Nanomaterials* **2021**, *11*, 1620. [CrossRef]
30. Tu, C.; Nagata, K.; Yan, S. Influence of melt-mixing processing sequence on electrical conductivity of polyethylene/polypropylene blends filled with graphene. *Polym. Bull.* **2017**, *74*, 1237–1252. [CrossRef]
31. Bouhfid, R.; Arrakhiz, F.; Qaiss, A. Effect of graphene nanosheets on the mechanical, electrical, and rheological properties of polyamide 6/Acrylonitrile–Butadiene–Styrene blends. *Polym. Compos.* **2016**, *37*, 998–1006. [CrossRef]
32. Wang, F.; Zhang, Y.; Zhang, B.; Hong, R.; Kumar, M.; Xie, C. Enhanced electrical conductivity and mechanical properties of ABS/EPDM composites filled with graphene. *Compos. Part B Eng.* **2015**, *83*, 66–74. [CrossRef]
33. Graziano, A.; Garcia, C.; Jaffer, S.; Tjong, J.; Yang, W.; Sain, M. Functionally tuned nanolayered graphene as reinforcement of polyethylene nanocomposites for lightweight transportation industry. *Carbon* **2020**, *169*, 99–110. [CrossRef]
34. Graziano, A.; Garcia, C.; Jaffer, S.; Tjong, J.; Sain, M. Novel functional graphene and its thermodynamic interfacial localization in biphasic polyolefin systems for advanced lightweight applications. *Compos. Sci. Technol.* **2020**, *188*, 107958. [CrossRef]
35. Zhong, Y.L.; Tian, Z.; Simon, G.P.; Li, D. Scalable production of graphene via wet chemistry: Progress and challenges. *Mater. Today* **2015**, *18*, 73–78. [CrossRef]
36. Moghimian, N.; Saeidlou, S.; Lentzakis, H.; Rosi, G.F.; Song, N.; David, E. Electrical conductivity of commercial graphene polyethylene nanocomposites. In Proceedings of the 2017 IEEE 17th International Conference on Nanotechnology (IEEE-NANO), Pittsburgh, PA, USA, 25–28 July 2017; pp. 757–761.
37. Aprianti, N.; Kismanto, A.; Supriatna, N.K.; Yarsono, S.; Nainggolan, L.M.T.; Purawiardi, R.I.; Fariza, O.; Ermada, F.J.; Zuldian, P.; Raksodewanto, A.A.; et al. Prospect and challenges of producing carbon black from oil palm biomass: A review. *Bioresour. Technol. Rep.* **2023**, *23*, 101587. [CrossRef]
38. Batista, N.L.; Helal, E.; Kurusu, R.S.; Moghimian, N.; David, E.; Demarquette, N.R.; Hubert, P. Mass-produced graphene—HDPE nanocomposites: Thermal, rheological, electrical, and mechanical properties. *Polym. Eng. Sci.* **2019**, *59*, 675–682. [CrossRef]
39. Singh, M.K.; Mohanty, A.K.; Misra, M. Upcycling of waste polyolefins in natural fiber and sustainable filler-based biocomposites: A study on recent developments and future perspectives. *Compos. Part B Eng.* **2023**, *263*, 110852. [CrossRef]
40. Karlsson, S. Recycled polyolefins. Material properties and means for quality determination. *Adv. Polym. Sci.* **2004**, *169*, 201–229. [CrossRef]
41. Vandebril, S.; Vermant, J.; Moldenaers, P. Efficiently suppressing coalescence in polymer blends using nanoparticles: Role of interfacial rheology. *Soft Matter* **2010**, *6*, 3353–3362. [CrossRef]
42. Thareja, P.; Moritz, K.; Velankar, S.S. Interfacially active particles in droplet/matrix blends of model immiscible homopolymers: Particles can increase or decrease drop size. *Rheol. Acta* **2010**, *49*, 285–298. [CrossRef]

43. Grace, H.P. Dispersion phenomena in high viscosity immiscible fluid systems and application of static mixers as dispersion devices in such systems. *Chem. Eng. Commun.* **1982**, *14*, 225–277. [CrossRef]
44. Tucker, C.L., III; Moldenaers, P. Microstructural evolution in polymer blends. *Annu. Rev. Fluid Mech.* **2002**, *34*, 177–210. [CrossRef]
45. Graziano, A.; Dias, O.A.T.; Garcia, C.; Jaffer, S.; Tjong, J.; Sain, M. Impact of Reduced Graphene Oxide on structure and properties of polyethylene rich binary systems for performance-based applications. *Polymer* **2020**, *202*, 122622. [CrossRef]
46. Strugova, D.; David, É.; Demarquette, N.R. Effect of steady shear deformation on electrically conductive PP/PS/MWCNT composites. *J. Rheol.* **2023**, *67*, 977–993. [CrossRef]
47. Guo, Y.; Zuo, X.; Xue, Y.; Tang, J.; Gouzman, M.; Fang, Y.; Zhou, Y.; Wang, L.; Yu, Y.; Rafailovich, M.H. Engineering thermally and electrically conductive biodegradable polymer nanocomposites. *Compos. Part B Eng.* **2020**, *189*, 107905. [CrossRef]
48. Parameswaranpillai, J.; Sanjay, M.; Varghese, S.A.; Siengchin, S.; Jose, S.; Salim, N.; Hameed, N.; Magueresse, A. Toughened PS/LDPE/SEBS/xGnP ternary composites: Morphology, mechanical and viscoelastic properties. *Int. J. Light. Mater. Manuf.* **2019**, *2*, 64–71. [CrossRef]
49. Juan, L. Simultaneous improvement in the tensile and impact strength of polypropylene reinforced by graphene. *J. Nanomater.* **2020**, *2020*, 7840802. [CrossRef]
50. Arjmand, M.; Apperley, T.; Okoniewski, M.; Sundararaj, U. Comparative study of electromagnetic interference shielding properties of injection molded versus compression molded multi-walled carbon nanotube/polystyrene composites. *Carbon* **2012**, *50*, 5126–5134. [CrossRef]
51. Karimi, S.; Helal, E.; Gutierrez, G.; Moghimian, N.; Madinehei, M.; David, E.; Samara, M.; Demarquette, N. A review on graphene's light stabilizing effects for reduced photodegradation of polymers. *Crystals* **2021**, *11*, 3. [CrossRef]
52. Gijsman, P.; Meijers, G.; Vitarelli, G. Comparison of the UV-degradation chemistry of polypropylene, polyethylene, polyamide 6 and polybutylene terephthalate. *Polym. Degrad. Stab.* **1999**, *65*, 433–441. [CrossRef]

crystals

MDPI

Article

Effect of Few-Layer Graphene on the Properties of Mixed Polyolefin Waste Stream

S. M. Nourin Sultana [1], Emna Helal [1,2,*], Giovanna Gutiérrez [2], Eric David [1], Nima Moghimian [2] and Nicole R. Demarquette [1,*]

[1] Mechanical Engineering Department, Ecole de Technologie Superieure, 1100 Notre-Dame Street West, Montreal, QC H3C 1K3, Canada
[2] NanoXplore Inc., 4500 Thimens Blvd, Montreal, QC H4R 2P2, Canada
* Correspondence: emna.helal@nanoxplore.ca (E.H.); nicoler.demarquette@etsmtl.ca (N.R.D.)

Abstract: This work demonstrates how the addition of few-layer graphene (FLG) influences the processability and mechanical properties of the mixed polyolefin waste stream (R-(PE/PP)). Three different types of compounds were investigated: (1) R-(PE/PP) with FLG; (2) blends of R-(PE/PP) with prime polyethylene (PE) or polypropylene (PP) or PP copolymer; and (3) R-(PE/PP) with both the prime polymer and FLG. The processability was assessed by measuring the torque during melt extrusion, the melt flow index (MFI), and viscosity of the compounds. Investigations of the processability and mechanical properties of the composites indicate that the presence of FLG can reinforce the composites without hindering the processability, an unusual but desired feature of rigid fillers. A maximum increase in tensile strength by 9%, flexural strength by 23%, but a reduction in impact strength were observed for the compounds containing R-(PE/PP), 4 wt.% FLG, and 9 wt.% prime PP. The addition of FLG concentrations higher than 4 wt.% in R-(PE/PP), however, resulted in higher tensile and flexural properties while preserving the impact strength. Remarkably, the addition of 10 wt.% FLG increased the impact strength of the composite by 9%. This increase in impact strength is attributed to the dominant resistance of the rigid FLG particles to crack propagation.

Keywords: graphene; mixed polyolefin; waste stream; processability; mechanical properties

Citation: Sultana, S.M.N.; Helal, E.; Gutiérrez, G.; David, E.; Moghimian, N.; Demarquette, N.R. Effect of Few-Layer Graphene on the Properties of Mixed Polyolefin Waste Stream. *Crystals* **2023**, *13*, 358. https://doi.org/10.3390/cryst13020358

Academic Editors: Justina Gaidukevic and Jurgis Barkauskas

Received: 31 January 2023
Revised: 13 February 2023
Accepted: 15 February 2023
Published: 19 February 2023

1. Introduction

Over the recent years, polymer materials have demonstrated a promising role in household as well as industrial applications. Consequently, plastic waste accumulation is soaring. According to a survey by Deloitte [1], over 11 Mt of plastics were wasted in Canada during 2016 only, and only 9% was recycled; around 4% was incinerated for energy recovery and 87% was dumped into the environment. The estimated loss was CAD 7.8 billion for Canada.

Polyolefins (different types of polyethylene (PE) and polypropylene (PP)) are widely used in packaging materials [2] and are heavily present in post-consumer plastic residues. However, because they have comparable densities, they cannot be easily separated using standard techniques available in most sorting facilities. The most straightforward solution to recycle them consists of blending them and using them as is. However, this is quite problematic as PE and PP are thermodynamically immiscible with each other, resulting in a material with weak mechanical properties. Hence, this type of mixed plastic waste is often directed to incineration and landfills [3]. Therefore, improving the properties of mixed polyolefin waste streams is highly desired to enable the recycling of a huge portion of polymer waste.

In general, the mechanical recycling of plastic waste involves mixing with a rigid filler, prime polymer, or both [4]. In this regard, several investigations have been reported on the influence of adding different copolymers [5–8], solid particles [9–12], and combinations of copolymers and solid particles [6,12–14] on the mechanical properties of prime polyolefin mixtures. However, these studies cannot be, necessarily, applied to mixed polyolefin waste due to many uncertainties such as polymer aging, presence of impurities, varied molecular structure, and molecular weight. Hence, more research work on the improvement of the mechanical properties of a mixed polyolefin waste stream is needed. A few studies are available on the upgrade of mixed polyolefin waste streams [15–18], involving a copolymer and a few rigid fillers. These additives in PE/PP blends mostly increase the impact strength (impact toughness) and elongation at break (tensile toughness or ductility), but deteriorate the strength and modulus (stiffness) compared to the unfilled blends. Generally, impact modifying copolymers increase toughness but hinder the strength of the original material. Unlike copolymers, inorganic rigid fillers act as stress concentrators and mostly increase stiffness and strength but reduce toughness. Therefore, different dosages of mechanically strong fillers and/or impact modifying copolymers need to be explored to obtain a better balance of stiffness, strength, and toughness in the composite.

Concomitantly, graphene is a 2D carbonaceous filler with lightweight and excellent mechanical strength. Moreover, commercially available few-layer graphene (FLG), mass-produced from natural graphite by a mechanochemical exfoliation process, allows for an economically viable use of graphene at an industrial scale. Although there have been several reports on the effect of adding modified and unmodified graphene on electrical properties [19–24], only a few have focused on the mechanical properties [19,25] of prime PE/PP blends. Furthermore, the effect of the addition of FLG to improve the properties of a mixed polyolefin waste stream, to our knowledge, has not been investigated. Therefore, this study was conducted to analyze the potential of commercially available FLG, which is less expensive than lab-grade graphene, to improve the properties of a mixed polyolefin waste stream.

In this work, a mixed polyolefin waste stream (designated as R-(PE/PP) mixture was melt blended with either FLG alone, a prime polymer alone (PE, PP homopolymer, or PP copolymer), or FLG in combination with a prime polymer. To prepare the composites containing R-(PE/PP) and FLG, the FLG concentration was varied from 1 to 10 wt.%. To prepare the blends of R-(PE/PP), FLG, and each of the three prime polymers, only a 4 wt.% loading was selected. Torque during extrusion, melt flow index (MFI), and viscosity of the samples were measured to evaluate the influence of adding FLG on the processability of the composites. In addition, the tensile, flexural, and impact properties of the composites of R-(PE/PP) containing FLG and/or the prime polymers were evaluated. Finally, the influence of adding FLG on the microstructure, and subsequently, the mechanical properties of the composites were investigated by scanning electron microscopy (SEM).

2. Materials and Methods

2.1. Materials

FLG powder (GrapheneBlack 3X) from NanoXplore Inc., Montreal, QC, Canada, produced through a water-based environmentally friendly mechano-chemical exfoliation process was used in this work. The average thickness of FLG is 6–10 atomic layers and the lateral primary particle size is 1 to 2 μm. Loose clusters of primary particles form secondary particles. The average lateral size of secondary particles is around 30 μm.

A R-(PE/PP) mixture sourced from a local recycler in Quebec was selected for this work. According to the supplier's information, the mixture contains around 30 to 40 wt.% PP with less than 5 wt.% contamination (dye, ink, or pigment, etc.). Differential scanning calorimetry (DSC) analysis of the R-(PE/PP) samples (Figure S1, provided in the Supplementary Materials) indirectly confirmed the composition to be about PE/PP-60/40.

Masterbatches loaded with 30 wt.% FLG in three different prime polymers (PE homopolymer, PP homopolymer, and PP copolymer), provided by NanoXplore Inc., Montreal, QC, Canada, were used in this work. In addition, a masterbatch obtained by adding 20 wt.% FLG to the R-(PE/PP) was also prepared and used.

The specifications of the masterbatch polymers used in the present work are summarized in Table 1.

Table 1. Identification, MFI, and density of the polymers used in this work.

Masterbatch Polymer	Commercial Name	MFI (g/10 min)	Density (g/cm^3)
R-(PE/PP)	N/A	≥4 (230 °C, 2.16 kg)	N/A
PE homopolymer	Alathon H5618	17 (190 °C, 2.16 kg)	0.955
PP homopolymer	Polypropylene 3720 WZ	20 (230 °C, 2.16 kg)	0.905
PP copolymer	Formolene 2620 A	20 (230 °C, 2.16 kg)	0.900

2.2. Processing

Three types of composites were obtained:

Type 1: R-(PE/PP)/FLG composites by blending the R-(PE/PP) with the R-(PE/PP)/FLG masterbatch to analyze the individual effect of adding FLG.

Type 2: R-(PE/PP)/prime polymer compounds by mixing the R-(PE/PP) with the prime polymer (PE, PP homopolymer, and PP copolymer) to analyze the effect of adding only the prime polymer.

Type 3: R-(PE/PP)/prime polymer/FLG composites by mixing the R-(PE/PP) with the prime polymer/FLG masterbatch to investigate the combined effect of adding the prime polymer and FLG.

All of the samples were processed in a HAAKE twin-screw extruder, (Rheomex OS PTW16/40) at 150 rpm, and 200 °C in all zones. Table 2 summarizes the nomenclature and corresponding compositions of the samples.

Table 2. Sample nomenclature and respective composition of the samples.

Type of Compounds	Samples Nomenclature	R-(PE/PP) Concentration (wt.%)	Prime Polymer Concentration (wt.%)	FLG Concentration (wt.%)
As received mixed polyolefin waste stream	R-(PE/PP)	100	0	0
Type 1	R-(PE/PP)/FLG = 99/1	99	0	1
	R-(PE/PP)/FLG = 96/4	96	0	4
	R-(PE/PP)/FLG = 93/7	93	0	7
	R-(PE/PP)/FLG = 90/10	90	0	10
Type 2	R-(PE/PP)/PE/FLG = 87/13	87	13	0
	R-(PE/PP)/PP/FLG = 87/13	87	13	0
	R-(PE/PP)/PPcop/FLG = 87/13	87	13	0
Type 3	R-(PE/PP)/PE/FLG = 87/9/4	87	9	4
	R-(PE/PP)/PP/FLG = 87/9/4	87	9	4
	R-(PE/PP)/PPcop/FLG = 87/9/4	87	9	4

In the case of the Type 3 composites, 4 wt.% concentration of FLG was selected for further investigation. This choice was based on the performance of Type 1 composites, which, after the addition of 4 wt.% FLG, started to exhibit observable differences in the mechanical properties of the composites. Therefore, 87 wt.% R-(PE/PP) was blended with the 13 wt.% prime polymer/FLG masterbatch, resulting in Type 3 composites with 4 wt.% FLG. The final composition resulted in the R-(PE/PP)/prime polymer/FLG = 87/9/4.

2.3. *Characterizations*

2.3.1. Melt Flow Index (MFI)

A melt flow index (MFI) tester (manufactured by International Equipments, Mumbai, India) was used to estimate the MFI of the samples according to ASTM D1238.

2.3.2. Rheological Properties

A capillary rheometer (model: Instron SR20) was used to measure the viscosity of the samples at 200 °C at a shear rate varying from 50 s^{-1} to 1000 s^{-1}. This range was selected because it includes the shear rate that the polymer undergoes during the extrusion or injection processes. The length and diameter of the capillary were 40 mm and 1 mm, respectively, resulting in an L/D ratio of 40.

2.3.3. Mechanical Properties

An MTS Alliance RF/200 tensile test device was used to study the tensile and flexural properties of the samples. Tensile properties were studied according to ASTM D638 at a crosshead speed of 50 mm/min to measure the tensile strength, tensile modulus, and elongation at break, at room temperature. Flexural strength and modulus were analyzed according to ASTM D790 with a 0.1 (mm/mm)/min outer fiber strain rate (Z). For each case, a 10 kN load cell was used.

An impact strength tester device (manufactured by International Equipments, Mumbai, India) was used to measure the notched impact strength of the samples. The impact strength was evaluated according to ASTM D256. A motorized notch cutter (manufactured by International Equipments, Mumbai, India) was used to make notches in the samples.

At least five specimens of each composite were tested at room temperature to report the tensile, flexural, and impact properties. The average value is reported along with the standard deviation.

2.3.4. Morphological Characterization

A SEM S3600-N Hitachi (Model: MEB-3600-N) was used to obtain scanning electron microscope (SEM) images of the samples' fractured surface. The FLG powder was attached to the double-sided carbon tape without any gold coating to obtain the SEM image. To obtain the SEM images of the composites, the samples were cryo-fractured to analyze the morphology. In addition, a few samples were fractured by applying an impact load. A Gold Sputter Coater (Model: K550X) was used to coat the surface of the compounds before taking the SEM images.

3. Results and Discussions

3.1. *Rheological Properties*

The MFI values (Figure S2, provided in the Supplementary Materials) of the neat R-(PE/PP) and FLG-filled composites (Type 1 and Type 3) were found to be independent of the studied additive concentrations and types. Similarly, the applied torque during extrusion remained at 48 ± 1 Nm, irrespective of the FLG concentration in the composites. Both observations indicate that unlike other solid particles [26,27], which adversely affect the processability and flow properties of the polymeric materials, FLG had very little or no effect on the flow behavior of the R-(PE/PP) when added up to 10 wt.%.

Furthermore, Figure 1 shows the viscosity as a function of shear rate at 200 °C (processing temperature) for a few of the studied compounds. It can be seen that the viscosity of the materials was not affected by the addition of graphene. This can be attributed to the lubrication effect of FLG [28]. Similar findings were reported by the co-authors in a previous study regarding prime high density polyethylene (HDPE)/recycled HDPE/FLG blend composites [4]. Another study [28] reported similar results for HDPE/graphite composites.

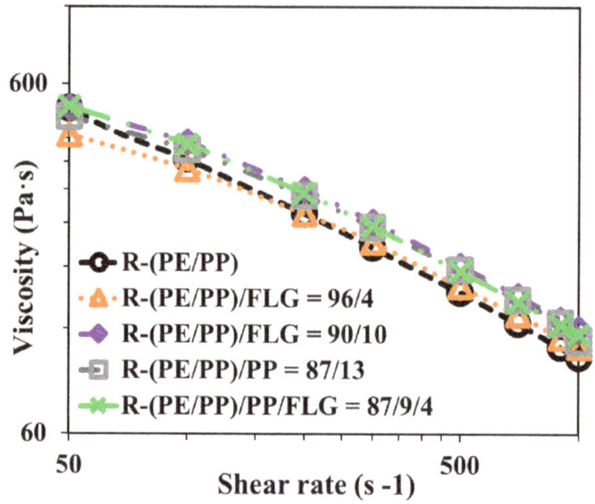

Figure 1. Apparent viscosity of the neat R-(PE/PP) and the additive (FLG, prime polymer) filled R-(PE/PP) samples at 200 °C as a function of shear rate.

3.2. Mechanical Properties

The tensile and flexural properties of the R-(PE/PP)/FLG composites (Type 1) as a function of FLG concentration were investigated. Figure 2a shows the tensile and flexural strengths, and Figure 2b shows the tensile and flexural moduli of the studied compounds. Figure 2c displays their impact strength, and Figure 2d indicates their tensile toughness and elongation at break.

From Figure 2a,b, it can be observed that the presence of FLG resulted in an increase in the strength and modulus (stiffness) of the composites when compared to the one of the unfilled R-(PE/PP). The most significant increases in the tensile, flexural, and impact properties were observed for the R-(PE/PP)/FLG = 90/10 composite. These increases in strength and stiffness can be attributed to the reinforcement effect of the FLG particles in the R-(PE/PP). Previously, a similar reinforcing behavior of carbon fiber [9], glass beads [10], short glass fiber [11], and wood flour [29] in a controlled PE/PP blend system has been reported in the literature.

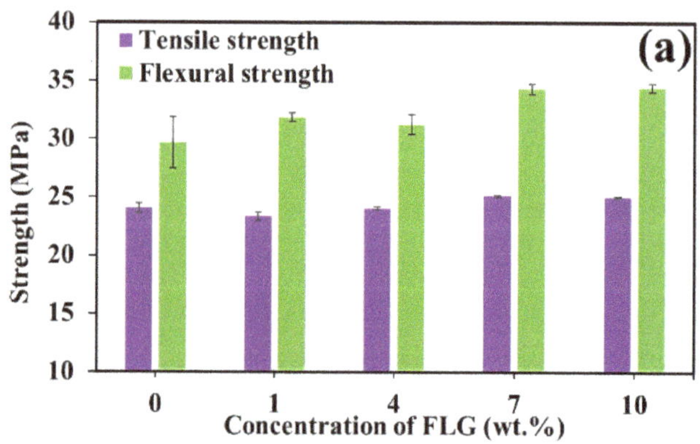

Figure 2. *Cont.*

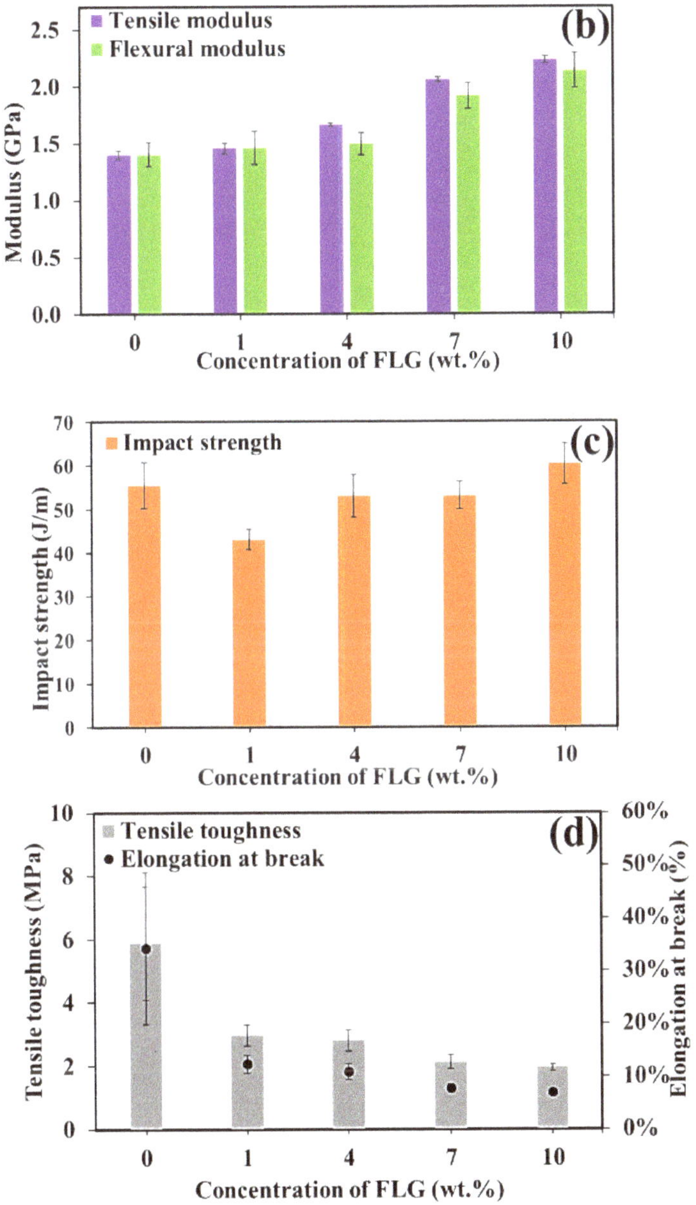

Figure 2. (**a**) Tensile strength and flexural strength, (**b**) tensile modulus and flexural modulus and (**c**) impact strength, and (**d**) tensile toughness and elongation at break of the R-(PE/PP)/FLG composites as a function of FLG concentration.

As illustrated in Figure 2c, the addition of 1 wt.% FLG resulted in a reduced impact strength while 4 and 7 wt.% FLG did not reduce the impact strength of the composites. However, an increase in the impact strength was observed in the R-(PE/PP)/FLG = 90/10 composite. This unusual trend in the impact strength by the addition of rigid fillers has only been reported in a few studies, and was attributed to the resistance of the rigid filler particles to crack propagation throughout the composite [30].

Both elongation at break and tensile toughness of the FLG-filled composites decreased as the FLG concentration increased, as can be seen in Figure 2d. This is a common effect of the addition of rigid fillers and has been reported in the literature [12,31].

It is worth noting that the increase in the FLG loading yielded the opposite effect on the tensile toughness and impact resistance of the Type 1 composites. Possible reasons explaining this difference are presented in the Discussion section.

Mechanical properties of the Type 2 (R-(PE/PP)/prime polymer = 87/13) and Type 3 (R-(PE/PP)/prime polymer/FLG = 87/9/4) composites are summarized in Figure 3. The tensile strength, flexural strength, tensile modulus, flexural modulus, elongation at break, tensile toughness (estimated by the calculating the area underneath the stress–strain curve), and impact strength are, respectively, presented in Figure 3a–g, as a function of the type of prime polymer present in the Type 2 and Type 3 composites.

As observed in Figure 3a–d, the Type 3 composite containing both the prime polymer (PE or PP) and FLG presented higher tensile and flexural properties than that of the neat R-(PE/PP) and respective Type 2 samples. For example, with 4 wt.% FLG and 9 wt.% PP, the flexural strength of Type 3 was 21% higher than the corresponding one of the Type 2 composite with 13 wt.% PP.

However, the Type 2 composites showed a higher elongation at break, tensile toughness, and impact strength than the corresponding ones of Type 3. A substantial increase in the impact strength and ductility was observed in the Type 2 composite containing the PP copolymer. Precisely, the addition of the impact-modifying PP copolymer resulted in a sharp increase in the fracture toughness by 132% for Type 2 and by 50% for Type 3 when compared to the neat R-(PE/PP). A similar effect resulting from the addition of copolymers has been reported [5–8].

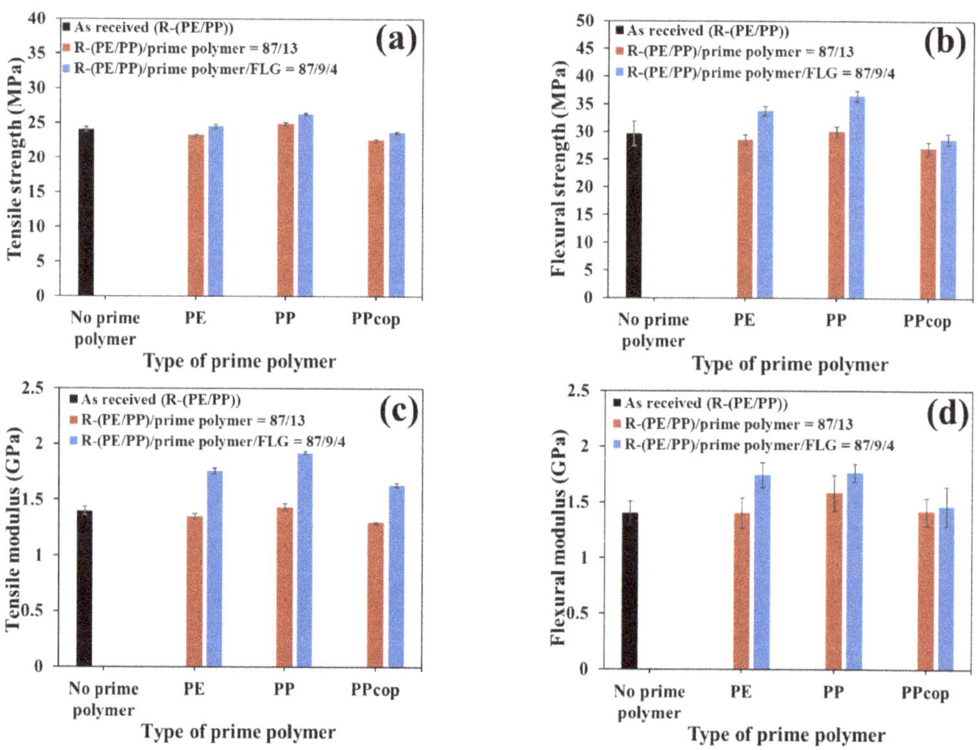

Figure 3. *Cont.*

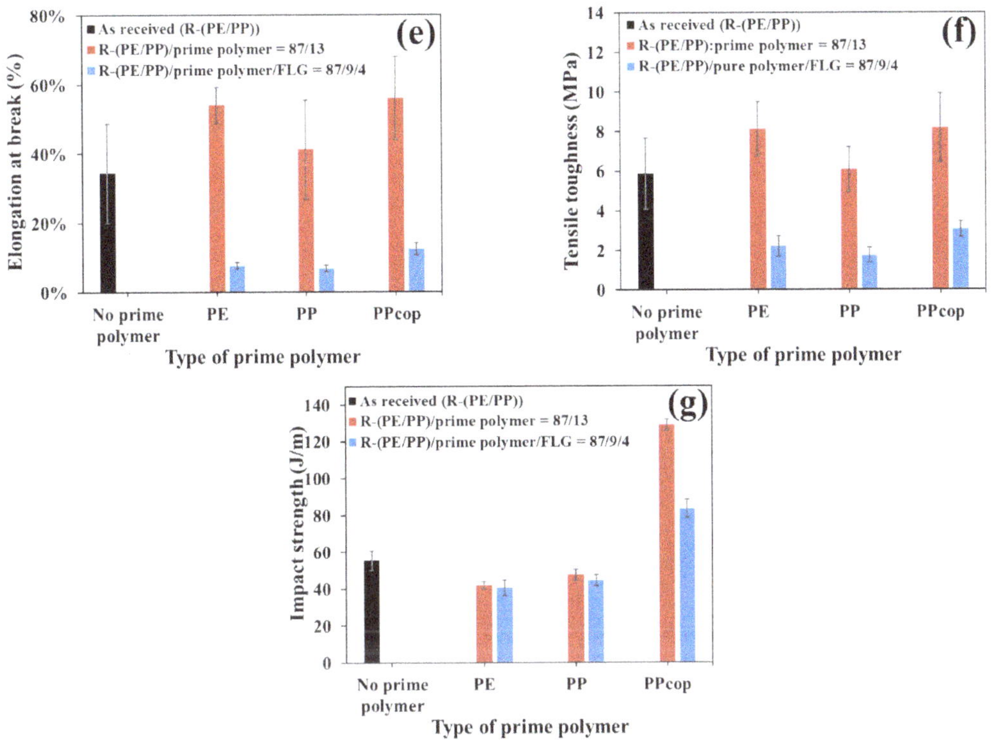

Figure 3. (**a**) Tensile strength, (**b**) flexural strength, (**c**) tensile modulus, (**d**) flexural modulus, (**e**) elongation at break, (**f**) tensile toughness, and (**g**) impact strength of the Type 2 and Type 3 samples as a function of the type of the used prime polymer. The properties of the as received R-(PE/PP) are presented by the black bars as a baseline.

3.3. Morphology

Figure 4 shows the SEM images of the cryo-fractured surface of (a) R-(PE/PP), (b) and (c) R-(PE/PP)/FLG = 96/4 and (d) FLG powder.

In Figure 4a,b, two different phases are visible. The rough phase corresponds to PE and the smooth phase to PP. PP has a much higher glass transition temperature than PE. During cryo-fracturing, PP dips below its glass transition temperature much easier than the case for PE. Consequently, the PE undergoes ductile fracture while the PP undergoes brittle fracture [18]. Although PE was the major component of the mixed polyolefin waste stream of this work, the area of the PE phase domain was smaller than that of the PP phase. In the case of an immiscible polymer blend, the morphology is directly dependent on the viscosity ratio of the constituting polymer components at the processing conditions. In the PE/PP immiscible blend systems, PP can form the matrix of the blend morphology with minor weight contribution, if the viscosity of PP is less than that of PE. A similar observation of creating a matrix by the minor component of the blend has been reported in the literature [18]. The authors showed this observation in the case of a PE/PP-65/35 blend where the viscosity of PE was 500 Pa·s and the viscosity of PP was 300 Pa·s.

Moreover, Figure 4c indicates the localization of FLG in the PE phase, which was similar to the observations by other authors [23]. The SEM image of the FLG powder on carbon tape is provided in Figure 4d as a reference and for the purpose of comparison. The preferential localization of FLG in the PE phase is attributed to lower interfacial tension between PE and graphene than that of PP and graphene [32]. Moreover, it is observed that the addition of FLG resulted in a finer or elongated domain size of the PE phase in the R-(PE/PP)/FLG = 96/4 composite compared to that of the neat R-(PE/PP). A similar refinement of morphology in the presence of FLG has been reported by other authors [4].

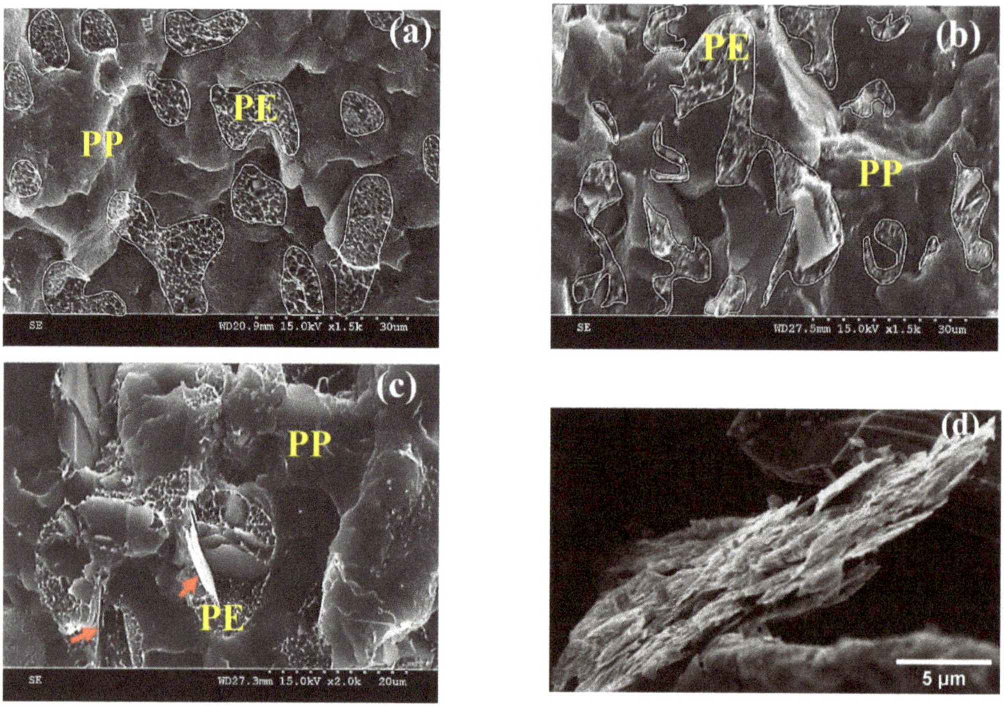

Figure 4. SEM images of (**a**) neat R-(PE/PP), (**b**,**c**) R-(PE/PP)/FLG = 96/4, and (**d**) FLG powder; in (**a**,**b**), the domain of the PE phase is marked by a white outline and in (**c**), the localization of FLG in the PE phase is marked by the red arrow to guide the readers' eyes.

3.4. Discussion

As shown in Figure 1, the viscosity of the FLG-filled composites remained independent of the loading of FLG. The observed torque during extrusion and the MFI values (provided in Figure S2 of Supplementary Materials) also support this finding. Therefore, it can be stated that FLG loadings up to 10 wt.% do not affect the processability of the composites. This may be the outcome of commensurable contacts of filler agglomerates, which maintain both the slippage and mobility of the molten polymer chain in the filled composites, similar to the unfilled compound (i.e., lubricant effect) [28].

It was observed that in the Type 1 composites, higher loadings of FLG increased both the tensile and flexural strength and modulus of the composites. This is attributed to the reinforcing effect of strong solid particles [9,11,29] such as FLG. Furthermore, decreasing trends of both elongation at break and tensile toughness were observed at increasing concentrations of FLG. In contrast, the impact strength (fracture toughness) behavior as a function of FLG loading showed a different trend. Low (1 wt.%), medium (4 and 7 wt.%), and high loadings (10 wt.%) of FLG resulted in reduced, similar, and increased impact strength, respectively, compared to the one of neat R-(PE/PP). In particular, the R-(PE/PP)/FLG = 90/10 composites showed an increase in the fracture toughness but a decrease in elongation at break and tensile toughness. In homogeneous or single-phased materials, both impact toughness and tensile toughness follow a similar trend of variation. More complicated trends might be at play in heterogeneous or multiphase composite systems. In fact, the elongation at break of a polymer composite depends mostly on the extent of the polymer chains' mobility while the impact strength depends on the rate of crack propagation throughout the compound. Solid particles reinforcing the polymer composites may have a positive or negative effect on the impact strength by acting either as defect centers [33] or crack resisting points [30], depending on their strength and stiffness. In other words, the overall impact strength of the solid particle filled blend composites is the outcome of the competition between the negative effect of a filler as a defect center and the positive effect of the crack resistance of the filler. A higher impact strength of the composite can be achieved if the loading of the rigid filler is within an optimum concentration range where the crack propagation resistance of the filler is the dominant factor. In this work, the reduction in the impact strength of the R-(PE/PP)/FLG = 99/1 composite indicates that 1 wt.% FLG was not within the desired optimum concentration range to overcome the negative effect of fillers as defect centers in the composite. On the other hand, the increased impact strength of the R-(PE/PP)/FLG = 90/10 composite suggests that 10 wt.% FLG was adequate and within the optimum concentration range to increase the resistance to crack propagation, and to overcome the negative effect of FLG as a defect center, on the impact toughness of the composite. For reference, the SEM images of the fractured surfaces of the neat R-(PE/PP), R-(PE/PP)/FLG = 96/4, and R-(PE/PP)/FLG = 90/10 composites are shown in Figure 5a–c, respectively.

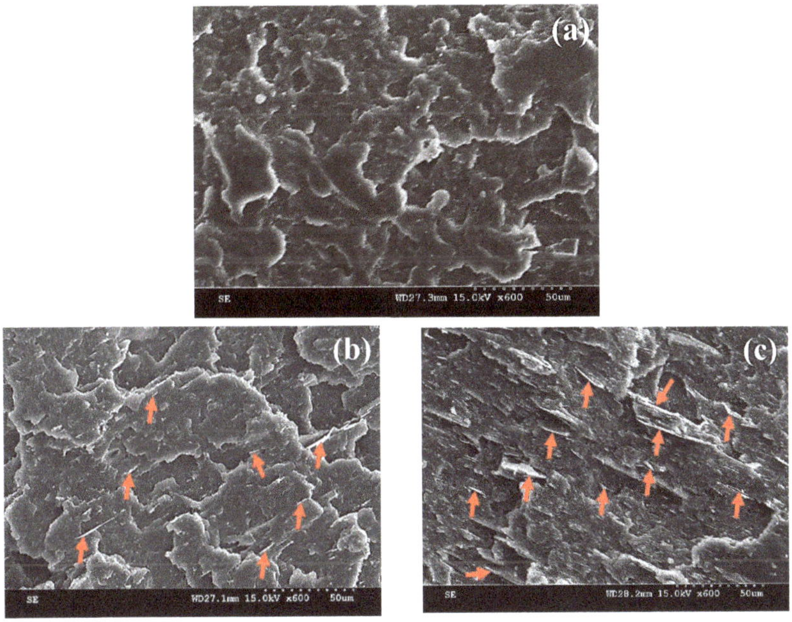

Figure 5. SEM images of the fractured surface of the (**a**) neat R-(PE/PP), (**b**) R-(PE/PP)/FLG = 96/4, and (**c**) R-(PE/PP)/FLG = 90/10. The arrows indicate a few of the FLGs, visible on the fractured surface of the R-(PE/PP)/FLG composites.

It can be seen in Figure 5b,c that more FLG was visible in the R-(PE/PP)/FLG = 90/10 composite than in the R-(PE/PP)/FLG = 96/4 composite. In other words, R-(PE/PP)/FLG-90/10 (shown in Figure 5c) had more impact crack resisting sites than the neat R-(PE/PP) and R-(PE/PP)/FLG = 96/4 (as can be seen in Figure 5b)). This phenomenon resulted in a higher impact strength for the R-(PE/PP)/FLG = 90/10 composite. However, the polymer chain mobility or slippage of the polymer chains in the solid state was restricted by strong and rigid FLG, which reduced the elongation at break of R-(PE/PP)/FLG = 90/10. Simultaneous increment in the impact toughness and reduction in the tensile toughness in the same composite was also observed by Antimo et al. [25] in a PE/PP-90/10 composite with 4 wt.% graphene derivatives.

Based on the analysis of the plots in Figures 2 and 3, it can be concluded that R-(PE/PP)/FLG = 90/10, R-(PE/PP)/PP/FLG = 87/9/4, and R-(PE/PP)/PPcopolymer = 87/13 compounds have shown significant improvement in different properties compared to that of the unmodified R-(PE/PP). Table 3 shows the variation in the mechanical properties of the best performing samples in this work.

Table 3. Mechanical property variation (%) in compounds showing a significant change in properties compared to the neat R-(PE/PP).

		Property Variation (%)		
Properties	**As Received R-(PE/PP)**	**Type 1 with 10 wt.% FLG**	**Type 2 with 13 wt.% PPcopolymer**	**Type 3 with 9 wt.% PP and 4 wt.% FLG**
Tensile strength	24 MPa	+4%	−6%	+9%
Tensile modulus	1.4 GPa	+60%	−7%	+37%
Flexural strength	30 MPa	+16%	−9%	+23%
Flexural modulus	0.9 GPa	+46%	+1%	+34%
Elongation at break	34%	−80%	+63%	−80%
Tensile toughness	6 MPa	−67%	+39%	−70%
Impact strength	55.5 J/m	+9%	+132%	−20%

Further analysis of the data presented in Table 3 suggests that, if the target application requires an overall increase in the strength, stiffness, and fracture toughness in the mixed polyolefin waste stream, then the addition of a higher loading of FLG is recommended. Since any reprocessing step or inclusion of any additive (fillers or nanofillers) will increase the cost of R-(PE/PP), the addition of a higher loading of FLG will increase the price of the R-(PE/PP)/FLG composite. However, it is worth mentioning that an industrial grade FLG was used in this work, which is more cost effective than other grades of carbon based nanofillers such as carbon nanotubes (CNT). In addition, FLG has the potential to reduce the photodegradation of polymer composites [34]. Therefore, it is expected that FLG-filled composites feature a higher strength and durability compared to those of the unmodified mixed polyolefin waste stream, at a cost of the added FLG. Finally, the presented approach is intended to recover mixed plastic waste streams that have little or no value at the base and are otherwise landfilled. For superior impact and tensile toughness, the strategy of the mixing of R-(PE/PP) with a copolymer would be more useful. For applications requiring higher strength and stiffness and less sensitivity to toughness, compounds containing both prime PP and FLG would be interesting.

4. Conclusions

We investigated the influence of adding FLG on the processability and mechanical properties of mixed polyolefin waste streams. FLG alone, the prime polymer (PE, PP, or PP copolymer) alone, and FLG in combination with different prime polymers were melt blended with the R-(PE/PP). MFI, effective torque during extrusion, and viscosity of the FLG-filled composites remained unchanged for concentrations of FLG up to 10 wt.%. Moreover, it was shown that the addition of 4 wt.% FLG in combination with prime PP increased the tensile strength by 9%, the tensile modulus by 37%, the flexural strength by 23%, and the flexural modulus by 34%, but reduced the impact strength by 20% compared to the respective properties of the neat R-(PE/PP). However, the addition of 10 wt.% of FLG increased the tensile strength by 4%, tensile modulus by 60%, flexural strength by 16%, flexural modulus by 46%, and the impact strength by 9% compared to that of the neat R-(PE/PP). This desired increase in the tensile and flexural properties is attributed to the reinforcing effect of FLG. The increase in the impact strength of R-(PE/PP)/FLG with a higher loading (10 wt.%) of FLG was attributed to the resistance effect of strong FLG particles to impact crack propagation.

Based on the experimental and morphological observations, this work reports that FLG enables the processability of R-(PE/PP) while improving the mechanical properties of FLG-filled R-(PE/PP) composites. This work outlines the possibility of channeling a mixed polyolefin waste stream back to various regular applications with more strength, without altering the processability. FLG-reinforced R-(PE/PP) composites can be used in outdoor furniture and in the construction industry. As a result, the demand for virgin polymers could be reduced, which would be a big step toward the protection of biodiversity and a reduction in harmful gas emissions. Moreover, this study suggests a useful application of commercially produced FLG.

Supplementary Materials: The following supporting information can be downloaded at: https://www.mdpi.com/article/10.3390/cryst13020358/s1, Figure S1: DSC analysis of unfilled R-(PE/PP), Figure S2: MFI of (**a**) R-(PE/PP)/FLG composites as a function of the concentration of FLG and (**b**) R-(PE/PP)/prime polymer/FLG composites as a function of the type of prime polymer.

Author Contributions: Conceptualization, S.M.N.S., E.H., G.G., E.D., N.M. and N.R.D.; Methodology, S.M.N.S., E.H., E.D. and N.R.D.; Validation, S.M.N.S., E.H., G.G., E.D., N.M. and N.R.D.; Formal analysis, S.M.N.S., E.H., G.G., E.D., N.M. and N.R.D.; Investigation, S.M.N.S., E.H., E.D. and N.R.D.; Resources, E.H., G.G., E.D., N.M. and N.R.D.; Writing—original draft, S.M.N.S.; Writing—review & editing, E.H., G.G., E.D., N.M. and N.R.D.; Visualization, S.M.N.S., E.H., E.D. and N.R.D.; Supervision, E.H., G.G., E.D., N.M. and N.R.D.; Project administration, E.H., G.G., E.D., N.M. and N.R.D.; Funding acquisition, S.M.N.S., E.D. and N.R.D. All authors have read and agreed to the published version of the manuscript.

Funding: This research was funded by NanoXplore Inc.; Natural Sciences and Engineering Research Council (grant number CRDPJ 538482—18); PRIMA Quebec (grant number R18-46-001); Fonds de Recherche du Québec-Nature et Technologies (grant number 318813). The APC was funded by discounts from Editorial Office, MDPI and IOAP of Ecole de Technologie Superieure (ETS) Montreal.

Institutional Review Board Statement: Not applicable.

Informed Consent Statement: Not applicable.

Data Availability Statement: The author confirms that the data supporting the reported results can be found within the article and its Supporting Materials. Raw data that supports the findings of this work can be inquired to the corresponding authors, upon reasonable request.

Acknowledgments: The authors would like to thank Mazen Samara for his helpful feedback to improve the language of this paper, Laure Castagna for preparing a few samples during her internship, and Jose Carlos Ferreira Junior for the help in obtaining the SEM image of the FLG powder.

Conflicts of Interest: The authors declare no conflict of interest.

References

1. Petigny, J.; Ménigault, C.; Luisce, T.; Harscoët, E.; David, A.; Mitsios, A.; Laberge, M.; Lysenko, D.; Moore, P.; Dimoff, A.; et al. Economic Study of the Canadian Plastic Industry, Markets and Waste: Summary Report to Environment and Climate Change Canada. 2019, pp. 1–43. Available online: https://publications.gc.ca/site/eng/9.871296/publication.html (accessed on 30 January 2023).
2. Jubinville, D.; Esmizadeh, E.; Saikrishnan, S.; Tzoganakis, C.; Mekonnen, T. A comprehensive review of global production and recycling methods of polyolefin (PO) based products and their post-recycling applications. *Sustain. Mater. Technol.* **2020**, *25*, e00188. [CrossRef]
3. Maris, J.; Bourdon, S.; Brossard, J.-M.; Cauret, L.; Fontaine, L.; Montembault, V. Mechanical recycling: Compatibilization of mixed thermoplastic wastes. *Polym. Degrad. Stab.* **2018**, *147*, 245–266. [CrossRef]
4. Diallo, A.K.; Helal, E.; Gutiérrez, G.; Madinehei, M.; David, E.; Demarquette, N.; Moghimian, N. Graphene: A multifunctional additive for sustainability. *Sustain. Mater. Technol.* **2022**, *33*, e00487. [CrossRef]

5. Vranjes, N.; Rek, V. Effect of EPDM on Morphology, Mechanical Properties, Crystallization Behavior and Viscoelastic Properties of iPP+HDPE Blends. *Macromol. Symp.* **2007**, *258*, 90–100. [CrossRef]
6. Parameswaranpillai, J.; Parambil, H.P.; Sanjay, M.R.; Siengchin, S. Polypropylene/high-density polyethylene based blends and nanocomposites with improved toughness. *Mater. Res. Express* **2019**, *6*, 075334. [CrossRef]
7. Blom, H.; Teh, J.; Rudin, A. I-PP/HDPE blends. III. Characterization and compatibilization at lower i-PP contents. *J. Appl. Polym. Sci.* **1996**, *61*, 959–968. [CrossRef]
8. Yang, M.; Wang, K.; Ye, L.; Mai, Y.-W.; Wu, J. Low density polyethylene-polypropylene blends: Part 2—Strengthening and toughening with copolymer. *Plast. Rubber Compos.* **2003**, *32*, 27–31. [CrossRef]
9. Hsieh, C.-T.; Pan, Y.-J.; Lin, J.-H. Polypropylene/high-density polyethylene/carbon fiber composites: Manufacturing techniques, mechanical properties, and electromagnetic interference shielding effectiveness. *Fibers Polym.* **2017**, *18*, 155–161. [CrossRef]
10. Qiu, G.-X.; Raue, F.; Ehrenstein, G.W. Mechanical properties and morphologies of PP/mPE/filler composites. *J. Appl. Polym. Sci.* **2002**, *83*, 3029–3035. [CrossRef]
11. Ramos, M.A.; Belmontes, F.A. Polypropylene/low density polyethylene blends with short glass fibers. II: Effect of compounding method on mechanical properties. *Polym. Compos.* **1991**, *12*, 1–6. [CrossRef]
12. Ghasemi, F.A.; Daneshpayeh, S.; Ghasemi, I. Multi-response optimization of impact strength and elongation at break of nanocomposites based on polypropylene/polyethylene binary polymer matrix in the presence of titanium dioxide nanofiller. *J. Elastomers Plast.* **2017**, *49*, 633–649. [CrossRef]
13. Parameswaranpillai, J.; Elamon, R.; Sanjay, M.R.; Siengchin, S. Synergistic effects of ethylene propylene diene copolymer and carbon nanofiber on the thermo-mechanical properties of polypropylene/high-density polyethylene composites. *Mater. Res. Express* **2019**, *6*, 085302. [CrossRef]
14. Mofokeng, T.G.; Ray, S.S.; Ojijo, V. Structure-property relationship in PP/LDPE blend composites: The role of nanoclay localization. *J. Appl. Polym. Sci.* **2018**, *135*, 46193. [CrossRef]
15. Fang, C.; Nie, L.; Liu, S.; Yu, R.; An, N.; Li, S. Characterization of polypropylene–polyethylene blends made of waste materials with compatibilizer and nano-filler. *Compos. Part B Eng.* **2013**, *55*, 498–505. [CrossRef]
16. Karaagac, E.; Koch, T.; Archodoulaki, V.-M. The effect of PP contamination in recycled high-density polyethylene (rPE-HD) from post-consumer bottle waste and their compatibilization with olefin block copolymer (OBC). *Waste Manag.* **2021**, *119*, 285–294. [CrossRef]
17. Najafi, S.K.; Hamidinia, E.; Tajvidi, M. Mechanical properties of composites from sawdust and recycled plastics. *J. Appl. Polym. Sci.* **2006**, *100*, 3641–3645. [CrossRef]
18. Kazemi, Y.; Kakroodi, A.R.; Rodrigue, D. Compatibilization efficiency in post-consumer recycled polyethylene/polypropylene blends: Effect of contamination. *Polym. Eng. Sci.* **2015**, *55*, 2368–2376. [CrossRef]
19. Al-Saleh, M.H. Electrical, EMI shielding and tensile properties of PP/PE blends filled with GNP:CNT hybrid nanofiller. *Synth. Met.* **2016**, *217*, 322–330. [CrossRef]
20. Du, B.X.; Hou, Z.H.; Li, J.; Li, Z.L. Effect of graphene nanoplatelets on space charge and breakdown strength of PP/ULDPE blends for HVDC cable insulation. *IEEE Trans. Dielectr. Electr. Insul.* **2018**, *25*, 2405–2412. [CrossRef]
21. Wei, Z.; Hou, Y.; Jiang, C.; Liu, H.; Chen, X.; Zhang, A.; Liu, Y. Graphene Enhanced Electrical Properties of Polyethylene Blends for High-Voltage Insulation. *Electron. Mater. Lett.* **2019**, *15*, 582–594. [CrossRef]
22. Mun, S.C.; Kim, M.J.; Cobos, M.; Gu, L.; Macosko, C.W. Strategies for interfacial localization of graphene/polyethylene-based cocontinuous blends for electrical percolation. *AIChE J.* **2019**, *65*, e16579. [CrossRef]
23. Tu, C.; Nagata, K.; Yan, S. Morphology and electrical conductivity of polyethylene/polypropylene blend filled with thermally reduced graphene oxide and surfactant exfoliated graphene. *Polym. Compos.* **2017**, *38*, 2098–2105. [CrossRef]
24. Tu, C.; Nagata, K.; Yan, S. Dependence of Electrical Conductivity on Phase Morphology for Graphene Selectively Located at the Interface of Polypropylene/Polyethylene Composites. *Nanomaterials* **2022**, *12*, 509. [CrossRef] [PubMed]
25. Graziano, A.; Garcia, C.; Jaffer, S.; Tjong, J.; Sain, M. Novel functional graphene and its thermodynamic interfacial localization in biphasic polyolefin systems for advanced lightweight applications. *Compos. Sci. Technol.* **2020**, *188*, 321–329. [CrossRef]
26. Lago, E.D.; Cagnin, E.; Boaretti, C.; Roso, M.; Lorenzetti, A.; Modesti, M. Influence of Different Carbon-Based Fillers on Electrical and Mechanical Properties of a PC/ABS Blend. *Polymers* **2020**, *12*, 29. [CrossRef]
27. Koodehi, A.V.; Koohi, A.D. Optimization of Thermal Stability of High-Density Polyethylene Composite Using Antioxidant, Carbon Black and Nanoclay Addition by a Central Composite Design Method: X-ray Diffraction and Rheological Characterization. *J. Macromol. Sci. Part B* **2018**, *57*, 660–678. [CrossRef]
28. Ferreira, E.H.C.; Andrade, R.J.E.; Fechine, G.J.M. The "Superlubricity State" of Carbonaceous Fillers on Polyethylene-Based Composites in a Molten State. *Macromolecules* **2019**, *52*, 9620–9631. [CrossRef]
29. Feng, Y.; Yuan, Z.; Sun, H.; He, H.; Zhang, G. Toughening and reinforcing wood flour/polypropylene composites with high molecular weight polyethylene under elongation flow. *Compos. Sci. Technol.* **2020**, *200*, 108395. [CrossRef]
30. Juan, L. Simultaneous Improvement in the Tensile and Impact Strength of Polypropylene Reinforced by Graphene. *J. Nanomater.* **2020**, *2020*, 7840802. [CrossRef]
31. Batista, N.L.; Helal, E.; Kurusu, R.S.; Moghimian, N.; David, E.; Demarquette, N.R.; Hubert, P. Mass-produced graphene—HDPE nanocomposites: Thermal, rheological, electrical, and mechanical properties. *Polym. Eng. Sci.* **2019**, *59*, 675–682. [CrossRef]

32. Tu, C.; Nagata, K.; Yan, S. Influence of melt-mixing processing sequence on electrical conductivity of polyethylene/polypropylene blends filled with graphene. *Polym. Bull.* **2017**, *74*, 1237–1252. [CrossRef]
33. Dikobe, D.; Luyt, A. Thermal and mechanical properties of PP/HDPE/wood powder and MAPP/HDPE/wood powder polymer blend composites. *Thermochim. Acta* **2017**, *654*, 40–50. [CrossRef]
34. Karimi, S.; Helal, E.; Gutierrez, G.; Moghimian, N.; Madinehei, M.; David, E.; Samara, M.; Demarquette, N. A Review on Graphene's Light Stabilizing Effects for Reduced Photodegradation of Polymers. *Crystals* **2021**, *11*, 3. [CrossRef]

crystals

MDPI

Article

Tailoring the Graphene Properties for Electronics by Dielectric Materials

Isaac Appiah Otoo [1], Aleksandr Saushin [1], Seth Owusu [1], Petri Karvinen [1], Sari Suvanto [2], Yuri Svirko [1], Polina Kuzhir [1] and Georgy Fedorov [1,*]

[1] Department of Physics and Mathematics, University of Eastern Finland, FI-80100 Joensuu, Finland; isaac.appiah.otoo@uef.fi (I.A.O.); aleksandr.saushin@uef.fi (A.S.); sowusu@student.uef.fi (S.O.); petri.karvinen@uef.fi (P.K.); yuri.svirko@uef.fi (Y.S.); polina.kuzhir@uef.fi (P.K.)

[2] Department of Chemistry, University of Eastern Finland, FI-80100 Joensuu, Finland; sari.suvanto@uef.fi

* Correspondence: georgy.fedorov@uef.fi

Abstract: Tunability of properties is one of the most important features of 2D materials, among which graphene is attracting the most attention due to wide variety of its possible applications. Here, we demonstrated that the carrier concentration in graphene can be efficiently tuned by the material of the dielectric substrate on which it resides. To this end, we fabricated samples of CVD-grown graphene transferred onto silicon wafers covered with alumina, titanium dioxide, and silicon dioxide. We measured the transmission spectra of these samples using a time-domain terahertz spectrometer and extracted the Drude frequency-dependent graphene conductivity. We found that the sheet resistance of graphene is strongly affected by the underlying dielectric material, while the carrier scattering time remains the same. The carrier concentration value was found to range from $7 \times 10^{11}/\text{cm}^2$ in the case of alumina and $4.5 \times 10^{12}/\text{cm}^2$ in the case of titanium dioxide. These estimations are consistent with what can be extracted from the position of the G-peak in the Raman spectra of graphene. Our results show a way to control the graphene doping level in applications where it does not have to be adjusted.

Keywords: graphene; synthesis; Raman; terahertz

Citation: Otoo, I.A.; Saushin, A.; Owusu, S.; Karvinen, P.; Suvanto, S.; Svirko, Y.; Kuzhir, P.; Fedorov, G. Tailoring the Graphene Properties for Electronics by Dielectric Materials. *Crystals* **2024**, *14*, 595. https://doi.org/10.3390/cryst14070595

Academic Editors: Justina Gaidukevic and Jurgis Barkauskas

Received: 3 June 2024
Revised: 18 June 2024
Accepted: 23 June 2024
Published: 27 June 2024

1. Introduction

Graphene has attracted a lot of attention during the last decades because of its unique electronic properties giving it enormous potential [1]. The everlasting interest graphene has received is due to its remarkable electrical, mechanical, and optical properties [2] that have a large promise for numerous applications [3–7]. These include an enormous thermal conductivity [8] (up to 5×10^3 W/mK in the ref. [9]), high mobility of electrons (2×10^5 cm^2 V^{-1} s^{-1} [10]), and ability to absorb 2.3% of incident radiation in a wide spectral range spanning from infrared to ultraviolet [6].

These properties of graphene are mainly governed by the concentration of charge carriers that can be controlled via electrostatic [11] or chemical [12] doping. Correspondingly, varying the carrier's concentration by electrostatic doping allows one to control its conductivity and absorptivity and to create graphene-based transistors [13], plasmonic interferometers [14], optical switches [15], and other optoelectronic devices. However, applications such as chemical sensing [16] often require graphene to be doped to a certain constant level. This can be achieved either by exploring electrostatic doping by adding a gate electrode to the device architecture or by chemical doping via graphene functionalization. The former approach essentially complicates the fabrication process, while the latter one introduces extra defects that may affect sensor performance.

It is known that underlying dielectric substrates are capable of providing constant doping of deposited graphene and carbon nanotubes [17,18]. However, the origin of this effect is still unknown. Recent work [19] reported on the substrate effect on the THz dynamic conductivity of graphene transferred onto bare silicon, silicon coated with Si_3N_4 and silicon coated with SiO_2. The graphene layer electromagnetic response was described within the Drude model and the DC surface conductivity as well as the scattering time were estimated based on the transmission spectra of the graphene-coated substrate. It has been shown that both the carrier scattering time and concentration depend on the substrate material. This approach allows for the non-invasive characterization of graphene's transport properties. The methodology of graphene AC conductivity reconstruction developed in [19] relies on the assumption that the THz response of graphene is not dependent on the THz dispersion of the used dielectric substrate, which leads to substantial blurring of the conductivity values reconstructed from the experimental data.

In this paper, we extended the approach developed in ref. [19] to show that the properties of the charge carriers in graphene can be efficiently tuned by the material of the dielectric substrate on which it resides. For this work, we have chosen three different dielectric substrates with thicknesses ensuring good Raman signals [20] in order to estimate the graphene doping level based on its Raman spectra [21]. Using THz time-domain spectroscopy (TTDS), we measured the transmission spectra of graphene deposited on a silicon wafer coated with SiO_2, TiO_2, or Al_2O_3. We compared the transmission spectra of the substrate with and without graphene and used the transfer matrix technique [22] to obtain the DC sheet resistance and the carrier scattering time of the graphene layer. We further improved this simple and nondestructive methodology of measuring the AC conductivity of supported graphene via THz transmission data analysis. We considered consistently the contribution of the substrate as is, and then reconstructed the constitutive parameters of the supported graphene from the THz spectra. It decreased the measurement error down to less than 20%. We thus achieved an estimation of the carrier concentration with an accuracy that is much better than that in ref [19]. Our results confirmed that the THz conductivity of graphene can be well described by the Drude model and allowed us to obtain the DC conductance and scattering time in our graphene samples. Using this improved methodology, we obtained an important result, i.e., we demonstrated that the scattering time is not sensitive to the substrate, indicating that the carrier's mean free path is determined by the defects introduced during the graphene synthesis and transfer rather than the substrate material. On the other hand, the DC sheet resistance varied from 350 Ω/sq for graphene deposited on TiO_2 to 900 Ω/sq for graphene deposited on Al_2O_3. The graphene carrier concentration varied from $7 \times 10^{11}/\text{cm}^2$ on the alumina substrate to $4.5 \times 10^{12}/\text{cm}^2$ on the titanium dioxide substrate. These estimations are consistent with those of the position of the G-peak in the Raman spectra of graphene [21]. We believe that our results provide an opportunity to achieve a constant concentration of the charge carriers in graphene without electrostatic and/or chemical doping.

The collected data of graphene doping due to the presence of conventional substrates are in good agreement with those measured via Raman scattering and give even less scattered results than those provided by the Raman technique. This approach allows for treating the developed THz spectroscopy method of determining the carrier concentrations and AC conductivity of supported graphene as a future standard methodology to support graphene characterization. Moreover, through the reconstruction of the doping level of supported graphene through its interaction with several dielectric substrates, we found the most conductive and less defective samples, i.e., graphene supported by Al_2O_3 and TiO_2.

2. Materials and Methods

In these experiments, we used commercial graphene samples (Graphenea) synthesized on copper foil using the chemical vapor deposition (CVD) method and covered with a 65 nm layer of poly-methyl methacrylate (PMMA) on one side (see Figure 1). Graphene was transferred onto silicon 275 μm wafers covered with a layer of Al_2O_3 (96 nm) or TiO_2 (150 nm) using atomic layer deposition (ALD). Additionally, we used a commercially available silicon wafer covered with a 3 μm thick layer of thermally grown SiO_2.

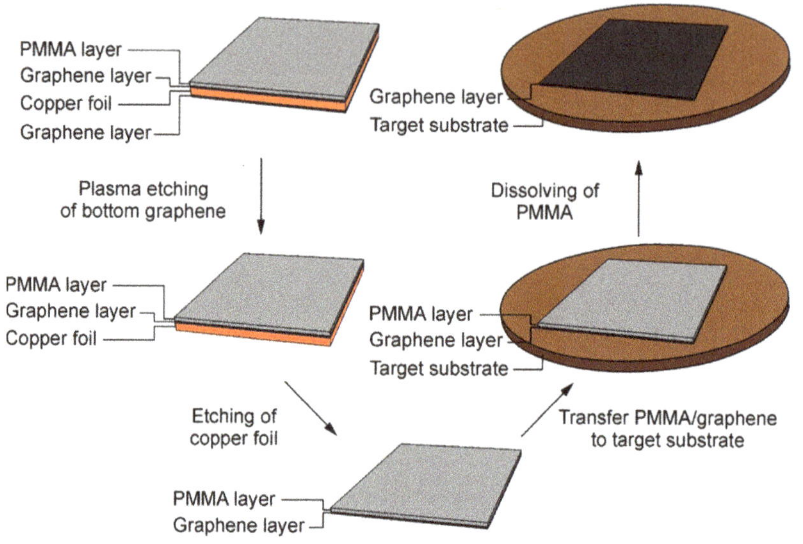

Figure 1. Schematic illustration of transferring graphene to substrate.

The transfer process started with etching graphene on the side of the copper foil not covered with PMMA using an Oxford Instruments PlasmaLab80 etcher. Then, to dissolve the copper foil and get the graphene ready for transferring, we left the copper with graphene overnight in an iron chloride ($FeCl_3$) solution consisting of 90 g of $FeCl_3$ and 200 mL deionized water. After that, we moved the graphene/PMMA layer to a beaker with deionized water to wash away the remains of the $FeCl_3$ solution. Next, we transferred the graphene to another beaker with new deionized water five times. Then we transferred the graphene with PMMA on the target substrates and let it dry overnight. We next removed the PMMA layer by keeping it in acetone for 30 min and dried the sample under air for 5 min. After that, each sample was washed in isopropanol for 5 min, dried under air, washed in deionized water and dried one more time.

The samples were characterized by measuring their Raman spectra in the 1100 to 3500 cm^{-1} range by using 514 and 785 nm excitation wavelengths at 25 and 150 mW power, respectively.

As the main characterization technique, we employed transmission spectroscopy. The transmission of the samples in the THz range was studied with a time-domain spectrometer (TDS) (TeTechS, Waterloo, ON, Canada) based on a femtosecond laser with 795 nm in the transmission geometry with an aperture of 3 mm wavelength.

3. Results and Discussions

Figure 2a shows the SEM image of the graphene deposited onto a dielectric substrate. The Raman spectra of the TiO_2, SiO_2, and Al_2O_3 substrates (blue curves) with the graphene on them (red curves) are shown in Figure 2b–d, respectively. One can see that the SEM image is typical for graphene and that the spectra are dominated by D, G and 2D peaks. The D-peak lying around 1360 cm^{-1} is low in comparison with both the G and 2D peaks for each sample, which means that the graphene does not have a lot of defects. The intensity I_{2D} of the 2D peak was more than two times that of the G peak (I_G) for all samples. It is an indication of an unbent monolayer graphene. The exact peaks' positions and I_{2D}/I_G ratios are presented in Table 1.

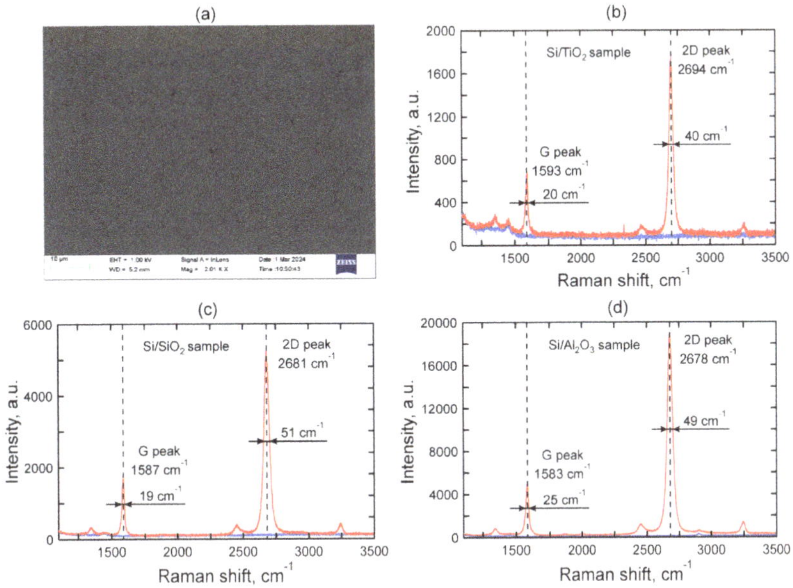

Figure 2. (**a**) SEM image and Raman spectra of graphene (red lines) and its substrates (blue lines), (**b**) graphene on TiO_2, (**c**) graphene on SiO_2, (**d**) graphene on Al_2O_3.

Table 1. Results of Raman characterization. G- and 2D-peaks positions and I_{2D}/I_G ratios were obtained directly from the measured spectra. The concentration of the carriers was evaluated based on the G-peak position following [21].

	Graphene on TiO_2	Graphene on SiO_2	Graphene on Al_2O_3
G-peak position (cm^{-1})	1593	1587	1583
FWHM of G-peak (cm^{-1})	20 ± 2	20 ± 2	25 ± 2
2D-peak position (cm^{-1})	2694	2681	2678
FWHM of 2D-peak (cm^{-1})	40 ± 3	50 ± 3	50 ± 3
I_{2D}/I_G	2.53	2.88	3.69
n_s, $10^{12}/cm^2$	5 ± 1	2 ± 0.5	± 1

THz transmission spectroscopy allows one to measure the graphene surface conductivity (inverse of the sheet resistance) non-invasively [23], avoiding the lithography and metallization process required for the four-probe sheet resistance measurements. This approach was proven to be efficient in previous works. In our work, we used the transfer matrix method [22] and the Drude model for dynamic dielectric permittivity to analyze the measured spectra. We first fit the measured and simulated spectra for the bare substrate to extract the Drude parameters (DC conductivity and scattering time) of the silicon wafer and then used the same approach for graphene by fitting the corresponding spectra of graphene-coated substrates.

The thickness and refractive index of the silicon wafer were obtained from the time traces measured with the TDS. Figure 3 demonstrates typical time traces of the signal recorded in the case of an empty aperture (blue curve) and the case of a Si/TiO_2 substrate.

The time trace of the signal in the case of the Si/TiO_2 substrate is delayed relative to the case of the empty aperture and has an additional echo because of reflections inside the substrate (see Figure 3, inset). Based on these two time traces we can calculate the refractive index n and thickness d_{calc} of the substrates according to the following equations:

$$n = \frac{t_3 - t_1}{t_3 - 3t_2 + 2t_1}, \tag{1}$$

$$d_{calc} = \frac{c(t_2 - t_1)}{n - 1}, \tag{2}$$

where c is the speed of light in a vacuum and t_1, t_2 and t_3 are illustrated in Figure 3. The results for all of the used substrates are summarized in Table 1. We note that the thicknesses of the silicon wafers we obtained using this method match the results of the direct measurements and the wafer thickness value provided by the producer. The calculated refractive indexes and thickness are presented in Table 2 along with the thicknesses d_{meas} measured with a micrometer and one can see that they approximately match.

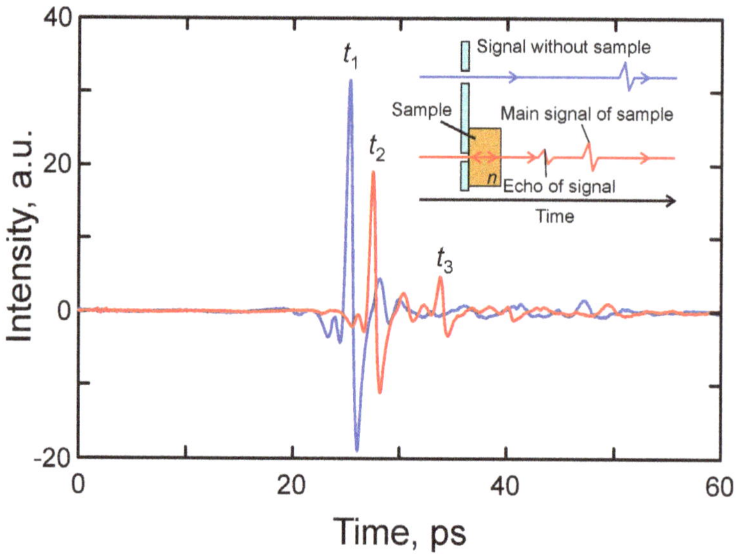

Figure 3. Example of experimental time traces of the THz pulse propagating through the empty aperture and through the Si/TiO_2 substrate.

Table 2. Refractive indexes: measured and calculated thicknesses of the samples.

	Si/TiO_2	Si/SiO_2	Si/Al_2O_3
t_1, ps	25.28953	25.30211	25.25176
t_2, ps	27.51762	27.49245	27.47986
t_3, ps	33.78651	33.66062	33.74874
n	3.46	3.45	3.46
d_{calc}, µm	271.7	267.9	271.7
d_{meas}, µm	275	275	275

The THz transmission spectra of the substrates covered with graphene and without it are illustrated in Figure 4a,b. The Fabri–Perot oscillations with a period depending on the substrate's thickness and refractive index are seen in each graph. We also note that in the case of the bare substrate, transmittance tends to grow with frequency. This already indicates that the Drude scattering time is comparable with the inverse of the characteristic frequency. Transmission of the graphene-coated substrates is lower than that of the bare ones in all frequency ranges due to the intrinsic conductance of the graphene. Figure 4c shows the THz transmission spectra of the silicon substrates with 150 nm and 100 nm layers of TiO_2. One can see that the spectra match well enough. This illustrates first the overall level of the reproducibility of the results. Secondly, it shows the lack of an effect of the dielectric layer thickness on the THz transmission spectra.

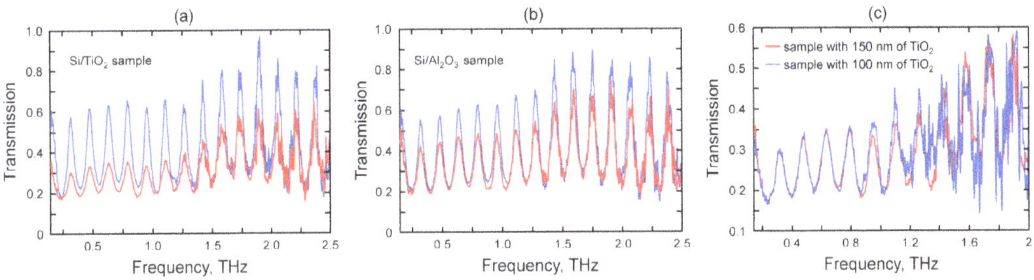

Figure 4. Experimental THz range transmission spectra of samples based on (**a**) Si/TiO_2, (**b**) Si/Al_2O_3 (blue lines—spectra of the substrate, red lines—spectra of the substrate covered with graphene) and (**c**) experimental THz transmission spectra of silicon covered with 150 nm (red line) and 100 nm (blue line) of TiO_2.

The propagation of the radiation through the sample can be described in terms of the transfer matrix method [22], which allows us to calculate the transmission, reflection, and absorption through a stack of parallel layers with known dielectric functions $\varepsilon(\omega)$ (the magnetic susceptibility is assumed to be the unit). The dielectric function of both the substrate and silicon are described within the Drude model, implying as follows:

$$\varepsilon(\omega) = \varepsilon_\infty + \frac{4\pi i \sigma(\omega)}{\omega}, \tag{3}$$

$$\sigma(\omega) = \frac{\sigma_0}{1 - i\omega\tau}, \tag{4}$$

where ω is the frequency of the incident THz radiation and σ_0 is the DC conductivity $\sigma(\omega = 0)$. We first fit the transmission of the bare substrates using the silicon DC conductivity and scattering time as fitting parameters. Importantly, the silicon refractive index thickness is defined directly from the time traces. The scattering time turns out to be about 100 fs, consistent with previous studies [24], while the conductivity ranges from 10 to 30 S/m. Importantly the thickness of the ALD grown dielectric is much smaller that the radiation wavelength, so taking it into account does not affect the results of the simulations. This justifies considering the substrate as a uniform wafer with the thickness and refracted index evaluated based on the time traces. The same applies to the silica-on-silicon substrates. Next, we used the obtained dielectric function of the wafer and fit the transmission spectrum of the graphene-coated substrate to evaluate the sheet resistance $R_{sh} = 1/\sigma_s$ and scattering for the graphene layer. In order to use Equations (3) and (4) for the graphene, we assume that the graphene sheet thickness is $d = 0.35$ nm. Still, the obtained value of the graphene sheet resistance does not depend on the graphene sheet thickness d, which is much smaller than both the wavelength and skin depth.

A comparison of the simulated and measured transmission spectra of the bare substrates is illustrated in Figure 5a and for the graphene-coated substrates in Figure 5b. The resulting characteristics of the graphene evaluated based on the transmission spectra are shown in Table 3.

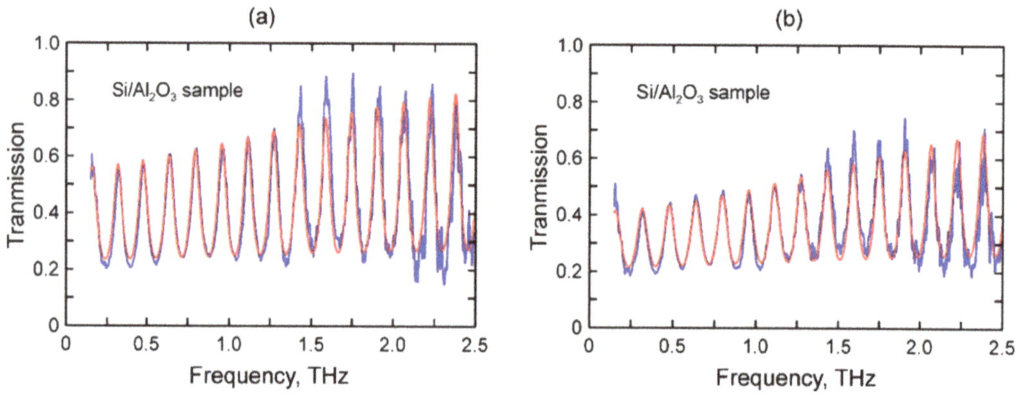

Figure 5. Experimental and theoretical THz spectra of Si/Al_2O_3 (**a**) substrate and (**b**) substrate covered with graphene, blue lines—experimental spectra, red lines—calculated spectra.

Table 3. Drude parameters of samples.

Drude's Parameter	Si/TiO_2	Si/SiO_2	Si/Al_2O_3
n	3.46	3.45	3.46
d, μm	271.7	267.9	271.7
τ_{sub}, fs	100 ± 20	100 ± 20	100 ± 20
σ_0, s/m	11 ± 0.5	29 ± 1	11 ± 0.5
τ_{graph}, fs	90 ± 5	90 ± 5	70 ± 5
R_{sheet}, Ω/sq	350 ± 50	550 ± 50	900 ± 50
n_s, $10^{12}/cm^2$	4.5 ± 0.25	1.5 ± 0.08	0.7 ± 0.05

The results obtained show that the sheet resistance of the graphene significantly depends on the substrate where it is placed. It varies from 350 Ω/sq for graphene on Si/TiO_2 to 900 Ω/sq for graphene on Si/Al_2O_3 with an intermediate value of 550 Ω/sq for graphene on Si/SiO_2. This allows us to suppose that the electronic properties of the substrate affect the Fermi level of the graphene due to the different levels of its doping. The scattering time of graphene derived from the calculated transmission spectra was about 80 ± 10 fs for all samples. Such a value corresponds to a mean free path of about 0.1 μm, which is typical for CVD-grown graphene [25]. The fact that the mean free path is the same for all graphene samples indicates that the concentration of defects does not depend on the substrate and is related only to the transfer of graphene and the CVD growth process. Correspondently, the difference in sheet resistance is determined mainly by the difference in the charge carriers' concentration n_s and we may describe its transport properties in terms of the semiclassical Boltzmann transport theory [26]. Since in our case $E_F >> k_BT$ (degenerate Fermi gas), we may write an equation for the conductivity of 2D materials using the relaxation time calculated as follows:

$$\sigma_S = \frac{e^2 v_f \tau}{\hbar}\sqrt{\frac{n_s}{\pi}}, \quad (5)$$

where $\sigma_S = 1/R_{sheet}$—sheet conductivity. Correspondingly,

$$n_s = \frac{\pi\hbar^2}{R_{sheet}^2 e^4 v_f^2 \tau^2}, \tag{6}$$

where $e = 1.6 \times 10^{-19}$—electron charge, $v_f = 10^6$ m/s—Fermi velocity. We found that $n_s = 4.5 \times 10^{12}$ cm^{-2} for Si/TiO$_2$, 1.8×10^{12} cm^{-2} for Si/SiO$_2$ and 0.7×10^{12} cm^{-2} for Si/Al$_2$O$_3$.

It is informative to relate the graphene carrier concentration evaluated based on the transmission spectra to the Raman spectra of the corresponding samples. The graphene on the Si/TiO$_2$ substrate had the highest charge concentration, and at the same time, its G and 2D peaks had the biggest Raman shift. The lowest Raman shifts were registered for graphene on Si/Al$_2$O$_3$, and the intermediate parameters belonged to graphene on the Si/SiO$_2$ substrate; in other words, the greater the charge carrier's concentration, the greater the Raman shifts of the G and 2D peaks. These results are consistent with the previous studies. According to [27], the charge concentration of graphene affects its Raman spectra. We used the results of ref [21] and the position of the G-peak in the measured Raman spectra to estimate the charge carrier concentration; see the last row of Table 1. The good match between the obtained concentration values provides solid ground for the reliability of our analysis. We note that the use of transmission spectroscopy allows for better accuracy in the evaluation of the carrier concentration, while the Raman data are used in our case for independent verification of these data.

Importantly, our data cannot be used to draw any conclusions on the carrier type. In other words, in no case can we tell whether graphene is *p*- or *n*- doped. Further studies will be needed to resolve this issue. Still, the absolute value of the charge carrier concentration is by itself a very important characteristic.

4. Conclusions

The dynamic charge transport of CVD graphene transferred onto Si/TiO$_2$, Si/SiO$_2$, and Si/Al$_2$O$_3$ substrates was investigated using time-domain terahertz range spectroscopy. Analysis of the transmission spectra unambiguously showed that graphene dynamic conductivity is well described by the Drude model with the DC surface conductance strongly dependent on the dielectric material in direct contact with the graphene. On the other hand, the carrier scattering time was found to be insensitive to it, meaning that the scattering time was limited by the intrinsic defects in the graphene and not by the substrate. Our data allow for a relatively accurate evaluation of carrier concentration. We found it to vary from 0.7×10^{12} cm^{-2} for graphene in contact with alumina to 4.5×10^{12} cm^{-2} for graphene in contact with titanium dioxide. These values are consistent with the carrier concentration estimations based on the positions of the G and 2D peaks of the measured Raman spectra.

We thus demonstrated that the graphene doping level can be efficiently engineered by choosing an appropriate substrate without a significant effect on the graphene quality (the mean free path). This approach may facilitate graphene applications for sensing and optoelectronic applications. The described methodology of the transmission spectra analysis may be extrapolated to use time-domain terahertz spectroscopy as a versatile tool for the characterization of 2D materials and thin films supported by silicon-based substrates.

Author Contributions: I.A.O. and P.K. (Petri Karvinen) fabricated the samples; A.S. and S.O. conducted the TDS measurements; S.S. measured the Raman spectra and analyzed the results; and G.F., P.K. (Polina Kuzhir) and Y.S. supervised the project and wrote the manuscript, with crucial contributions from all authors. All authors have read and agreed to the published version of the manuscript.

Funding: This study was accomplished with the financial support of subproject H-Cube of EU ATTRACT phase 2 Research infrastructure H2020 (Project No 101004462); the work is part of the Research Council of Finland Flagship Program, Photonics Research and Innovation (PREIN), decision number 346518; authors also acknowledge support of the Horizon 2020 RISE TERASSE Project (Project No 823878) and Research Council of Finland project (decision number 343393).

Data Availability Statement: The data that support the findings of this study are available from the corresponding authors upon reasonable request.

Conflicts of Interest: The authors declare no conflicts of interest.

References

1. Geim, A.K. Graphene: Status and Prospects. *Science* **2009**, *324*, 1530–1534. [CrossRef]
2. Castriota, M.; Politano, G.G.; Vena, C.; De Santo, M.P.; Desiderio, G.; Davoli, M.; Cazzanelli, E.; Versace, C. Variable Angle Spectroscopic Ellipsometry investigation of CVD-grown monolayer graphene. *Appl. Surf. Sci.* **2019**, *467–468*, 213–220. [CrossRef]
3. Xia, Y.; Gao, W.; Gao, C. A Review on Graphene-Based Electromagnetic Functional Materials: Electromagnetic Wave Shielding and Absorption. *Adv. Funct. Mater.* **2022**, *32*, 2204591. [CrossRef]
4. Lin, C.-H.; Chen, Y.-S.; Lin, J.-T.; Wu, H.C.; Kuo, H.-T.; Lin, C.-F.; Chen, P.; Wu, P.C. Automatic Inverse Design of High-Performance Beam-Steering Metasurfaces via Genetic-type Tree Optimization. *Nano Lett.* **2021**, *21*, 4981–4989. [CrossRef]
5. Liao, C.; Li, Y.; Tjong, S.C. Graphene Nanomaterials: Synthesis, Biocompatibility, and Cytotoxicity. *Int. J. Mol. Sci.* **2018**, *19*, 3564. [CrossRef]
6. Nair, R.R.; Blake, P.; Grigorenko, A.N.; Novoselov, K.S.; Booth, T.J.; Stauber, T.; Peres, N.M.R.; Geim, A.K. Fine Structure Constant Defines Visual Transparency of Graphene. *Science* **2008**, *320*, 1308. [CrossRef]
7. Narayan, J.; Bezborah, K. Author response for "Recent advances in the functionalization, substitutional doping and applications of graphene/graphene composite nanomaterials". *RSC Adv.* **2024**, *14*, 13413–13444. [CrossRef]
8. Döğüşcü, D.K.; Sarı, A.; Hekimoğlu, G. Effects of graphene doping on shape stabilization, thermal energy storage and thermal conductivity properties of PolyHIPE/PEG composites. *J. Energy Storage* **2024**, *76*, 109804. [CrossRef]
9. Balandin, A.A.; Ghosh, S.; Bao, W.; Calizo, I.; Teweldebrhan, D.; Miao, F.; Lau, C.N. Superior Thermal Conductivity of Single-Layer Graphene. *Nano Lett.* **2008**, *8*, 902–907. [CrossRef]
10. Bolotin, K.; Sikes, K.; Jiang, Z.; Klima, M.; Fudenberg, G.; Hone, J.; Kim, P.; Stormer, H. Ultrahigh electron mobility in suspended graphene. *Solid State Commun.* **2008**, *146*, 351–355. [CrossRef]
11. Özyilmaz, B.; Jarillo-Herrero, P.; Efetov, D.; Kim, P. Electronic transport in locally gated graphene nanoconstrictions. *Appl. Phys. Lett.* **2007**, *91*, 192107. [CrossRef]
12. Lherbier, A.; Blase, X.; Niquet, Y.-M.; Triozon, F.; Roche, S. Charge transport in chemically doped 2D graphene. *Phys. Rev. Lett.* **2008**, *101*, 036808. [CrossRef]
13. Berger, C.; Song, Z.M.; Li, T.B.; Li, X.B.; Ogbazghi, A.Y.; Feng, R.; Dai, Z.N.; Marchenkov, A.N.; Conrad, E.H.; First, P.N.; et al. Ultrathin epitaxial graphite: 2D electron gas properties and a route toward graphene-based nanoelectronics. *J. Phys. Chem. B* **2004**, *108*, 19912–19916. [CrossRef]
14. Bandurin, D.A.; Svintsov, D.; Gayduchenko, I.; Xu, S.G.; Principi, A.; Moskotin, M.; Tretyakov, I.; Yagodkin, D.; Zhukov, S.; Taniguchi, T.; et al. Resonant terahertz detection using graphene plasmons. *Nat. Commun.* **2018**, *9*, 5392. [CrossRef]
15. Wang, F.; Zhang, Y.; Tian, C.; Girit, C.; Zettl, A.; Crommie, M.; Shen, Y.R. Gate-Variable Optical Transitions in Graphene. *Science* **2008**, *320*, 206–209. [CrossRef]
16. Shao, Y.; Wang, J.; Wu, H.; Liu, J.; Aksay, I.A.; Lin, Y. Graphene based electrochemical sensors and biosensors: A review. *Electroanal. Int. J. Devoted Fundam. Pract. Asp. Electroanal.* **2010**, *22*, 1027–1036. [CrossRef]
17. Minot, E.D.; Yaish, Y.; Sazonova, V.; McEuen, P.L. Determination of electron orbital magnetic moments in carbon nanotubes. *Nature* **2004**, *428*, 536–539. [CrossRef]
18. Shi, Y.; Dong, X.; Chen, P.; Wang, J.; Li, L.J. Effective doping of single-layer graphene from underlying SiO_2 substrates. *Phys. Rev. B* **2009**, *79*, 115402. [CrossRef]
19. Scarfe, S.; Cui, W.; Luican-Mayer, A.; Ménard, J.-M. Systematic THz study of the substrate effect in limiting the mobility of graphene. *Sci. Rep.* **2021**, *11*, 8729. [CrossRef]
20. Van Velson, N.; Zobeiri, H.; Wang, X. Rigorous prediction of Raman intensity from multi-layer films. *Opt. Express* **2020**, *28*, 35272–35283. [CrossRef]
21. Bruna, M.; Ott, A.K.; Ijäs, M.; Yoon, D.; Sassi, U.; Ferrari, A.C. Doping dependence of the raman spectrum of defected graphene. *ACS Nano* **2014**, *8*, 7432–7441. [CrossRef] [PubMed]
22. Mackay, T.G.; Lakhtakia, A. *The Transfer-Matrix Method in Electromagnetics and Optics*; Springer Nature: Berlin/Heidelberg, Germany, 2022.
23. Tomaino, J.L.; Jameson, A.D.; Kevek, J.W.; Paul, M.J.; Van Der Zande, A.M.; Barton, R.A.; Lee, Y.S. Terahertz imaging and spectroscopy of large-area single-layer graphene. *Opt. Express* **2011**, *19*, 141–146. [CrossRef] [PubMed]
24. van Exter, M.; Grischkowsky, D. Carrier dynamics of electrons and holes in moderately doped silicon. *Phys. Rev. B* **1990**, *41*, 12140. [CrossRef] [PubMed]
25. Gayduchenko, I.A.; Fedorov, G.E.; Moskotin, M.V.; Yagodkin, D.I.; Seliverstov, S.V.; Goltsman, G.N.; Kuntsevich, A.Y.; Rybin, M.G.; Obraztsova, E.D.; Leiman, V.G.; et al. Manifestation of plasmonic response in the detection of sub-terahertz radiation by graphene-based devices. *Nanotechnology* **2018**, *29*, 245204. [CrossRef] [PubMed]

26. Das Sarma, S.; Adam, S.; Hwang, E.H.; Rossi, E. Electronic transport in two-dimensional graphene. *Rev. Mod. Phys.* **2011**, *83*, 407–470. [CrossRef]
27. Bukowska, H.; Meinerzhagen, F.; Akcöltekin, S.; Ochedowski, O.; Neubert, M.; Buck, V.; Schleberger, M. Raman spectra of graphene exfoliated on insulating crystalline substrates. *New J. Phys.* **2011**, *13*, 063018. [CrossRef]

crystals

Article

Mechanical Properties of Small Quasi-Square Graphene Nanoflakes

Andrés Serna-Gutiérrez [1] and Nicolás A. Cordero [1,2,3,*]

1 Departamento de Física, Universidad de Burgos, 09001 Burgos, Spain; aserna@ubu.es
2 International Research Center in Critical Raw Materials for Advanced Industrial Technologies (ICCRAM), Universidad de Burgos, 09001 Burgos, Spain
3 Institute Carlos I for Theoretical and Computational Physics (IC1), 18016 Granada, Spain
* Correspondence: ncordero@ubu.es

Abstract: The rise of straintronics—the possibility of fine-tuning the electronic properties of nanosystems by applying strain to them—has enhanced the interest in characterizing the mechanical properties of these systems when they are subjected to tensile (or compressive), shear and torsion strains. Four parameters are customarily used to describe the mechanical behavior of a macroscopic solid within the elastic regime: Young's and shear moduli, the torsion constant and Poisson's ratio. There are some relations among these quantities valid for elastic continuous isotropic systems that are being used for 2D nanocrystals without taking into account the non-continuous anisotropic nature of these systems. We present in this work computational results on the mechanical properties of six small quasi-square (aspect ratio between 0.9 and 1.1) graphene nanocrystals using the PM7 semiempirical method. We use the results obtained to test the validity of two relations derived for macroscopic homogeneous isotropic systems and sometimes applied to 2D systems. We show they are not suitable for these nanostructures and pinpoint the origin of some discrepancies in the elastic properties and effective thicknesses reported in the literature. In an attempt to recover one of these formulas, we introduce an effective torsional thickness for graphene analogous to the effective bending thickness found in the literature. Our results could be useful for fitting interatomic potentials in molecular mechanics or molecular dynamics models for finite carbon nanostructures, especially near their edges and for twisted systems.

Keywords: graphene; mechanical properties; stretching; shear; torsion; straintronics

Citation: Serna-Gutiérrez, A.; Cordero, N.A. Mechanical Properties of Small Quasi-Square Graphene Nanoflakes. *Crystals* **2024**, *14*, 314. https://doi.org/10.3390/cryst14040314

Academic Editor: Lin Gan

Received: 28 February 2024
Revised: 24 March 2024
Accepted: 27 March 2024
Published: 28 March 2024

1. Introduction

Strain engineering or "straintronics" has emerged as a way to change the behavior of materials, in particular, 2D nanomaterials, especially graphene [1–9]. In this case, straintronics has been considered in recent years as a way to fine-tune, among other properties, its band gap in order to use it in many different applications [10–19]. In fact, strained graphene nanobubbles have been recently proposed as qubits for quantum computing [20]. It is also possible to take advantage of the changes in properties caused by strain in sensing devices [21]. For instance, a graphene strain sensor that can detect various types of strain induced via stretching, bending and torsion [22] and a graphene-based torsion balance [23] have been constructed. Graphene's properties make it suitable to be used as an actuator to build artificial muscles [24] and in many other biomedical applications [25].

Material simulation methods can be classified according to, among other criteria, two distinct features. The first one uses classical or quantum mechanics, and the second considers materials as continuous media or takes into account their discrete atomistic nature. Therefore, we can consider four different approaches in the study of carbon nanostructures:

1. Quantum atomistic calculations, which explicitly treat materials as atoms obeying the rules of quantum mechanics. This category includes methods that solve either

full (like quantum chemistry and Density Functional Theory approaches) or approximate quantum equations (like tight-binding and semiempirical methods). See, for instance, [14,26];
2. Quantum continuous models, in which approximate quantum equation solutions are extrapolated to the continuum limit. See, as an example, [27,28];
3. Classical atomistic simulations, which consider matter as being built of particles that interact following the equations of classical mechanics with potentials (force fields) that try to mimic experimental properties (or those calculated by quantum methods) using empirical parameters. This category includes static calculations called molecular mechanics (MM), also known as molecular structural mechanics (MSM) or nanoscale continuum modelling (NCM)—in which molecular bonds are considered as springs or beams—as well as classical molecular dynamics (MD). See, for instance, [29,30];
4. Classical continuous models, traditionally used in engineering. See, as an example, [31,32].

Present computational power precludes using quantum atomistic calculations for very large systems, and multi-scale approaches are employed. In these models, macroscopic parameters for approach 4 are obtained from classical atomistic calculations (approach 3) that, in turn, fit their microscopic parameters using quantum mechanical results (approach 1). Therefore, these multi-scale models need a hierarchy of calculations. Jumping from approach 1 to 3 in the case of deformed graphene is still an open question, and better comprehension of the atomistic mechanisms at work in graphene deformations is needed [33].

There are many studies on the different mechanical properties of both infinite graphene and finite graphene nanostructures [30,34–44]. Nevertheless, there is a lack of a comprehensive study on the three different deformations a graphene nanoflake can be subjected to—stretching, shear and torsion (especially the last one)—and on the validity of the use of macroscopic formulas for isotropic materials [45,46] for this clearly anisotropic system.

We have studied the behavior of six different nearly square graphene nanoflakes of various sizes subjected to the three aforementioned deformations along their two edges (zigzag and armchair) and calculated for each of them Young's and shear moduli and the torsion constant, as well as Poisson's ratio. Our results could be useful for fitting interatomic potentials in MM/MSM/NCM or MD models for finite carbon nanostructures, especially near their edges and for twisted carbon systems. We have also tested how well commonly used formulas for calculating the torsion constant for a thin slab, as well as Poisson's ratio from Young's and shear moduli, perform.

The structure of the paper is as follows. We present in Section 2 (Materials and Methods) first the theoretical model employed, then the systems studied and, finally, the computational method used. Our results are presented in Section 3 in four different subsections: Young's Modulus, Shear Modulus, Torsion Constant and Poisson's Ratio. Finally, the conclusions of the study are summarized in Section 4.

2. Materials and Methods

2.1. Theoretical Model

Hooke's law states that when a material is loaded within the elastic limit, stress T is proportional to strain q:

$$T = -\kappa q \,. \tag{1}$$

The proportionality constant κ is the so-called elastic constant and can be calculated as the stress/strain ratio associated to a given deformation within the elastic limit.

From a computational point of view, it is easier to adopt a different approach and calculate this constant from the expression for the deformation energy (taking the solid in the absence of any external load as the energy origin) as follows:

$$U = \frac{1}{2}\kappa q^2 \,. \tag{2}$$

Using this energy approach yields values closer to the experimental results than the stress–strain approach since the latter requires an additional process, the differentiation of the energy, which introduces some inaccuracies [47].

Let us consider the elastic homogeneous and isotropic rectangular parallelepiped solid with length l, width w and height h depicted in Figure 1.

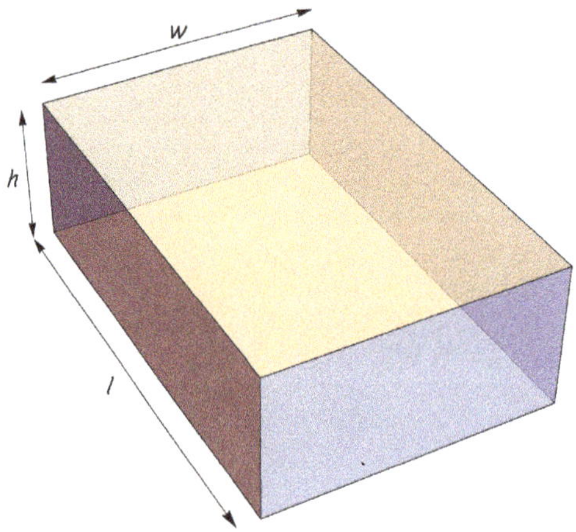

Figure 1. An elastic homogeneous and isotropic rectangular parallelepiped solid with length l, width w and height h.

We can load the solid in three different ways in order to obtain three small deformations: stretching, shear and torsion. In the case of uniaxial normal stress (either tensile or compressive), the ratio of the stress (force per unit area) applied to the solid, as depicted in Figure 2 ($\sigma = \frac{F}{wh}$), and the resulting axial strain ($\varepsilon = \frac{x}{l}$) is called the (tensile) elastic modulus or Young's modulus and can be calculated as follows:

$$E = \frac{\sigma}{\varepsilon} = \frac{Fl}{whx}\,. \tag{3}$$

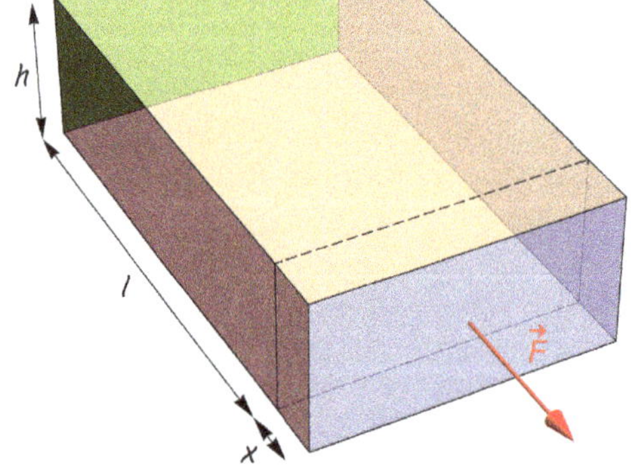

Figure 2. The solid in Figure 1 subjected to a tensile stress caused by fixing the rear face (colored in green) and pulling with a normal force $\vec{F}$ uniformly distributed over the front face.

It is easy to prove that the elastic energy of the stretched system is given by using a particular case of Equation (2),

$$U = \frac{1}{2}EV\varepsilon^2\,, \tag{4}$$

where $V = whl$ is the volume of the solid.

In the case of uniaxial tangential stress, the ratio of the shear stress (force per unit area) applied to the solid, as depicted in Figure 3 ($\tau = \frac{F}{wl}$), and the resulting shear strain ($\gamma = \phi \simeq \tan\phi = \frac{x}{h}$) is called the rigidity modulus or shear modulus and can be calculated as follows:

$$G = \frac{\tau}{\gamma} \simeq \frac{Fh}{wlx}\,. \tag{5}$$

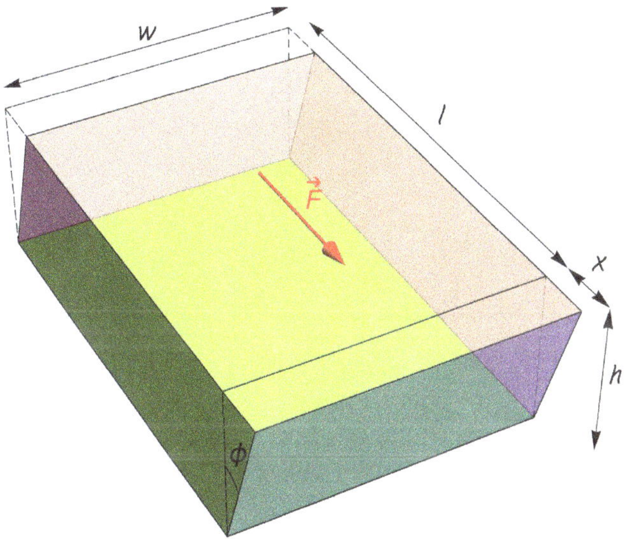

Figure 3. The solid in Figure 1 subjected to a shear stress caused by fixing the lower face (colored in green) and tearing with a tangential force $\vec{F}$ uniformly distributed over the upper face.

It is straightforward to prove that the elastic energy of the deformed system is given by using another particular case of Equation (2):

$$U = \frac{1}{2}GV\gamma^2 \simeq \frac{1}{2}GV\phi^2\,. \tag{6}$$

Finally, in the case of torsion, shown in Figure 4, the ratio of the applied torque M to the twist angle θ is known as torsional stiffness, calculated as follows:

$$k = \frac{M}{\theta}\,. \tag{7}$$

Once again, it is easy to prove that the elastic energy of the twisted system is given by using a third particular case of Equation (2):

$$U = \frac{1}{2}k\theta^2\,. \tag{8}$$

Torsion can be seen as a special case of shear, and there exists a relationship between their elastic constants. Taking into account this [48,49],

$$\theta = \frac{Ml}{GJ}, \tag{9}$$

where J is the so-called torsion constant, torsional constant or polar second moment of area (also known as polar moment of inertia). Using this expression and Equation (7),

$$k = \frac{GJ}{l}. \tag{10}$$

J can be calculated by reordering this equation as follows:

$$J = \frac{kl}{G}. \tag{11}$$

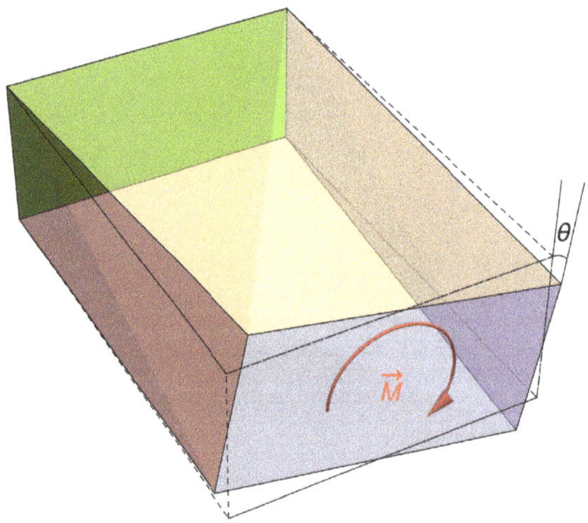

Figure 4. The solid in Figure 1 subjected to a torsional stress caused by fixing the rear face (colored in green) and twisting with torque $\vec{M}$ on the front face.

Graphene is considered as a 2D system. It is, in fact, extremely thin, but it has a small thickness due to the electron cloud surrounding the honeycomb structure of carbon nuclei. The accepted value for this thickness is 3.4 Å [50] (though other values are used by different studies, especially in the case of graphene sheets rolled up to build carbon nanotubes [43,51–54]). Therefore, in principle, rectangular graphene nanoflakes could be considered as elastic solids of the type shown in Figure 1 with $h = 3.4$ Å. When the thickness of a homogeneous isotropic solid is much smaller that its other two dimensions, there is a formula for calculating the torsion constant [46] that, in our notation, reads:

$$J = \frac{1}{3}h^3 w \left[1 - \frac{192}{\pi^5}\frac{h}{w}\sum_{n=1,3,5,\ldots}^{\infty}\frac{1}{n^5}\tanh\frac{n\pi w}{2h}\right], \tag{12}$$

where, in practice, going up to $n = 7$ in the series is enough.

Another consideration to take into account is that the situation depicted in Figure 2 is only a first approximation. When a solid is stretched, its perpendicular cross section $w \times h$ usually changes. In most cases, it decreases. This phenomenon can be measured by using the so-called Poisson's ratio (denoted by ν), which is the opposite of the ratio of transverse strain to axial strain. For small deformations, ν can be calculated as the transversal compression divided by the axial elongation. In our case, for small deformations, the height of the system (i.e., the thickness of the nanoflake) does not change since the breadth of the electron cloud with an appreciable density practically remains constant around the central carbon plane. Therefore,

$$\nu = -\frac{\text{transverse strain}}{\text{axial strain}} = \frac{\text{transverse compression}}{\text{axial elongation}} = -\frac{\Delta w}{x} \,. \tag{13}$$

For homogeneous systems, the following expression links Poisson's ratio to Young's and shear moduli:

$$\nu = \frac{E}{2G} - 1 \,. \tag{14}$$

2.2. Systems Studied

In order to test whether all the equations valid for homogeneous isotropic systems hold for graphene, which is clearly an anisotropic system, we have selected six small quasi-square hydrogen-passivated carbon nanoflakes. The selection criteria were that the aspect ratio $\frac{w}{l}$ must lie between 0.9 and 1.1 and that the electronic ground state has to be a spin singlet. The reason for these rules was the need to be able to analyze the effect of the kind of border (zigzag or armchair) without contamination from changes in aspect ratio or spin state. In order to label the nanoflakes, we have used n $\times$ m, where n (m) is the number of hexagons along the zigzag (armchair) edge. Figure 5 presents as an example the 6 $\times$ 7 case, while Figure 6 shows all the nanoflakes studied.

The six stresses considered in this study are shown in Figure 7 for the 3 $\times$ 3 nanoflake.

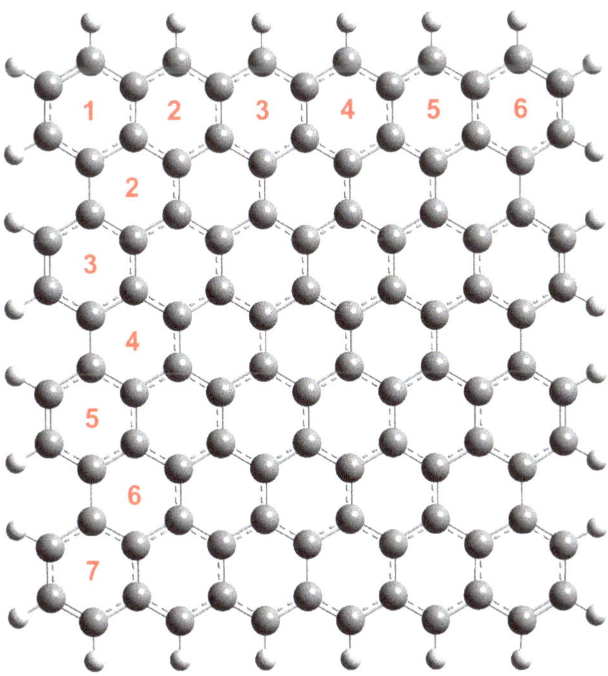

Figure 5. A 6 $\times$ 7 nanoflake with the hexagons along zigzag and armchair edges numbered.

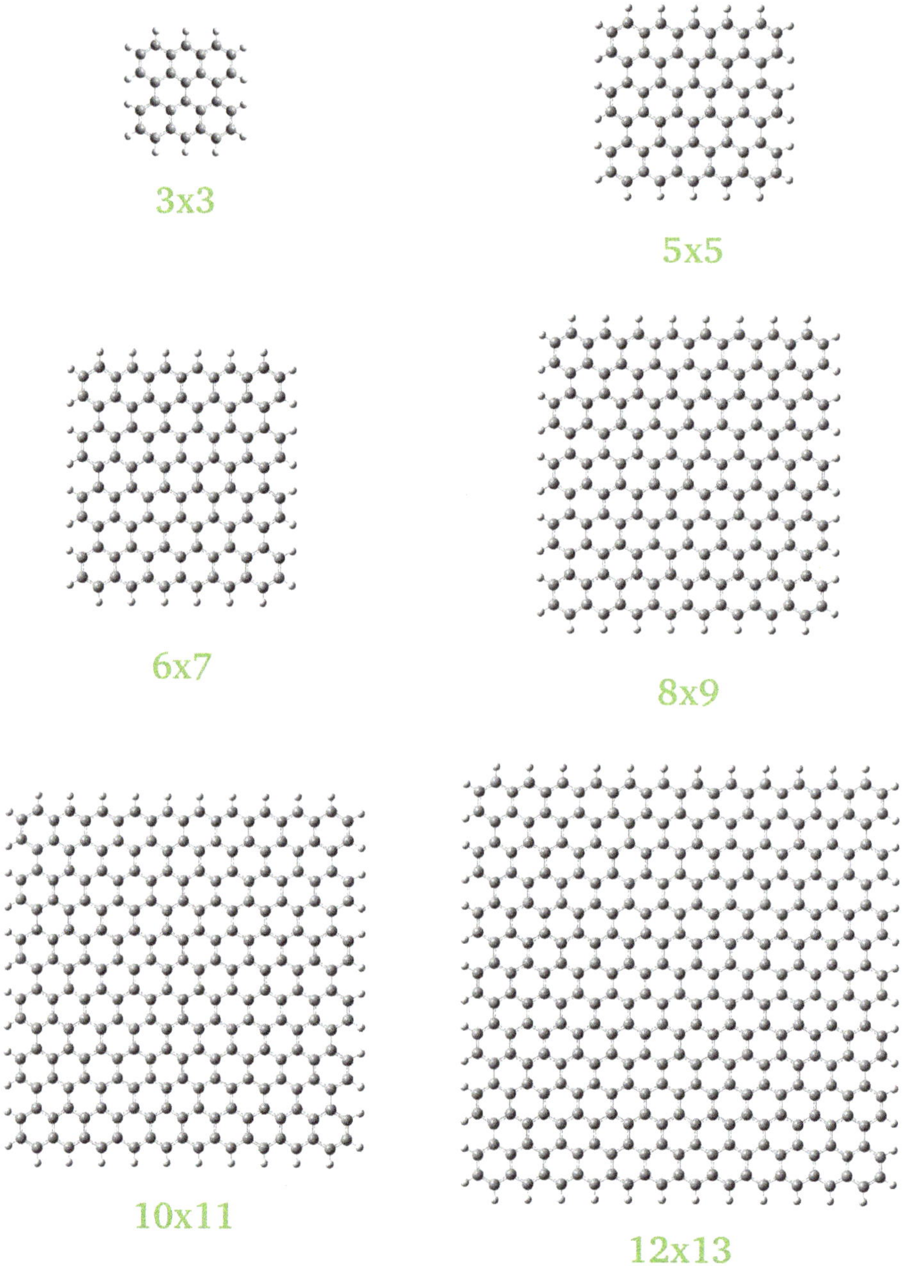

Figure 6. The six nanoflakes considered in this study.

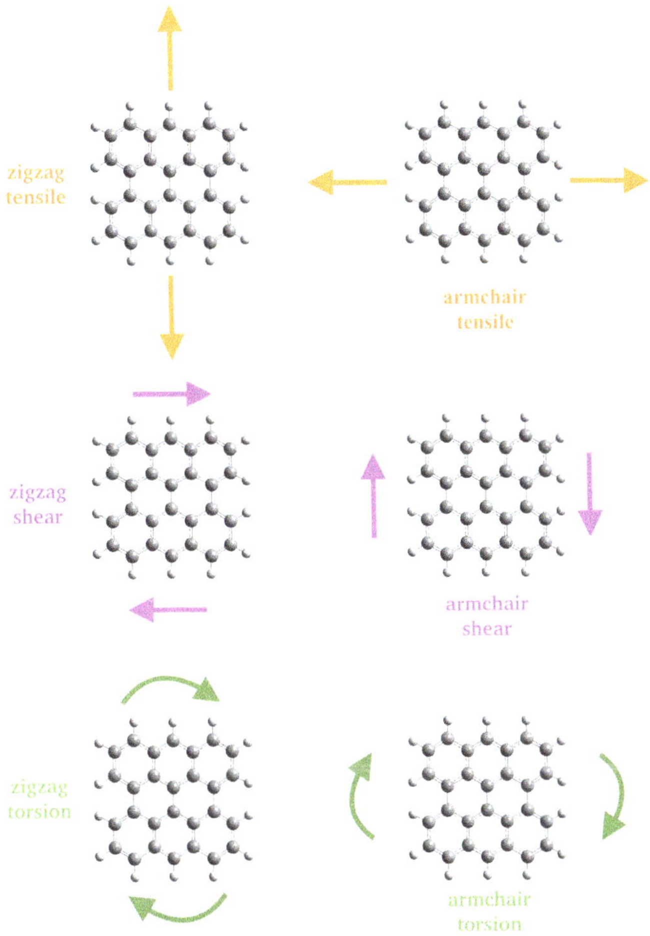

Figure 7. The six different stresses considered applied on a 3 × 3 nanoflake: tensile (**top row**), shear (**middle row**), torsion (**bottom row**); zigzag (**left column**), armchair (**right column**).

2.3. Computational Method

As we pointed out in the Introduction, quantum calculations are needed in order to fit the interatomic potentials used in MM/MSM/NCM and MD methods. Taking into account that these methods only deal with mechanical properties and do not try to explain electronic properties, in order to provide data useful for potential fitting, it is not necessary to use full ab initio quantum mechanics calculations (which would be unavoidable when, for instance, establishing the minimum size for antiferromagnetic ordering to appear in finite graphene nanostructures), and semiempirical approaches (much faster than first-principles calculations) are enough for our purpose.

We have selected the semiempirical PM7 method [55], as implemented in Gaussian16 [56] (i.e., PM7R8 [57]), because it is a fast quantum model that explicitly includes dispersion as well as hydrogen bonding corrections and has been successfully used to analyze interactions in graphene nanostructures [58–65].

We have fully relaxed all atomic positions to obtain the different initial geometries. We have calculated all vibrational frequencies to check for the absence of imaginary values and thus are sure that our initial geometries correspond to minima in the potential energy surface. We have then moved and fixed some boundary atoms in order to deform the nanoflakes and relaxed the positions of the rest of the atoms using the same algorithm to obtain deformation energies. As an example, we present in Figure 8 the final optimized geometries for the six stresses applied to the 3 × 3 nanoflake. The carbon atoms fixed in each case to achieve the desired deformation are highlighted in blue.

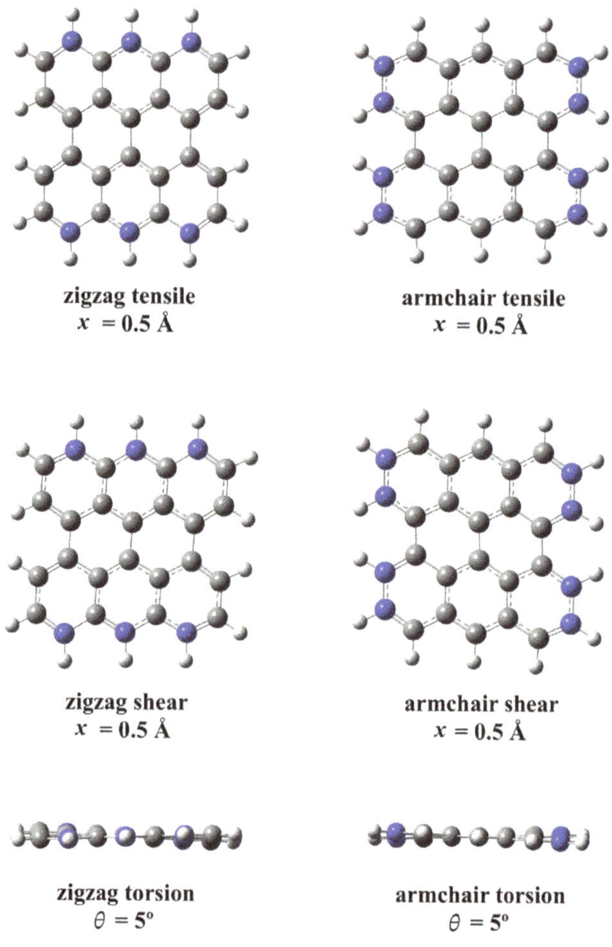

Figure 8. The six different optimized geometries corresponding to the maximum deformations applied to the 3 × 3 nanoflake: tensile (**top row**), shear (**middle row**), torsion (**bottom row**); zigzag (**left column**), armchair (**right column**). The carbon atoms fixed to achieve the desired deformation are highlighted in blue.

3. Results

3.1. Young's Modulus

We have displaced the carbon atoms on one of the edges along the stretching direction between 0.1 Å and 0.5 Å in 0.1 Å steps, as shown in the upper row of Figure 7, relaxed the rest of the atoms until the new equilibrium geometry depicted in the upper row of Figure 8 is reached and calculated the corresponding strain energies. We have fitted the results to a second-order polynomial and used the quadratic coefficient to calculate E.

As an example of this procedure, we show in Figure 9 the stretching energies corresponding to both armchair and zigzag edges for the 10 × 11 nanoflake. The graphs for other nanoflakes are similar.

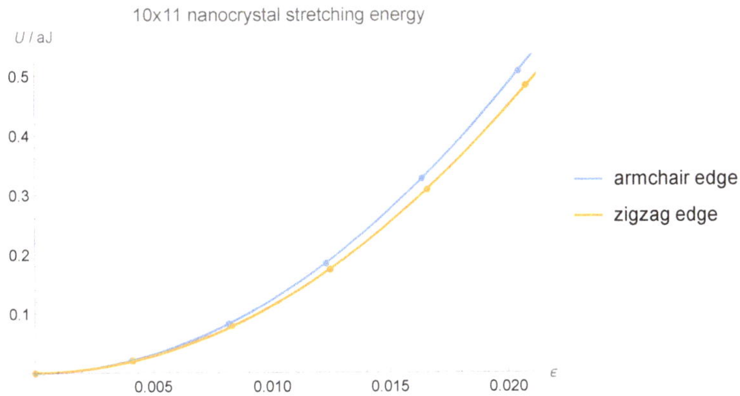

Figure 9. Stretching energies U as a function of strain ε for the 10 × 11 nanoflake. Points correspond to calculated energies, while lines show a second-order polynomial fit.

We present in Table 1 the geometrical characteristics of the fully relaxed nanoflakes (the lengths of the two carbon atoms' edges and the aspect ratio) as well as our results for their Young's moduli, obtained using Equation (4).

Table 1. Young's modulus of square carbon nanoflakes.

Nanoflake	Aspect Ratio	Edge	Edge Length/Å	E/TPa
3 × 3	1.03	zigzag armchair	7.099 7.288	0.927 1.224
5 × 5	1.07	zigzag armchair	11.368 12.191	0.869 1.229
6 × 7	0.94	zigzag armchair	15.618 14.667	1.061 1.200
8 × 9	0.98	zigzag armchair	19.879 19.580	1.105 1.220
10 × 11	1.01	zigzag armchair	24.143 24.484	1.217 1.134
12 × 13	1.04	zigzag armchair	28.388 29.409	1.127 1.205

For small nanoflakes, the zigzag results are smaller than the armchair ones, but they increase (while armchair Young's moduli remain practically constant), and both edges show similar E values, for large nanoflakes. Results from other calculations (in ascending order) are presented in Table 2. It can be seen that our results are in the upper part of the interval defined by all of them, excluding Shi et al.'s value, which is very far from the rest. The table also includes the thicknesses (assumed or calculated) when available.

Table 2. Young's moduli (E) and thicknesses (h) of graphene found in the literature.

Source	E/TPa	Method	h/Å
Reddy et al. [30]	0.671	interatomic potential	3.4
Lebedeva et al. [66]	0.7058–1.343 (depending on the potential used)	interatomic potential	3.34
Giannopoulos [67]	0.745208 for zigzag graphene nanoribons	spring-based structural mechanics	N/A
Giannopoulos [67]	0.745204 for armchair graphene nanoribbons	spring-based structural mechanics	N/A
Scarpa et al. [68]	0.762–1.000 (depending on the potential used)	cellular material mechanics theory	0.74–0.84
Polyakova et al. [69]	0.820	molecular dynamics	N/A
Tsai and Tu [38]	0.912	molecular dynamics	3.4
Tzeng and Tsai [70]	0.912	molecular dynamics	3.4
Zhang et al. [71]	0.985	spring finite element model	N/A
Sakhaee-Pour [37]	1.040 for zigzag graphene	interatomic potential	3.4
Sakhaee-Pour [37]	1.042 for armchair graphene	interatomic potential	3.4
Sha'bani and Rash-Ahmadi [72]	1.05	molecular dynamics	N/A
Zaeri et al. [73]	1.040	molecular structural mechanics finite element method	3.4
Tapia et al. [74]	1.042	atomistic finite element method	3.4
Anastasi et al. [34]	1.061 for zigzag graphene	molecular dynamics	3.35
Anastasi et al. [34]	1.035 for armchair graphene	molecular dynamics	3.35
Chandra et al. [40]	1.082	atomistic finite element method	1.46
Tahani and Safarian [75]	1.13	homogenization composite shell model	N/A
Cho et al. [35]	1.153 for graphite	molecular mechanics	3.35
Shi et al. [51]	2.81	atomic interaction based continuum model	1.27

The experimental value for the (infinite) graphene Young's modulus is 1.0 TPa [76], very similar to the in-plane Young's modulus in graphite (1.020 TPa [77]). Our results for small nanoflakes are, globally, slightly bigger. This is not surprising since Young's modulus tends to decrease when the size of the system grows from a few tens to a few hundreds of angstroms [78]. The reason is the edge effect on C–C distances. In infinite graphene, all distances are equal. This is not the case for graphene nanoflakes. To show this, we present in Table 3 a comparison between average C–C distances in the central ring (we have selected the nanoflakes with a central hexagon) and on the edge. C–C distances in the center ring are very close to the experimental infinite graphene value (1.42 Å), but they slightly decrease on the edges. This means C–C bonds are a little bit stronger on the edges of carbon nanoflakes than in graphene, and this translates into an increase in Young's modulus. As can be seen in the table, the percentage of edge carbon atoms, logically, decreases with the size of the system, but, for our nanoflakes, it is high. For our biggest nanoflake (12 × 13, not shown in the table because it has no central C ring), it is 28%. This explains our values for E being bigger than that for infinite graphene.

Table 3. Average C–C distance in square carbon nanoflakes and percentage of edge C atoms.

Nanoflake	Average C–C Distance Central Ring	Edge	Percentage of Edge C Atoms
5 × 5	1.420	1.399	58 %
6 × 7	1.418	1.402	48 %
10 × 11	1.418	1.403	33 %

As we have pointed out, graphene is clearly an anisotropic system because the properties in the direction perpendicular to the honeycomb plane are completely different to those in plane. Nevertheless, infinite graphene is transversely isotropic, i.e., its properties are symmetric about an axis normal to the honeycomb plane [79]. This means zigzag and armchair Young's moduli should be equal. In order to test how far our nanoflakes are from this transverse isotropy, we define their transverse isotropicity TI as the ratio of their zigzag and armchair Young's moduli as follows:

$$TI = \frac{E(\text{zigzag})}{E(\text{armchair})}.$$

With this definition, transverse isotropy corresponds to $TI = 1$, and anisotropy can be measured as the absolute deviation from this value $|\delta| = |1 - TI|$. Using the data in Table 1, we have calculated these two quantities for the six nanoflakes studied. The results are presented in Figure 10. For very small flakes (3 × 3 and 5 × 5), aspect ratio is more important than nanoflake size, and the anisotropy increases with size, but, for the rest of the flakes, as expected, the bigger the flake, the smaller the anisotropy.

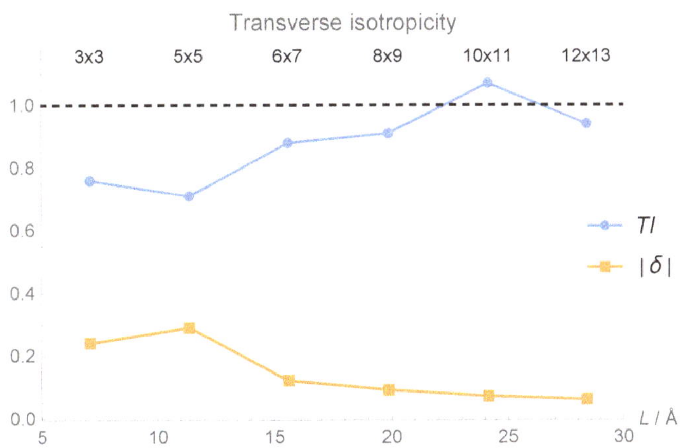

Figure 10. Transverse isotropicities of the Young's moduli TI of the six nanoflakes studied and their anisotropies (absolute deviations from unity) $|\delta|$ as a function of the zigzag edge length L.

3.2. Shear Modulus

We have displaced the carbon atoms on one of the edges along the shear direction between 0.1 Å and 0.5 Å in 0.1 Å steps, as shown in the middle row of Figure 7, relaxed the rest of the atoms until the new equilibrium geometry depicted in the middle row of Figure 8 is reached and calculated the corresponding strain energies. We have fitted the results to a second-order polynomial and used the quadratic coefficient to calculate G.

As an example of this procedure, we show in Figure 11 the shear energies corresponding to both armchair and zigzag edges for the 8 × 9 nanoflake. The graphs for other nanoflakes are similar.

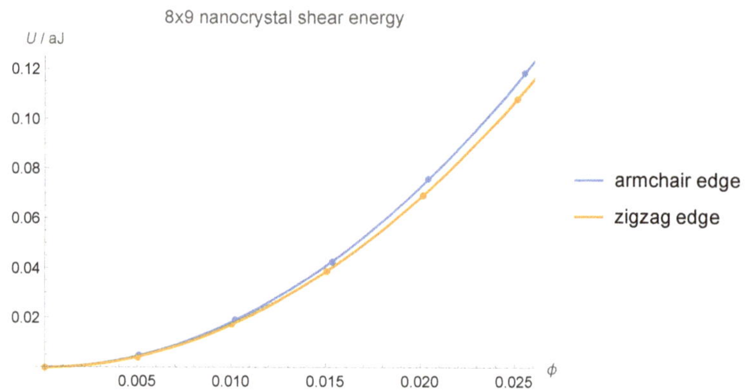

Figure 11. Shear energies U as a function of the shear angle ϕ (in radians) for the 8 × 9 nanoflake. Points correspond to calculated energies, while lines show a second-order polynomial fit.

Table 4 presents our results for the shear modulus of the nanoflakes calculated using Equation (6).

Table 4. Shear modulus of square carbon nanoflakes.

Nanoflake	Edge	G/GPa
3 × 3	zigzag	221
	armchair	255
5 × 5	zigzag	226
	armchair	259
6 × 7	zigzag	189
	armchair	288
8 × 9	zigzag	263
	armchair	279
10 × 11	zigzag	222
	armchair	240
12 × 13	zigzag	229
	armchair	250

Armchair values for G are bigger than zizgzag ones, in agreement with Sakhaee-Pour [37] and Min and Aluru [36].

Results from other calculations (in ascending order) are presented in Table 5. Our results agree very well with most of them and are similar to the experimental value for graphene, which is 280 GPa [80]. It is worth noting that the shear modulus for graphite is 440 GPa [77]. This is proof of the fact that using in-plane graphite properties for isolated graphene is not always valid.

Table 5. Shear moduli (G) and thicknesses (h) of graphene found in the literature.

Source	G/GPa	Method	h/Å
Mukhopadhyay et al. [81]	125.4	molecular mechanics	3.4
Scarpa et al. [68]	202–270 (depending on the potential used)	cellular material mechanics theory	0.74–0.84
Tahani and Safarian [75]	212	homogenization composite shell model	N/A
Sakhaee-Pour [37]	213 for zigzag graphene	interatomic potential	3.4
Sakhaee-Pour [37]	228 for armchair graphene	interatomic potential	3.4
Tapia et al. [74]	213	atomistic finite element method	3.4
Zhang et al. [71]	242	spring-based finite element model	N/A
Georgantzinos et al. [82]	280	spring-based finite element model	3.4
Polyakova et al. [69]	302	molecular dynamics	N/A
Reddy et al. [30]	384	interatomic potential	3.4
Tsai and Tu [38]	358	molecular dynamics	3.4
Zheng et al. [83]	434	beam finite element method	3.4
Zakharchenko et al. [39]	445	atomistic Monte Carlo based on empirical bond order potential	N/A
Min and Aluru [36]	≈460 for zigzag graphene	molecular dynamics	3.335
Min and Aluru [36]	≈360 for armchair graphene	molecular dynamics	3.335
Cho et al. [35]	482 for graphite	molecular mechanics	3.35
Zaeri et al. [73]	490	molecular structural mechanics finite element method	3.4
Chandra et al. [40]	606	atomistic finite element method	1.46

3.3. Torsion Constant

We present in Table 6 our results for the torsion constant of the nanoflakes calculated using Equations (11) and (12). In the first case, we have twisted one of the edges between 1° and 5° in 1° steps, as shown in the lower row of Figure 7, relaxed the rest of the atoms until the new equilibrium geometry depicted in the lower row of Figure 8 is reached and calculated the corresponding strain energies. We have fitted the results to a second-order polynomial and used the quadratic coefficient to calculate the torsional stiffness according to Equation (8). We have then used our values for G and l and employed Equation (11) to determine J.

As an example of this procedure, we show in Figure 12 the torsion energies corresponding to both armchair and zigzag edges for the 6×7 nanoflake. The graphs for other nanoflakes are similar.

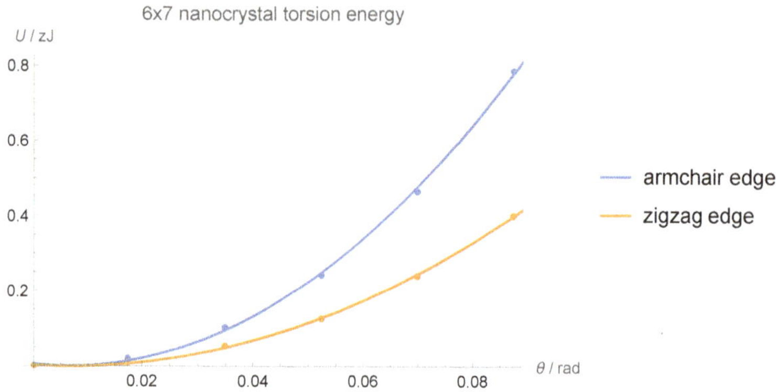

Figure 12. Torsion energies U as a function of the torsion angle θ for the 6×7 nanoflake. Points correspond to calculated energies, while lines show a second-order polynomial fit.

Table 6. Torsion constant and effective torsional thickness of square carbon nanoflakes.

Nanoflake	Edge	$J/(10^{-40}\ \mathrm{m}^4)$ Equation (11)	$J/(10^{-40}\ \mathrm{m}^4)$ Equation (12)	$\tilde{h}$/Å
3×3	zigzag	3.89	64.9	0.71
	armchair	7.02	62.6	0.98
5×5	zigzag	10.0	126	0.88
	armchair	6.76	116	0.75
6×7	zigzag	10.1	157	0.80
	armchair	12.2	169	0.85
8×9	zigzag	7.08	219	0.58
	armchair	12.6	223	0.76
10×11	zigzag	16.6	280	0.79
	armchair	22.2	276	0.92
12×13	zigzag	28.7	342	0.94
	armchair	30.1	329	0.99

The first thing to notice is that the direct use of Equation (12) with the standard value for graphene thickness h is not valid. This expression for homogeneous isotropic systems overestimates the value of J by an order of magnitude. The situation is similar to what happens for graphene bending rigidity B. Using continuous mechanics expressions for thin plates leads to a value for B one order of magnitude larger than B values obtained based on atomic-level calculations [84–87]. In fact, as we have already pointed out, there are several proposals for graphene thickness, and some of them arise precisely from trying to apply equations valid for homogeneous isotropic solids to 2D systems. Values of h determined from calculations on bent graphene sheets and carbon nanotubes (which can be seen as a particular case of bent carbon nanoribbons) are very different from the standard value $h = 3.4$ Å used in stretched (or compressed) planar graphene (the so-called "Yakobson's paradox"). Using molecular mechanics calculations, Duan et al. reported 0.52 Å [88], Scarpa et al. 0.84–1 Å [68] and Shi et al. 1.27 Å [51], while Yakobson et al. obtained 0.66 Å using molecular dynamics simulations [29]. From ab initio calculations, Kudin et al. [89] obtained 0.894 Å, Wang et al. 0.665 Å [90] and Shenderova et al. 0.9 Å [91]. Following this idea of an effective bending thickness, we report in the last column of the table the effective torsional thickness $\tilde{h}$ defined as the value of h that has to be used in Equation (12) in order to obtain the same value for J as the one obtained directly from torsion data and Equation (11). Our values for the effective torsional thicknesses are similar to these effective bending thicknesses and compatible with the upper limit for the effective single-walled carbon nanotube thickness determined by Pine et al., $h \leq 1$ Å [92].

Focusing on the results obtained using Equation 11, with the exception of in the 5×5 case, the torsion constant is greater when the nanoflake is twisted around the armchair edge, and, globally, J increases with the size of the system. Both behaviors are in agreement with results obtained in previous studies [93–95]. In this case, we cannot compare our numerical results to those in the literature because we have found neither calculations nor experiments dealing with any of our nanoflakes, and neither torsional stiffness nor torsion constant are intrinsic material properties but depend on the exact geometry of the system. Results for twisting are by far less abundant than those for other deformations, and our results could be useful for fitting interatomic potentials (especially for torsion and out-of-plane sp^2 hybridization terms in MM/MSM/NCM and MD models) for finite carbon nanostructures.

3.4. Poisson's Ratio

Table 7 includes our results for the Poison's ratio of the nanoflakes calculated using Equations (13) and (14). In the first case, we have used the optimized stretched geometries obtained for determining Young's moduli, measured the transverse compression at the center of the nanoflake and computed ν.

Table 7. Poisson's ratio of square carbon nanoflakes.

Nanoflake	Edge	ν	
		Equation (13)	Equation (14)
3×3	zigzag	0.33	1.09
	armchair	0.43	1.40
5×5	zigzag	0.30	0.93
	armchair	0.39	1.38
6×7	zigzag	0.33	1.81
	armchair	0.35	1.09
8×9	zigzag	0.31	1.10
	armchair	0.34	1.19
10×11	zigzag	0.31	1.74
	armchair	0.36	1.36
12×13	zigzag	0.31	1.46
	armchair	0.33	1.41

It is clear that Equation (14) overestimates the value of ν and cannot be used for these systems. This fact has been previously found for boron nitride nanoflakes [96,97], which also have a honeycomb structure. Results obtained by applying the definition of Poisson's ratio given by Equation (13) indicate that ν is bigger for stretching and pulling on the armchair edges (in agreement with Sakhaee-Pour [37]) but that the difference decreases as the size of the nanoflakes increases.

Results from other calculations are presented (in ascending order) in Table 8. We do not include thicknesses in this case because h is not needed in order to calculate ν. Poisson's ratio is strongly dependent on the strain applied (it can even become negative for large strains) [98,99]. In the table, we have selected the values corresponding to the smallest strain possible, but not all cases correspond to the so-called zero-strain limit. That is the reason for the variety of results reported, except for in the last case; Sakhaee-Pour's results [37] are much higher that those from other calculations because he uses Equation (14) to determine ν.

Table 8. Poisson's ratios (ν) of graphene found in the literature.

Source	ν	Method
Tapia et al. [74]	0.072	atomistic finite element method
Zakharchenko et al. [39]	0.16 ± 0.03	atomistic Monte Carlo based on empirical bond order potential
Shodja et al. [47]	0.19–0.20	Density Functional Theory
Cho et al. [35]	0.195 for graphite	molecular mechanics
Scarpa et al. [68]	0.211–0.848 (depending on the potential used)	cellular material mechanics theory
Lebedeva et al. [66]	0.221–0.987 (depending on the potential used)	interatomic potential
Tsai and Tu [38]	0.26	molecular dynamics
Caillerie et al. [100]	0.26	interatomic potential
Huang et al. [33]	0.28–0.30	bond-orbital tight-binding
Jiang et al. [98]	0.3	molecular mechanics
Cadelano et al. [101]	0.31	tight-binding
Tahani and Safarian [75]	0.333	homogenization composite shell model
Wang et al. [99]	0.35	molecular dynamics
Polyakova et al. [69]	0.36	molecular dynamics
Zhang et al. [71]	0.366	spring-based finite element model
Huang and Hwang [54]	0.397	interatomic potential
Lu and Huang [102]	0.398	molecular mechanics
Reddy et al. [30]	0.428	interatomic potential
Zheng et al. [83]	0.46	beam finite element method
Koberidze [103]	0.51	density-functional tight-binding
Georgantzinos et al. [82]	0.603	spring-based finite element model
Chandra et al. [40]	0.62	atomistic finite element method
Sakhaee-Pour [37]	1.285 for zigzag graphene *	interatomic potential
Sakhaee-Pour [37]	1.441 for armchair graphene *	interatomic potential

* taking into account that his naming convention is the opposite to ours.

Our results should be compared to those from Jiang et al. [98] and Wang et al. [99], which correspond to the zero-strain limit. The agreement with those results is complete. Regarding experiments, the experimental value of ν for graphite is 0.160 [77] and, for graphene, 0.19 [104], but those papers do not indicate the strain, and it is not possible to know how far they are from the zero-strain limit.

4. Conclusions

We have studied the mechanical properties of six small nearly square graphene nanoflakes using the semiempirical PM7 model. We have considered stretching, shear and torsion deformations along zigzag and armchair edges, calculating Young's and shear moduli as well as the torsion constant and Poisson's ratio. Results obtained could be useful for fitting interatomic potentials (especially for torsion and out-of-plane sp^2 hybridization terms and for near-the-edge bonds in MM/MSM/NCM and MD models) for finite carbon nanostructures. Our results are close to other calculations and to experimental values corresponding to infinite graphene, when these are available. Additionally, by making use of these results, we have tested two formulas valid for macroscopic homogeneous isotropic systems that are sometimes used for 2D nanostructures. We have shown that they do not hold for graphene and explained some strange results reported in the literature. In an attempt to recover one of these formulas, we have introduced an effective torsional thickness for 2D crystals analogous to the effective bending thickness found in the literature. The fast methodology developed in this work can be extended to two other dimensional nanostructures—fully planar and buckled—providing valuable results in order to fit interatomic potentials for faster and more reliable molecular mechanics and molecular dynamics calculations in these systems.

Author Contributions: Conceptualization, N.A.C.; methodology, N.A.C.; software, A.S.-G. and N.A.C.; validation, A.S.-G. and N.A.C.; formal analysis, A.S.-G. and N.A.C.; investigation, A.S.-G.; writing—original draft preparation, N.A.C.; writing—review and editing, A.S.-G. and N.A.C.; visualization, A.S.-G. and N.A.C. All authors have read and agreed to the published version of the manuscript.

Funding: This work was supported by the European Union NextGenerationEU (PRTR C17.I1).

Institutional Review Board Statement: Not applicable.

Informed Consent Statement: Not applicable.

Data Availability Statement: The original contributions presented in the study are included in the article, further inquiries can be directed to the corresponding author.

Conflicts of Interest: The authors declare no conflicts of interest. The funders had no role in the design of the study; in the collection, analyses or interpretation of data; in the writing of the manuscript; or in the decision to publish the results.

References

1. Bissett, M.A.; Tsuji, M.; Ago, H. Strain engineering the properties of graphene and other two-dimensional crystals. *Phys. Chem. Chem. Phys.* **2014**, *16*, 11124–11138. [CrossRef]
2. Deng, S.; Sumant, A.V.; Berry, V. Strain engineering in two-dimensional nanomaterials beyond graphene. *Nano Today* **2018**, *22*, 14–35. [CrossRef]
3. Peng, Z.; Chen, X.; Fan, Y.; Srolovitz, D.J.; Lei, D. Strain engineering of 2D semiconductors and graphene: From strain fields to band-structure tuning and photonic applications. *Light. Sci. Appl.* **2020**, *9*, 190. [CrossRef] [PubMed]
4. Ge, X.; Zhou, X.; Sun, D.; Chen, X. First-Principles Study of Structural and Electronic Properties of Monolayer PtX2 and Janus PtXY (X, Y = S, Se, and Te) via Strain Engineering. *ACS Omega* **2023**, *8*, 5715–5721. [CrossRef]
5. Qi, Y.; Sadi, M.A.; Hu, D.; Zheng, M.; Wu, Z.; Jiang, Y.; Chen, Y.P. Recent Progress in Strain Engineering on Van der Waals 2D Materials: Tunable Electrical, Electrochemical, Magnetic, and Optical Properties. *Adv. Mater.* **2023**, *35*, 2205714. [CrossRef]
6. Hasani, N.; Rajabi-Maram, A.; Touski, S.B. Strain engineering of electronic and spin properties in SnX (X = P, As, Sb, Bi) monolayers. *J. Phys. Chem. Solids* **2023**, *174*, 111131. [CrossRef]
7. Kim, J.M.; Haque, M.F.; Hsieh, E.Y.; Nahid, S.M.; Zarin, I.; Jeong, K.Y.; So, J.P.; Park, H.G.; Nam, S. Strain Engineering of Low-Dimensional Materials for Emerging Quantum Phenomena and Functionalities. *Adv. Mater.* **2023**, *35*, 2107362. [CrossRef] [PubMed]
8. Pandey, M.; Pandey, C.; Ahuja, R.; Kumar, R. Straining techniques for strain engineering of 2D materials towards flexible straintronic applications. *Nano Energy* **2023**, *109*, 108278. [CrossRef]
9. Fan, W.; Winands, T.; Doltsinis, N.L.; Li, Y.; Wang, Z. A Decatwistacene with an Overall 170° Torsion. *Angew. Chem. Int. Ed.* **2017**, *56*, 15373–15377. [CrossRef]
10. Liu, H.; Lee, M.; Šiškins, M.; van der Zant, H.S.J.; Steeneken, P.G.; Verbiest, G.J. Tuning heat transport in graphene by tension. *Phys. Rev. B* **2023**, *108*, L081401. [CrossRef]
11. He, S.; Zhang, Y.; Gao, J.; Nag, A.; Rahaman, A. Integration of Different Graphene Nanostructures with PDMS to Form Wearable Sensors. *Nanomaterials* **2022**, *12*, 950. [CrossRef]
12. Srimaneepong, V.; Skallevold, H.E.; Khurshid, Z.; Zafar, M.S.; Rokaya, D.; Sapkota, J. Graphene for Antimicrobial and Coating Application. *Int. J. Mol. Sci.* **2022**, *23*, 499. [CrossRef]

13. Zhang, Y.; Jin, Y.; Liu, J.; Ren, Q.; Chen, Z.; Zhao, Y.; Zhao, P. Strain engineering of graphene on rigid substrates. *Nanoscale Adv.* **2022**, *4*, 5056–5061. [CrossRef] [PubMed]
14. de-la Huerta-Sainz, S.; Ballesteros, A.; Cordero, N.A. Quantum Revivals in Curved Graphene Nanoflakes. *Nanomaterials* **2022**, *12*, 1953. [CrossRef]
15. Yuan, X.; Zhu, S. Compressive strain engineering of strong and sensitive pseudomagnetic fields in buckled graphene nanobubbles. *Phys. Rev. B* **2023**, *107*, 195417. [CrossRef]
16. De Beule, C.; Phong, V.T.; Mele, E.J. Network model for periodically strained graphene. *Phys. Rev. B* **2023**, *107*, 045405. [CrossRef]
17. Mahmud, M.T.; Zhai, D.; Sandler, N. Topological Flat Bands in Strained Graphene: Substrate Engineering and Optical Control. *Nano Lett.* **2023**, *23*, 7725–7732. [CrossRef]
18. Dawood, O.M.; Thameel, M.N.; Mohammad, J. Unveiling a new Raman active mode in graphene and its implications for band gap formation: A DFT study. *Phys. E Low-Dimens. Syst. Nanostruct.* **2023**, *154*, 115810. [CrossRef]
19. Tepliakov, N.V.; Lischner, J.; Kaxiras, E.; Mostofi, A.A.; Pizzochero, M. Unveiling and Manipulating Hidden Symmetries in Graphene Nanoribbons. *Phys. Rev. Lett.* **2023**, *130*, 026401. [CrossRef]
20. Park, H.C.; Han, J.; Myoung, N. A strain-engineered graphene qubit in a nanobubble. *Quantum Sci. Technol.* **2023**, *8*, 025012. [CrossRef]
21. Lee, H.C.; Hsieh, E.Y.; Yong, K.; Nam, S. Multiaxially-stretchable kirigami-patterned mesh design for graphene sensor devices. *Nano Res.* **2020**, *13*, 1406–1412. [CrossRef]
22. Chun, S.; Choi, Y.; Park, W. All-graphene strain sensor on soft substrate. *Carbon* **2017**, *116*, 753–759. [CrossRef]
23. Cong, L.; Yuan, Z.; Bai, Z.; Wang, X.; Zhao, W.; Gao, X.; Hu, X.; Liu, P.; Guo, W.; Li, Q.; et al. On-chip torsion balances with femtonewton force resolution at room temperature enabled by carbon nanotube and graphene. *Sci. Adv.* **2021**, *7*, eabd2358. [CrossRef]
24. Foroughi, J.; Spinks, G. Carbon nanotube and graphene fiber artificial muscles. *Nanoscale Adv.* **2019**, *1*, 4592–4614. [CrossRef] [PubMed]
25. Malisz, K.; Świeczko Żurek, B. Graphene Production and Biomedical Applications: A Review. *Crystals* **2023**, *13*, 1413. [CrossRef]
26. Cordero, N.A.; Alonso, J.A. Interaction of Surfactants Containing a Sulfuric Group with a (5,5) Carbon Nanotube. *J. Phys. Chem.* **2010**, *114*, 17249–17256. [CrossRef]
27. García, T.; Rodríguez-Bolívar, S.; Cordero, N.A.; Romera, E. Wavepacket revivals in monolayer and bilayer graphene rings. *J. Phys. Condens. Matter* **2013**, *25*, 235301. [CrossRef]
28. García, T.; Cordero, N.A.; Romera, E. Zitterbewegung and quantum revivals in monolayer graphene quantum dots in magnetic fields. *Phys. Rev. B* **2014**, *89*, 075416. [CrossRef]
29. Yakobson, B.I.; Brabec, C.J.; Bernholc, J. Nanomechanics of Carbon Tubes: Instabilities beyond Linear Response. *Phys. Rev. Lett.* **1996**, *76*, 2511–2514. [CrossRef]
30. Reddy, C.D.; Rajendran, S.; Liew, K.M. Equilibrium configuration and continuum elastic properties of finite sized graphene. *Nanotechnology* **2006**, *17*, 864. [CrossRef]
31. Ru, C.Q. Effect of van der Waals forces on axial buckling of a double-walled carbon nanotube. *J. Appl. Phys.* **2000**, *87*, 7227–7231. [CrossRef]
32. El Kass, D.; Monneau, R. Atomic to continuum passage for nanotubes: A discrete Saint-Venant principle and error estimates. *Arch. Ration. Mech. Anal.* **2014**, *213*, 25–128. [CrossRef]
33. Huang, K.; Yin, Y.; Qu, B. Tight-binding theory of graphene mechanical properties. *Microsyst. Technol.* **2021**, *27*, 3851–3858. [CrossRef]
34. Anastasi, A.A.; Ritos, K.; Cassar, G.; Borg, M.K. Mechanical properties of pristine and nanoporous graphene. *Mol. Simul.* **2016**, *42*, 1502–1511. [CrossRef]
35. Cho, J.; Luo, J.; Daniel, I. Mechanical characterization of graphite/epoxy nanocomposites by multi-scale analysis. *Compos. Sci. Technol.* **2007**, *67*, 2399–2407. [CrossRef]
36. Min, K.; Aluru, N.R. Mechanical properties of graphene under shear deformation. *Appl. Phys. Lett.* **2011**, *98*, 013113. [CrossRef]
37. Sakhaee-Pour, A. Elastic properties of single-layered graphene sheet. *Solid State Commun.* **2009**, *149*, 91–95. [CrossRef]
38. Tsai, J.L.; Tu, J.F. Characterizing mechanical properties of graphite using molecular dynamics simulation. *Mater. Des.* **2010**, *31*, 194–199. [CrossRef]
39. Zakharchenko, K.V.; Katsnelson, M.I.; Fasolino, A. Finite Temperature Lattice Properties of Graphene beyond the Quasiharmonic Approximation. *Phys. Rev. Lett.* **2009**, *102*, 046808. [CrossRef]
40. Chandra, Y.; Adhikari, S.; Mukherjee, S.; Mukhopadhyay, T. Unfolding the mechanical properties of buckypaper composites: nano- to macro-scale coupled atomistic-continuum simulations. *Eng. Comput.* **2022**, *38*, 5199–5229. [CrossRef]
41. Savin, A.; Korznikova, E.; Dmitriev, S. Improving bending rigidity of graphene nanoribbons by twisting. *Mech. Mater.* **2019**, *137*, 103123. [CrossRef]
42. Thakur, R.; Ahluwalia, P.; Kumar, A.; Sharma, M.; Sharma, R. Twisted helical armchair graphene nanoribbons: Mechanical and electronic properties. *Eur. Phys. J.* **2021**, *94*, 99. [CrossRef]
43. Sun, Y.W.; Papageorgiou, D.G.; Humphreys, C.J.; Dunstan, D.J.; Puech, P.; Proctor, J.E.; Bousige, C.; Machon, D.; San-Miguel, A. Mechanical properties of graphene. *Appl. Phys. Rev.* **2021**, *8*, 021310. [CrossRef]

44. Ragab, T.; McDonald, J.; Basaran, C. Aspect ratio effect on shear modulus and ultimate shear strength of graphene nanoribbons. *Diam. Relat. Mater.* **2017**, *74*, 9–15. [CrossRef]
45. Landau, L.; Lifshitz, E. *Theory of Elasticity*; Pergamon Press: Oxford, UK, 1970.
46. Timoshenko, S.; Goodier, J. *Theory of Elasticity*; McGRAW-HILL: New York, NY, USA, 1951.
47. Shodja, H.M.; Ojaghnezhad, F.; Etehadieh, A.; Tabatabaei, M. Elastic moduli tensors, ideal strength, and morphology of stanene based on an enhanced continuum model and first principles. *Mech. Mater.* **2017**, *110*, 1–15. [CrossRef]
48. Guo, C.; Papangelis, J. Torsion of beams with corrugated webs. In Proceedings of the Ninth International Conference on Advances in Steel Structures (ICASS 2018), Hong Kong, China, 5–7 December 2018; pp. 5–7. [CrossRef]
49. Trahair, N.; Bradford, M.; Nethercot, D.; Gardner, L. *The Behaviour and Design of Steel Structures to EC3*, 4th ed.; Taylor & Francis: New York, NY, USA, 2008.
50. Xia, D.; Li, Q.; Xue, Q.; Liang, C.; Dong, M. Super flexibility and stability of graphene nanoribbons under severe twist. *Phys. Chem. Chem. Phys.* **2016**, *18*, 18406–18413. [CrossRef]
51. Shi, J.X.; Natsuki, T.; Lei, X.W.; Ni, Q.Q. Equivalent Young's modulus and thickness of graphene sheets for the continuum mechanical models. *Appl. Phys. Lett.* **2014**, *104*, 223101. [CrossRef]
52. Lee, D.H.; Lee, M.J.; Kim, Y.S.; Choi, J.S.; Park, B.H. Sample rotation angle dependence of graphene thickness measured using atomic force microscope. *Carbon* **2015**, *81*, 210–215. [CrossRef]
53. Delfani, M.R.; Shodja, H. Dual ideal shear strengths for chiral single-walled carbon nanotubes. *Int. J.-Non-Linear Mech.* **2020**, *120*, 103382. [CrossRef]
54. Huang, Y.; Wu, J.; Hwang, K.C. Thickness of graphene and single-wall carbon nanotubes. *Phys. Rev. B* **2006**, *74*, 245413. [CrossRef]
55. Stewart, J. Optimization of parameters for semiempirical methods VI: More modifications to the NDDO approximations and re-optimization of parameters. *J. Mol. Model.* **2013**, *19*, 1–32. [CrossRef] [PubMed]
56. Frisch, M.J.; Trucks, G.W.; Schlegel, H.B.; Scuseria, G.E.; Robb, M.A.; Cheeseman, J.R.; Scalmani, G.; Barone, V.; Petersson, G.A.; Nakatsuji, H.; et al. *Gaussian 16*; Revision C.01; Gaussian Inc.: Wallingford, CT, USA, 2019.
57. Throssel, K. Evaluating and Improving Approximate LCAO-MO Theory with Restored Overlap and Bond Order Bond Energy Corrections. Ph.D. Thesis, Wesleyan University, Middletown, CT, USA, 2017. Available online: https://digitalcollections.wesleyan.edu/object/ir-2308 (accessed on 24 March 2024).
58. Jurkiewicz, K.; Hawelek, L.; Balin, K.; Szade, J.; Braghiroli, F.L.; Fierro, V.; Celzard, A.; Burian, A. Conversion of Natural Tannin to Hydrothermal and Graphene-Like Carbons Studied by Wide-Angle X-ray Scattering. *J. Phys. Chem. A* **2015**, *119*, 8692–8701. [CrossRef] [PubMed]
59. Jaiyong, P.; Bryce, R.A. Approximate quantum chemical methods for modelling carbohydrate conformation and aromatic interactions: β-cyclodextrin and its adsorption on a single-layer graphene sheet. *Phys. Chem. Chem. Phys.* **2017**, *19*, 15346–15355. [CrossRef] [PubMed]
60. Vorontsov, A.V.; Tretyakov, E.V. Determination of graphene's edge energy using hexagonal graphene quantum dots and PM7 method. *Phys. Chem. Chem. Phys.* **2018**, *20*, 14740–14752. [CrossRef] [PubMed]
61. Hariharan, S.; Majumder, M.; Edel, R.; Grabnic, T.; Sibener, S.J.; Hase, W.L. Exploratory Direct Dynamics Simulations of 3O2 Reaction with Graphene at High Temperatures. *J. Phys. Chem. C* **2018**, *122*, 29368–29379. [CrossRef]
62. Bertorelle, F.; Basu, S.; Fakhouri, H.; Bakulic, M.P.; Mignon, P.; Russier-Antoine, I.; Brevet, P.F.; Thomas, S.; Kalarikkal, N.; Antoine, R. Covalent anchoring of atomically precise glutathione-protected gold nanoclusters on graphene oxide nanosheets. *Nano Express* **2020**, *1*, 030005. [CrossRef]
63. Gun'ko, V. Quantum-chemically computed integral characteristics of complex nanomaterials. *Chem. Phys. Technol. Surf./Khimiya Fiz. Tekhnologiya Poverhni* **2021**, *12*, 157–167. Available online: https://www.cpts.com.ua/index.php/cpts/article/download/590/598 (accessed on 24 March 2024). [CrossRef]
64. Chi, S.C.; Lee, C.L.; Chang, C.M. Adsorption of Pesticides, Antibiotics and Microcystin-LR by Graphene and Hexagonal Boron Nitride Nano-Systems: A Semiempirical PM7 and Theoretical HSAB Study. *Crystals* **2022**, *12*, 1068. [CrossRef]
65. Tromer, R.M.; Felix, I.M.; Pereira, L.F.C.; da Luz, M.G.E.; Junior, L.A.R.; Galvão, D.S. Lattice thermal conductivity of 2D nanomaterials: A simple semi-empirical approach. *Phys. Chem. Chem. Phys.* **2023**, *25*, 28703–28715. [CrossRef]
66. Lebedeva, I.V.; Minkin, A.S.; Popov, A.M.; Knizhnik, A.A. Elastic constants of graphene: Comparison of empirical potentials and DFT calculations. *Phys. E Low-Dimens. Syst. Nanostruct.* **2019**, *108*, 326–338. [CrossRef]
67. Giannopoulos, G.I. Elastic buckling and flexural rigidity of graphene nanoribbons by using a unique translational spring element per interatomic interaction. *Comput. Mater. Sci.* **2012**, *53*, 388–395. [CrossRef]
68. Scarpa, F.; Adhikari, S.; Phani, A.S. Effective elastic mechanical properties of single layer graphene sheets. *Nanotechnology* **2009**, *20*, 065709. [CrossRef] [PubMed]
69. Polyakova, P.; Galiakhmetova, L.K.; Murzaev, R.; Lisovenko, D.; Baimova, J. Elastic properties of diamane. *Lett. Mater.* **2023**, *13*, 171–176. [CrossRef]
70. Tzeng, S.H.; Tsai, J.L. Characterizing the Mechanical Properties of Graphene and Single Walled Carbon Nanotubes. *J. Mech.* **2011**, *27*, 461–467. [CrossRef]
71. Zhang, Y.; Huang, L.; Huang, J. A modified spring finite element model for graphene elastic properties study. *Mater. Today Commun.* **2023**, *34*, 105158. [CrossRef]

72. Sha'bani, F.; Rash-Ahmadi, S. Length scale effect on the buckling behavior of a graphene sheets using modified couple stress theory and molecular dynamics method. *Acta Mech.* **2022**, *233*, 943–960. [CrossRef]
73. Zaeri, M.; Ziaei-Rad, S.; Vahedi, A.; Karimzadeh, F. Mechanical modelling of carbon nanomaterials from nanotubes to buckypaper. *Carbon* **2010**, *48*, 3916–3930. [CrossRef]
74. Tapia, A.; Peón-Escalante, R.; Villanueva, C.; Avilés, F. Influence of vacancies on the elastic properties of a graphene sheet. *Comput. Mater. Sci.* **2012**, *55*, 255–262. [CrossRef]
75. Tahani, M.; Safarian, S. Determination of rigidities, stiffness coefficients and elastic constants of multi-layer graphene sheets by an asymptotic homogenization method. *J. Braz. Soc. Mech. Sci. Eng.* **2019**, *41*, 1–16. [CrossRef]
76. Lee, C.; Wei, X.; Kysar, J.W.; Hone, J. Measurement of the Elastic Properties and Intrinsic Strength of Monolayer Graphene. *Science* **2008**, *321*, 385–388. [CrossRef]
77. Blakslee, O.L.; Proctor, D.G.; Seldin, E.J.; Spence, G.B.; Weng, T. Elastic Constants of Compression-Annealed Pyrolytic Graphite. *J. Appl. Phys.* **2003**, *41*, 3373–3382. [CrossRef]
78. Lengvarský, P.; Bocko, J. Mechanical Properties and Eigenfrequencies of Graphene Sheets. *Acta Mech. Slovaca* **2017**, *21*, 26–32. [CrossRef]
79. Thomas, S.; Ajith, K.; Lee, S.U.; Valsakumar, M.C. Assessment of the mechanical properties of monolayer graphene using the energy and strain-fluctuation methods. *RSC Adv.* **2018**, *8*, 27283–27292. [CrossRef] [PubMed]
80. Liu, X.; Metcalf, T.H.; Robinson, J.T.; Houston, B.H.; Scarpa, F. Shear Modulus of Monolayer Graphene Prepared by Chemical Vapor Deposition. *Nano Lett.* **2012**, *12*, 1013–1017. [CrossRef]
81. Mukhopadhyay, T.; Mahata, A.; Adhikari, S.; Asle Zaeem, M. Probing the shear modulus of two-dimensional multiplanar nanostructures and heterostructures. *Nanoscale* **2018**, *10*, 5280–5294. [CrossRef]
82. Georgantzinos, S.; Giannopoulos, G.; Anifantis, N. Numerical investigation of elastic mechanical properties of graphene structures. *Mater. Des.* **2010**, *31*, 4646–4654. [CrossRef]
83. Zheng, W.; Guo, Z.; Wang, Z.; Boay, C.G.; Wang, X.; Li, Y. Modeling analysis of elastic properties of graphene-carbon nanotube (G-C) reinforced composites. *Polym. Compos.* **2022**, *43*, 3136–3149. [CrossRef]
84. Lindahl, N.; Midtvedt, D.; Svensson, J.; Nerushev, O.A.; Lindvall, N.; Isacsson, A.; Campbell, E.E.B. Determination of the Bending Rigidity of Graphene via Electrostatic Actuation of Buckled Membranes. *Nano Lett.* **2012**, *12*, 3526–3531. [CrossRef]
85. Androulidakis, C.; Zhang, K.; Robertson, M.; Tawfick, S. Tailoring the mechanical properties of 2D materials and heterostructures. *2D Mater.* **2018**, *5*, 032005. [CrossRef]
86. Androulidakis, C.; Koukaras, E.; Hadjinicolaou, M.; Galiotis, C. Non-Eulerian behavior of graphitic materials under compression. *Carbon* **2018**, *138*, 227–233. [CrossRef]
87. Androulidakis, C.; Koukaras, E.N.; Sampathkumar, K.; Rahova, J.; Galiotis, C.; Frank, O. Hierarchy of nanoscale graphene wrinkles on compliant substrate: Theory and experiment. *Extrem. Mech. Lett.* **2020**, *40*, 100948. [CrossRef]
88. Duan, W.H.; Wang, C.M. Nonlinear bending and stretching of a circular graphene sheet under a central point load. *Nanotechnology* **2009**, *20*, 075702. [CrossRef] [PubMed]
89. Kudin, K.N.; Scuseria, G.E.; Yakobson, B.I. C_2F, BN, and C nanoshell elasticity from ab initio computations. *Phys. Rev. B* **2001**, *64*, 235406. [CrossRef]
90. Wang, L.; Zheng, Q.; Liu, J.Z.; Jiang, Q. Size Dependence of the Thin-Shell Model for Carbon Nanotubes. *Phys. Rev. Lett.* **2005**, *95*, 105501. [CrossRef] [PubMed]
91. Shenderova, O.A.; Zhirnov, V.V.; Brenner, D.W. Carbon Nanostructures. *Crit. Rev. Solid State Mater. Sci.* **2002**, *27*, 227–356. [CrossRef]
92. Pine, P.; Yaish, Y.E.; Adler, J. Vibrational analysis of thermal oscillations of single-walled carbon nanotubes under axial strain. *Phys. Rev. B* **2014**, *89*, 115405. [CrossRef]
93. Shen, H. Mechanical properties and thermal conductivity of the twisted graphene nanoribbons. *Mol. Phys.* **2014**, *112*, 2614–2620. [CrossRef]
94. Chatzidakis, G.D.; Kalosakas, G.; Fthenakis, Z.G.; Lathiotakis, N.N. A torsional potential for graphene derived from fitting to DFT results. *Eur. Phys. J. B* **2018**, *91*, 11. [CrossRef]
95. Yoneyama, K.; Yamanaka, A.; Okada, S. Energetics and electronic structure of graphene nanoribbons under uniaxial torsional strain. *Jpn. J. Appl. Phys.* **2019**, *58*, SDDD05. [CrossRef]
96. Tapia, A.; Cab, C.; Hernández-Pérez, A.; Villanueva, C.; Peñuñuri, F.; Avilés, F. The bond force constants and elastic properties of boron nitride nanosheets and nanoribbons using a hierarchical modeling approach. *Phys. E Low-Dimens. Syst. Nanostruct.* **2017**, *89*, 183–193. [CrossRef]
97. Sakharova, N.A.; Pereira, A.F.G.; Antunes, J.M. A Study of the Mechanical Behaviour of Boron Nitride Nanosheets Using Numerical Simulation. *Nanomaterials* **2023**, *13*, 2759. [CrossRef]
98. Jiang, J.W.; Chang, T.; Guo, X.; Park, H.S. Intrinsic negative Poisson's ratio for single-layer graphene. *Nano Lett.* **2016**, *16*, 5286–5290. [CrossRef] [PubMed]
99. Wang, Y.; Lei, J.; Liu, Z. Molecular dynamics study on the anisotropic Poisson's ratio of the graphene. *Diam. Relat. Mater.* **2019**, *93*, 66–74. [CrossRef]
100. Caillerie, D.; Mourad, A.; Raoult, A. Discrete homogenization in graphene sheet modeling. *J. Elast.* **2006**, *84*, 33–68. [CrossRef]

101. Cadelano, E.; Palla, P.L.; Giordano, S.; Colombo, L. Nonlinear Elasticity of Monolayer Graphene. *Phys. Rev. Lett.* **2009**, *102*, 235502. [CrossRef] [PubMed]
102. Lu, Q.; Huang, R. Nonlinear mechanics of single-atomic-layer graphene sheets. *Int. J. Appl. Mech.* **2009**, *1*, 443–467. [CrossRef]
103. Koberidze, M. Computational Studies of Torsional Properties of Single-Walled Carbon Nanotubes. Master's Thesis, University of Jyväskylä, Jyväskylä, Findland, 2010. https://jyx.jyu.fi/handle/123456789/37243 (accessed on 24 March 2024).
104. Politano, A.; Marino, A.R.; Campi, D.; Farías, D.; Miranda, R.; Chiarello, G. Elastic properties of a macroscopic graphene sample from phonon dispersion measurements. *Carbon* **2012**, *50*, 4903–4910. [CrossRef]

crystals

MDPI

Article

Effect of Polar Faces of SiC on the Epitaxial Growth of Graphene: Growth Mechanism and Its Implications for Structural and Electrical Properties

Stefan A. Pitsch [1,2] and R. Radhakrishnan Sumathi [1,*]

1 Applied Crystallography and Materials Science Section, Department of Earth and Environmental Sciences, Ludwig-Maximilians-University (LMU), Theresienstrasse 41, D-80333 Munich, Germany
2 Presently at Institute for Geochemistry and Petrology, ETH Zürich, Clausiusstrasse 25, 8092 Zürich, Switzerland
* Correspondence: sumathi@lrz.uni-muenchen.de

Abstract: In this study, epitaxial graphene layers of cm^2 sizes were grown on silicon carbide (SiC) substrates by high-temperature sublimation. The behavior of the two crystallographic SiC-polar faces and its effect on the growth mechanism of graphene layers and their properties were investigated. Crystallographic structural differences observed in AFM studies were shown to cause disparities in the electrical conductivity of the grown layers. On the silicon-polar (Si-polar) face of SiC, the graphene formation occurred in spike-like structures that originated orthogonally from atomic steps of the substrate and grew outwards in the form of 2D nucleation with a fairly good surface coverage over time. On the carbon-polar (C-polar) face, a hexagonal structure already formed at the beginning of the growth process. On both polar faces, the known process of step-bunching promoted the formation of nm-scale structural obstacles. Such a step-bunching effect was found to be more pronounced on the C-polar face. These 2D-obstacles account for a low probability of a complete nano-sheet formation, but favor 2D-structures, comparable to graphene nanoribbons. The resulting direction-dependent anisotropic behavior in electrical conductivity measured by four-point probe method mainly depends on the height and spacing between these structural-obstacles. The anisotropy becomes less prudent as and when more graphene layers are synthesized.

Keywords: 2D materials; graphene; silicon carbide; sublimation growth; surface morphologies; step-bunching; electrical conductivity

Citation: Pitsch, S.A.; Sumathi, R.R. Effect of Polar Faces of SiC on the Epitaxial Growth of Graphene: Growth Mechanism and Its Implications for Structural and Electrical Properties. *Crystals* **2023**, *13*, 189. https://doi.org/10.3390/cryst13020189

Academic Editors: Justina Gaidukevic and Jurgis Barkauskas

Received: 13 December 2022
Revised: 15 January 2023
Accepted: 18 January 2023
Published: 21 January 2023

1. Introduction

Graphene has attracted much attention for its outstanding material properties such as strength, chemical durability, electrical conductivity and many more and is therefore extensively studied. Since its discovery in 2004, an exponential increase in the graphene material research could be observed. The reason is the material's high potential for multiple areas of application. Prominent examples are highly durable and strong mechanical components for the aviation and automobile industries, a new generation of solar cells, batteries and transistors, conductive inks, sensors [1], material for bio-medical applications [2], etc.

Graphene, ideally, is a single-layer material, solely composed of carbon atoms and arranged in a planar, hexagonal honeycomb lattice that accounts for many of its outstanding properties. Each carbon atom is bonded via sp^2 hybridized bonds to three other carbon atoms with a C-C distance of 1.42 Å [3]. Although graphene's properties largely depend on it being a single layer and free standing, i.e., chemically un-bound to other substrates, the term graphene is often used for bi-, tri-, and multilayers as well. As many as 10 layers have been reported to be graphene or to show graphene-like behavior [4].

The special bonding situation in graphene and its high lattice symmetry make it resistant to covalent modification and therefore chemically durable [4], highly efficient in electron and heat transport and result in graphene to be the first zero bandgap semiconductor material that is known [5]. The electrical conductivity of intrinsic graphene is quite low in the order of the conductance quantum $\sigma \sim e^2/h$ because at the Dirac points, the density of states is zero. Via doping, either through an electric field or adsorption, the conductivity changes are to be quite high, even outperforming copper at room temperature [5]. Even strain in a graphene sheet, which has been shown never to be a completely flat surface [4], acts as a bandgap modifying parameter [5,6] and must be considered when measuring graphene's electrical properties.

Multiple ways of fabrication are known, with each individual one suffering from setbacks. Chemical vapor deposition is mostly stated when large-areas of graphene are grown with both high quality and uniformity, mostly on Cu/Ni and Si/SiO_2 substrates. Another formation mechanism is the formation out of SiC substrates. Extensive work has been carried out to fabricate good quality, large-area graphene [7–10] resulting in many different growth techniques that are being improved continuously. However, a well-controlled process which yields a defined (low) number of layers, even better, single crystals, remains a challenge [11,12]. Goals are a better control of surface morphologies, individual crystal sizes and homogeneity.

The epitaxial growth of graphene on silicon carbide (SiC), a process which is also known as the sublimation technique, relies on SiC as a substrate. Due to SiC's crystal structure, with a distance of the carbon plane to the neighboring Si plane having a ratio of 1:3 in relation to the C-C interplanar distance, two polar faces can be subclassified: The Si-polar face and the C-polar face [13]. For the case of the resulting Si-polar face ([0001]), Si atoms occupy the top positions, while in the case of the C-face ([000-1]), C atoms form the top respectively. Although there is a relatively high lattice mismatch of about 0.62 Å (3.073 Å for SiC versus 2.46 Å for graphene) between graphene and the used substrate, it has been shown that 6H-SiC (a specific polymorph of SiC) acts as a very good surface for the sublimation synthesis of graphene [1]. The growth process is driven by the preferred Si sublimation compared to its stoichiometric counterpart composing the substrate. For the growth of one layer of graphene at least 3 layers of SiC bilayers are necessary [14]. Decomposition of SiC starts preferentially where the binding energy within the substrate is lowest, which is mainly at defect sites and atomic terrace edges of the SiC substrate [15]. A detailed review of the growth process can be found in G. R. Yazdi [3].

Surface reconstructions and growth kinetics differ for each polar surface. This results in different graphene growth rates, morphologies, and electronic properties for the graphene on the two polar faces [3]. The Si-face of SiC has the advantage that growth of graphene occurs much more slowly than on its counterpart and is therefore a lot easier to control by the growth parameters such as temperature and time alone. However, the Si-face also has worse electrical properties in comparison with the C-face due to a buffer layer that is formed before graphene formation occurs [16]. Generally, graphene on the Si-face is produced as a multilayer stack. The layers within interact with each other as they do in graphite, accounting for more and more graphite-like behavior the thicker the stacks become [17]. The Si-face produces domains that are only about a third of the size of graphene domains that were grown on the C-face. Under an argon atmosphere, domains could typically reach sizes of about 50 μm [1].

The growth of graphene on the C-face is faster, a lot harder to control and does not produce a buffer layer. From the first produced graphene layer onwards, no strong interaction with the substrate occurs any longer which results from a relatively large distance to the substrate (3.2 Å) and the absence of covalent bonds to it. Multilayers, in contrast to the Si-face, show a large amount of rotational disorder. Every single graphene layer in a stack of many thus behaves like a monolayer of its own because of the very weak interaction between the layers [3]. Therefore, graphene that is grown on the C-face generally shows better electrical properties and is able to retain those for quite thick stacks [17]. The intrinsic mobility of Si-face graphene is about two orders of magnitude lower than optimally grown graphene on the C-face due to its graphite-like behavior [14]. C-face graphene was measured to reach 10,000 to 30,000 $cm^2.V^{-1}$ s^{-1} under room temperature while Si-face graphene yielded 500 $cm^2.V^{-1}s^{-1}$ to 2000 cm^2. $V^{-1}s^{-1}$ at the same conditions [1]. Resistivity on SiC-grown graphene (grown on the [0001]-face) has been reported to be in the range of 1.2×10^{-4} Ω·cm to 7.4×10^{-5} Ω·cm [18].

Although the sublimation epitaxy, over time, received less attention than, for example, the predominant CVD growth method, sublimation growth of graphene is still a potential candidate for future use for its comparative ease in handling and investigating after the growth, compared with other techniques that require a transfer process off a substrate before consecutive processing. Special applications, among them the work on confined 2D materials, benefit from the sublimation method and an improvement of the thereby formed graphene (e.g., [19,20]). The problem in this technique, however, is the lack of control over the resulting physical attributes that result out of the formed graphene's size, quality and thickness and can be influenced by growth temperature, duration, substrate treatment and its crystallographic orientation, carbon supply, pressure in the growth chamber, heating rate, etc. [3,20]. An improvement on the control of these factors has been reported on 4H-SiC by face-to-face growth in ultra-high vacuum [21] and by polymer-assisted growth [20]. The crucial factor of the degree of substrate miscut in obtaining good and reproducible results was outlined in [22].

The main aim of this work is to contribute to the understanding of the growth mechanism on different polar faces of SiC substrates. One relevant process on the substrate's surface during the growth process is that due to different decomposition kinetics of structural units, atomic steps tend to combine in higher ones, which is called step bunching. This growth process does not yet allow for complete control in terms of both the number of layers that are produced and the morphology of the graphene sheet. In contrast to other methods that do allow for a control on those attributes such as CVD, this method's advantage lies in no need to further treat the produced graphene once it is prepared on semi-insulating SiC, which can remain pristine on the SiC substrate and will therefore not be damaged in necessary consecutive preparation procedures. Also, we report here for the first time, to the best of our knowledge, the interrelation in the characteristics of growth-related structural and electrical properties resulting from two different growth mechanisms on the two polar faces of SiC. It has already been reported, that the structure of graphene has big impacts on its properties [23] and that graphene which is formed on one of SiC's two polar faces varies greatly depending on the face it was grown on [3,24]. However, a difference in the growth morphology on the two polar faces has not been reported so far. The resulting growth-direction-dependent anisotropy of the electrical resistivity is investigated and discussed.

2. Materials and Methods

The graphene layers were grown on the 6H polytype of SiC (on-axis) which had semi-insulating behavior and thus allowed resistivity measurements of the graphitized surface on top without major influence of the substrate. SiC wafers were cut into small squares of 1 to 2 cm^2 and cleaned by immersing them subsequently in boiling isopropanol and acetone. The samples were put onto a graphite sample holder with the polar face of the substrate that was later to be investigated facing downwards. An excavation was adjustable there both in height and diameter to account for a local increase of Si vapor pressure to interact with the surface. The sublimation rate at the surface of the sample could thus be controlled. The samples were put into a furnace, which was heated under an argon atmosphere of 800 mbar, accounting for a retardation of the sublimation process which starts at approximately 1500 °C under an argon pressure in comparison to about 1150 °C under ultra-high vacuum [1]. To reach growth temperatures as high as 2070 °C, a certain routine (program) was kept throughout the experiments to account for constant conditions: After reaching 1000 °C under a high vacuum, argon flow was enabled, and the temperature was increased to 1400 °C. By reaching these starting conditions, a fast up-ramping of 40 °C/min was performed to the intended growth temperature, where it was kept constant for the set duration of the experiment. The down-ramping of temperature mimicked the up-ramping process down to 1400 °C, where in theory, no Si sublimation from the substrate was taking place any longer [1]. The two principal parameters to be varied in these experiments were temperature and duration of growth. Argon-pressure in the sample chamber was kept constant. The growth temperatures were chosen between the high end of the spectrum of most cited studies (1900 °C) and 2070 °C both to extend the available spectrum of data and because models and personal observations have indicated that higher temperatures are favorable to grow larger, single-crystal graphene areas [25].

After the growth, the samples were analyzed by means of (1) Raman spectroscopy and (2) Atomic Force Microscopy (AFM) to be able to allow for an estimation of the numbers of graphene layers that were produced on the substrate and to observe the surface morphology of the samples. The (3) four-point probe measurement was conducted to be able to relate the observations to the electrical properties of formed material.

(1) Raman spectroscopy was used to qualitatively determine whether the growth of graphene on the SiC surface was successful. Graphene and graphite in Raman spectroscopy showed 3 major peaks that could be used to differentiate between both materials, namely the D, G and 2D peaks. In this study, the ratios of the 2D (2700 cm^{-1}) and the G peak (1582 cm^{-1}) were used for the differentiation. This ratio,

$$IR = I_{2D}/I_G, \tag{1}$$

where IR is the aforementioned intensity ratio, I_G is the intensity of the G peak, and I_2D is the intensity of the 2D peak, was used to identify single, free-standing graphene layers, which are indicated by an IR value of ~2. Ratios between 1 and 2 have been reported to show monolayer graphene, however, still attached to the substrate [26]. This ratio decreases with an increasing number of layers, eventually resulting in a ratio of about 0.25 for graphite. To evaluate the thickness of the graphene layers further, an additional intensity ratio (IS) is used, taking into account the Raman signal of the underlying SiC substrate (1518 cm^{-1}). This ratio is calculated as the following:

$$IS = I_{substrate}/I_G \tag{2}$$

As in Raman spectroscopy, only the topmost atomic layers contribute to most of the signal intensity, this ratio can be used to qualitatively quantify the thickness of the graphene stack on top of the substrate. If IS is high, the underlying substrate has a high influence in the measurement which points towards a thin layer of graphene coverage on the surface and vice versa.

The full width at half maximum (FWHM) of the 2D peak allows a distinction of mono- and multi-layered graphene, which is 24 cm^{-1} for a free-standing single layer of graphene. Double, triple and multi layers show significantly higher FWHMs [27] and non-free-standing monolayers have been reported to show a FWHM of around 40 cm^{-1} [26]. The 2D peak allows for additional information about strain on the graphene lattice, which derives from both a slight mismatch between the crystal lattices of graphene and SiC about 0.62 Å [1] and from having inversely behaving thermal expansion coefficients [28,29]. It has been shown that during cooling, compressive strain of up to 0.8% accumulates in the graphene sheets, which is measurable by a corresponding blue-shift in G and 2D bands of about 22 cm^{-1} [29]. Additional stress would result in even higher shifts of the peak positions.

The selected wavelength for the laser was 532 nm with 100% intensity reaching the sample. The error of the measured Raman shift that derives from the instrument is given as a maximum shift of 1.5 cm^{-1} into both directions. To evaluate the samples, both mappings with an area of 180 × 360 μm^2 (with 100 single, equally spaced measurement points) as well as the arbitrarily chosen single-point measurements were conducted. To allow a direct comparison of overall quality between the samples, an arbitrary space was mapped in the middle of all the samples. The single-point measurements were used to search for the biggest area of the desired single layer-free-standing graphene.

(2) AFM was used to produce surface images of the samples and to determine topographic differences between graphene grown on the two polar faces of SiC. AFM was conducted in constant force mode. The typical step height of a single graphene layer was reported to be 0.275 ± 0.001 nm, but the friction and the effect of impurities and physisorbed water could affect the measured topography considerably [30]. The linear correction mode was used for the images, which is the reason why profile lines in the figures show terraces that make the impression that they are inclined.

(3) The four-point probe technique, which is a system of 2 separate current- and voltage-sensing electrodes, used to measure extremely low resistance values, was used to measure the electrical resistivity ρ of the samples. The spacing between the probes is 1 mm. For thin films, the equation

$$\rho = \Pi / \ln 2 \times t \times V / I \tag{3}$$

was used, with I being the applied electrical current, V the measured voltage and t the thickness of a conducting layer. From this formula, it is evident, that one has to know the thickness and thus the number of graphene layers that have been grown on the substrate. This was assumed to range between 2 and 10 layers for the samples of this study and was individually set to a value in this range from the observations from the other two measurement techniques. For all samples, 4 directions (two diagonal and 2 orthogonal) were conducted. Each directional value represents the mean of 10 individual measurements of the same orientation on the sample within ±0.5 mm at the same spot. Following this procedure, for all samples at least 6 individual values in all 4 directions were produced. The minimum value of each direction is reported as an average of all individual measurement cycles of different spots on the sample, evaluated at the same orientation. The highest values, which correspond to a measurement direction perpendicular to the lowest values are also reported.

3. Results

3.1. Raman Spectroscopy

As described in previous sections, the shape, position, and height of peaks can be used to determine few and single-layer graphene from Raman spectra [31]. A typical spectrum of a mapping (one hundred equally spaced point measurements) on a C-polar face of SiC from our experiments is depicted in Figure 1 with the theoretical isolated spectrum of a monolayer graphene. Table 1 summarizes most important information about the samples relevant for this study as well as the results obtained from Raman and AFM measurements. Differences in observed characteristics at similar growth conditions (StGr-3 and 7) may be explained by slight technical deviations from the on-axis cutting angle of the industrially produced SiC substrate, which has been shown to have large effects on the resulting growth kinetics and developing structure [22].

3.1.1. Si-Face

Generally, the Si-face produces graphene with an IS ratio of 2.5 or above, which indicates a strong influence of the SiC substrate and therefore points to a reasonably thin graphitized surface. It can thus be said that for all samples, graphene-like material was formed, not graphite. The coverage of the graphitized surface ranges from an estimated complete (100%) down to 53%. It can be noted that a complete coverage of the graphitized surface is not necessarily a positive thing as these samples often showed very thick stacks of graphene or even spots of graphite in some places, in extreme cases. The maximum area of monolayer graphene which was detected is approximately 30 × 40 μm^2. It must be noted, however, that these areas do not correspond to single-crystal graphene. Single-layer graphene, here, corresponds to an area, which was found to be formed predominantly by single layer graphene, with multi-layers or holes as possible smaller textural aberrations within the monolayer. The determined blue shift of the 2D peak of about 20 cm^{-1} indicates that the graphene is strained in a compressive manner [29]. Generally, for all Si-faced samples it was assumed that few-layer graphene was grown over the largest part of the surface area. When comparing the characteristic shapes of the 2D peak, all shapes that were described in the literature [1] are present in all samples. However, 5 to 7-layer stacks (estimated from IR ratio and shape) seem to be most abundant. The repeating FWHM of the 2D peak of 50 to 60 cm^{-1} corresponds to multilayers and substrate-attached stacks as well [29].

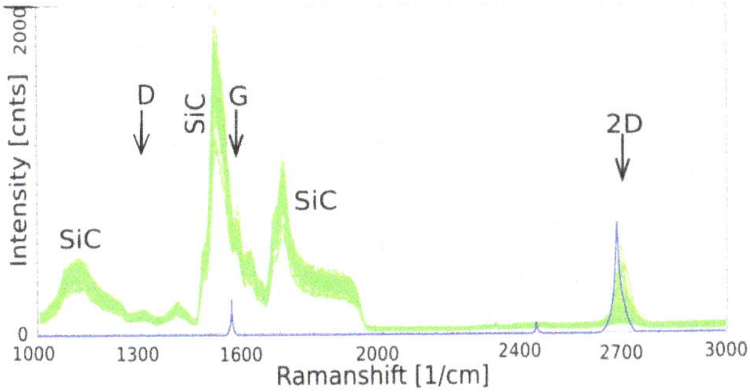

Figure 1. Measured Raman spectra of a characteristic sample with the signature of a 6H-SiC substrate (marked as SiC) with characteristic graphene peaks (marked as D, G and 2D). The green curves represent mapping spectra of 100 equally spaced points within the area of a typical Raman mapping ($100 \times 100\ \mu m^2$). Each individual spectrum corresponds to graphene grown on the C-polar face of SiC. For comparison, the typical spectrum of a monolayer graphene without a signature of a SiC substrate beneath the layer is plotted in blue as a reference. These reference data were replotted from [1].

3.1.2. C-Face

The C-face showed IS ratios between 2.5 and 3 for all samples, which indicates a thinner graphitization coverage compared with the Si-faced samples. The highest estimated percentage of monolayer coverage was found to be about 10% and the maximum detected area of monolayer graphene measured about $300 \times 30\ \mu m^2$. Again, like Si-face, it must be noted here that these areas do not correspond to single-crystal graphene and single-layer graphene, but corresponds to an area, which might contain multi-layers or holes as possible smaller textural aberrations within the monolayer. Apart from sample StGr-3, all these samples showed a characteristic bulk FWHM of the 2D peak of around 45 cm^{-1} which is both characteristic of attached monolayers as well as few-layer graphene [26]. All the samples showed a small blue shift in the 2D peak, indicating compressive strain in the material probably resulting from reverse thermal expansion behavior of substrate and graphene.

3.1.3. Reproducibility

In order to determine the reproducibility of graphene grown on the same SiC-polar faces, more samples would have been needed to make detailed observations about the growth conditions and the effect of their change on formed graphene. The main objective of this work however, was finding the difference of the growth characteristics between different SiC-polar faces that can be seen and reproduced to a high degree. These are (1) Generally lower 2D peak widths on the C-polar face compared with the Si-polar face, (2) constant negative 2D peak shifts on the C-polar face, while the Si-polar face shows both negative and positive shifts, (3) larger, yet more irregular and elongated (eccentric) forms of single-layer areas grown on the C-polar face at the same growth conditions (time, duration), while Si-polar face grown graphene areas are less eccentric in nature, (4) Higher graphene surface coverages on the Si-face at the same growth conditions, (5) Higher observed step-heights on the C-polar face with larger ones forming at lower growth temperatures.

Table 1. List of all the samples, showing the information about the basic growth parameters as well as the most important results. The abbreviation ND means that there are no data available. For the samples where numbers are replaced by lines, there was either no signal detected or the condition 2D > G was untrue.

	Samples (Numbered w.r.t. Growth Experiments)									
Parameters	**StGr-1**	**StGr-2**	**StGr-3**	**StGr-4**	**StGr-5**	**StGr-6**	**StGr-7**	**StGr-8**	**AGr -12**	**AGr-13**
T of growth (°C)	2000	2000	2000	1900	1900	1900	2000	1900	2050	2070
duration of growth (s)	300	150	150	150	150	150	150	150	300	300
Heating rate after 1400 °C (°C/min)	40	40	40	40	40	40	40	40	40	40
Polar face graphene was grown on	Si	Si	C	C	C	Si	C	C	Si	Si
2D average peak width (cm^{-1})	50	60	65	45	40	70	40	-	60	60
average shift of 2D (cm^{-1})	−20	10	−15	−10	−10	20	−10	-	−10	0
2D average peak width where 2D/G > 1 (cm^{-1})	37	57	63	45	44	60	42	-	-	ND
shift of 2D where 2D/G > 1 (cm^{-1})	−10	−5	−5	−7	−8	−10	0	-	-	ND
maximum area of monolayer graphene found (μm^2)	30 × 40	30 × 30	30 × 40	90 × 60	80 × 60	20 × 40	300 × 30	0	0	0
coverage of graphene-like surface (%)	93	69	90	75	61	53	81	0	100	ND
IR ratio (average 2D/G) without substrate only points	0.33	0.25	0.33	0.41	0.23	0.18	0.42	0.16	0.33	ND
IS ratio (average substrate/G)	2.36	2.61	2.52	3.70	2.93	3.29	2.40	2.81	2.50	ND
Maximum step height (nm)	13	12	25	30	29	13	26	35	10	10
Hexagonal network	Yes	Yes	Yes	Yes	Yes	No	Yes	No	No	Yes
Interpretation (formation features) MM = Multi and monolayers	MM	MM	Thick graphene and graphite	MM	MM	Started, yet incom-plete growth	MM	Almost no graphene present	MM	MM

3.2. *Atomic Force Microscopy*

3.2.1. Si-Face

Generally, all Si-face grown samples that went through the sublimation process displayed similar surface features, only varying in how severe these features were developed depending on the temperature and the time of growth. The former smooth, polished surface of the substrate became much rougher. Bigger steps of up to 9.5 nm vertical displacement dominate the surface with large terraces in between them. Using lower magnifications, it could be seen that these steps form a superstructure with a certain repetitive order (Figure 2a). In-between higher steps (in the following described as bunched steps), smaller ones can be found in a sub-parallel order with vertical step heights of about 0.2 to 1.5 nm. The smaller steps often display spike-shaped fronts that are orthogonal to the step structure. The spikes were measured to have typical step heights of around 2.5 to 3.5 Å(Figure 2b,c). These spikes are interpreted as the growth front of a new graphene layer/film on the surface. The height of around 3 Årepresents the typical height of one monolayer graphene. Further, these spike shapes could optimally be related to a starting growth front of one new layer of graphene.

Often, the surface showed a sub-hexagonal network of bright lines that are slightly elevated relative to all other features (Figure 3a). These lines span the whole surface and run over both small steps and bunched steps alike. Typical heights of the lines relative to the structure beneath them were measured to range between 0.5 and 1.3 nm. In some instances, the hexagons are replaced by pentagons or heptagons, displaying the presence of defects. Usually, networks of very pronounced hexagonal shapes showed a second, very faint network beneath them (Figure 3b). The bottom network is oriented like the more pronounced one, however is an independent structure. The maximum step-heights of the underlying network were determined to be typically between 1.5 and 3 Å(small peaks in Figure 3c) and are thus difficult to measure and may represent one layer of graphene.

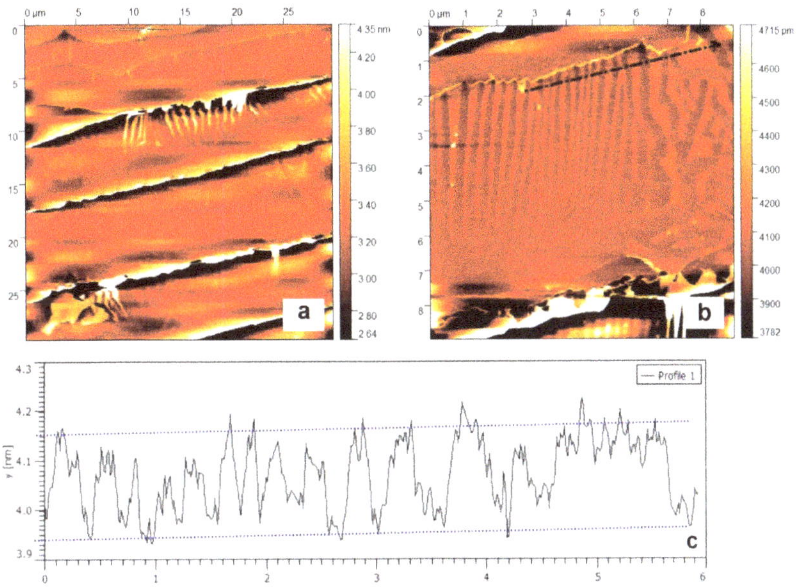

Figure 2. (**a**) 30 × 30 μm^2 of the graphitized surface of sample StGr-6. Bigger, parallel steps of about 4.5 nm in height are visible, running approximately from left to right in the picture. On the terraces between the steps, slightly brighter, slim fingers grow in the direction of the next step that is oriented higher than the point they originate from; (**b**) 9 × 9 μm^2 close-up of these bright fingers. The dashed black arrow indicates the trace of the height-profile depicted at the bottom (**c**); (**c**) 6 μm long profile across the surface of a terrace where a characteristic growth front of a sample grown on the Si-polar face of SiC. Dashed blue lines indicate the minimal height detected of the growth front; and the approximate mean of the valleys in between them respectively.

The elevated lines are seen as borderlines or overlapping regions of two adjacent hexagonal flakes of graphene—hence the elevation. Here it is interpreted that the elevated borderline of the lower graphene network deforms the topmost thin flakes growing above them. This causes the border zone to be traceable through the topmost layer. The stiff, yet bendable graphene on top smooths the surface expression. This is the reason why overlapping areas of individual flakes of the top layer (about 0.3 nm step height) are easily distinguished from the same structures one layer beneath (about half that height) that are not as pronounced and are cut by the bigger flake boundaries. The presence of an underlying structure directly confirms the Raman measurements, which predicted that predominantly multilayers were grown. Often, the surface displays small elevations that follow a random shape without interrupting it in a significant way. It is not trivial to differentiate between those and the crystal boundaries of the graphene, which are clearly hexagonally shaped and of smaller dimensions on the Si-faced substrate. The other elevations, however, often die out, branch out or show a zig-zag appearance. Even those can follow a pseudo hexagonal structure in some areas but have higher characteristic elevations. They can be interpreted as wrinkles, deformations of the surface that are usually formed during the cooling down process and are a direct effect of the different thermal expansion coefficients between SiC and graphene. One example of these wrinkles can be seen in Figure 3a.

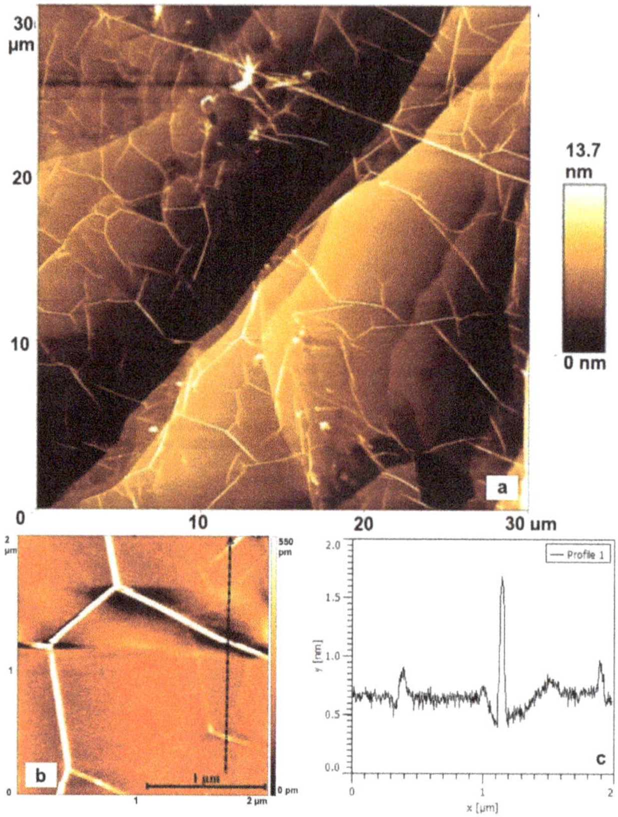

Figure 3. (**a**) 30 × 30 µm² of the graphitized surface of sample StGr-1. Visible are bigger steps of about 12 nm height where they coalesce, running diagonally from bottom left to top-right of the picture. On top of the surface, bright lines that create a pseudo hexagonal network. They are taken as wrinkles on the surface; (**b**) 2.1 × 2.1 µm² close-up of these bright lines. The dashed black arrow indicates the trace of the height-profile depicted on the right, (**c**); (**c**) 2 µm long profile across the surface. The trace of the profile-line can be viewed on the left (**b**).

3.2.2. C-Face

As in the case of the Si-polar face, the C-face displays a superstructure consisting of bunched steps. These bunched steps occur in a periodic arrangement, usually parallel to each other. Distances between them range from about 20 to 30 µm. This superstructure shows a medium-range structure on top of it, which is composed of steps as well. Medium-range steps (in the following m-steps) do not necessarily have to be parallel to each other and seem more randomly oriented compared to the bunched steps and may split up and reconnect again. The resulting are medium-range terrace islands that have shapes of lenses or stripes depending on the random orientation of the m-steps. Typically, one bunched step has five to seven m-steps in between. This accounts for step heights of roughly 3 to 4 nm and medium-range terrace widths of up to 9 µm. It is important to note that there is a variation in the distances of these m-steps to each other that does not seem random. The closer they are located to the lowest point of a terrace next to a bunched step, the closer the m-steps are situated to each other (Figure 4a–c). On smaller scale, a set of lines, like those found on the Si-face can be found and are again considered as the representation of overlapping grain boundaries of graphene on the surface, as well as wrinkles of graphene on the 2D structure. Some very distinctive small-range features that were found to a varying degree on every sample were hexagonal fronts at small-range steps. Typically, they are composed of a mixture of halves of hexagons arranged together to produce an armchair-shaped growth front of individual graphene layers. The difference in height across such an armchair-shaped front was characteristically in the range of 3 Å (Figure 5a–c). The growth fronts are typically found where many small-range steps are concentrating in the vicinity of an m-step, from which the growth front originates. They exclusively start growing from the bottom of such an m-step which clearly shows the formation of a new layer of graphene.

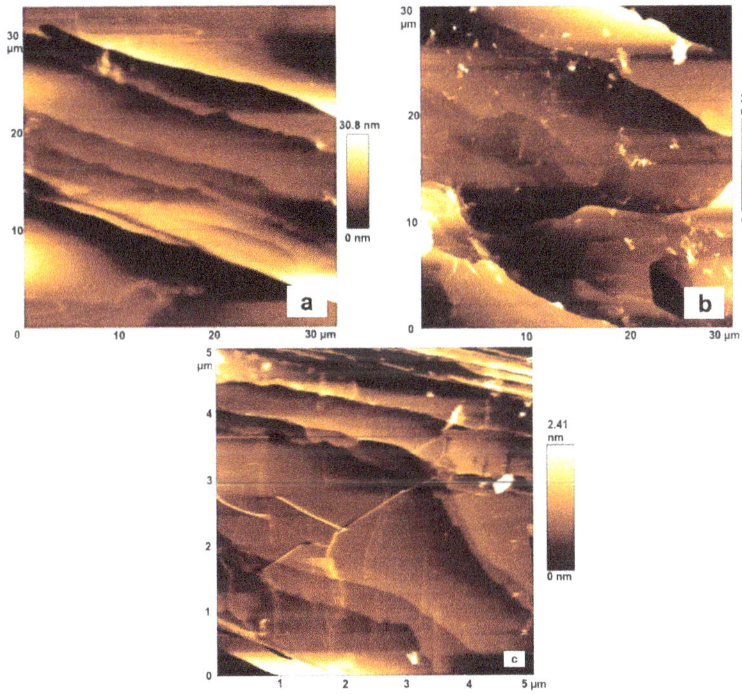

Figure 4. (**a**) Overview of the superstructure in the sample StGr-5. Bunched steps occur in periodic distances of about 15–20 µm to each other with step heights up to 20 nm. Smaller steps (medium range steps) do have approximately one fifth of both the step height and the distance compared to the superstructure. For yet smaller features, a higher magnification is necessary; (**b**) The same superstructure is visible in the other sample, StGr-7, but the medium range structure follows a more random order here. On top of the medium-range steps, small-range structures are faintly visible; (**c**) 5 × 5 µm^2 image showing a section of the graphitized surface of the sample StGr-5. Several steps can be seen with several growth fronts that all face towards the bottom-left of the picture. This structure is the fine-structure that can be found in between the bunched steps and m-steps which are displayed in (**a**,**b**). Over the visible terraces, two kinds of separations may be visible. The first is a distinctive set of lines, standing out about 1 nm from the mean height of the terrace. These lines often form 120° angles and cross small-range and medium-range steps in an undisturbed manner while being typically oriented at high angles to them. The second set of lines is less pronounced and never shows more than 0.6 nm difference in height to the surroundings. Typically, they are at very high angles close to 90° to small-range steps, where they often stop. These lines never cross medium-range steps.

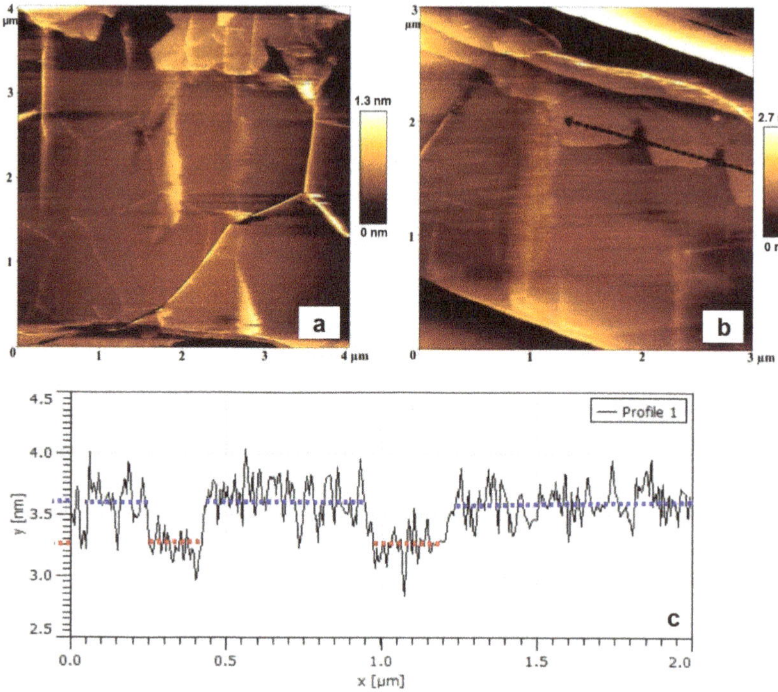

Figure 5. (**a**) Image of 4 × 4 µm^2 surface of the sample StGr-5 showing hexagonal structures. Both sets of lines that were described above, in Figure 4, are forming a hexagonal network over the surface of a terrace. Hexagons are replaced by pentagons or heptagons in some places as structural defects; (**b**) image of a 3 × 3 µm^2 area, both showing a hexagonal network as well as a growth front with a newly forming sub-hexagonal structure. The arrow indicates the trace of the profile-line that is shown below; (**c**) profile line taken over the boarder-zone of a newly forming hexagonal growth front of a sample grown on the C-polar face of SiC. The red and blue dashed lines indicate the average height of the terrace and the overgrowing growth-front.

4. Discussion

Because of different growth temperatures and growth durations, all observed features occurred to a varying degree in all samples. Exceptions to this observation (StGr 3 and 7) can be explained by a technical (slight) offset of the on-axis substrate which have a strong effect on the experimental result [22]. No two batches of wafers are completely identical. In consequence, a substrate that has slightly more atomic steps on its cut surface may produce a different result. As noted before, a general distinguishable trend between samples grown on Si and C-polar faces of SiC can be observed even though exact results cannot be reproduced between individual experiments that are grown on similar polar faces under the same growth conditions.

It was evident that samples which experienced the highest growth-temperature and duration also showed the highest amount of change in appearance as compared with the reference substrate. The spike morphologies on the Si-faced samples, with their terrace heights in the range of 3 Åcan be seen as the direct expression of a new graphene layer starting to form, as can be the half-hexagons on the C-face, respectively. Resulting from the described slower growth rate on the Si-polar face in the literature [3,16], one would expect the complete opposite growth geometry of a newly forming graphene surface. However, spike-shaped structures, as observed on the Si-polar face samples, are similar to skeletal or cellular growths in 3D crystal growth and thus have fast growth kinetics. An almost idiomorphic growth of the graphene, indicated by the hexagonal growth front on the samples that were grown on the C-polar faces, is indeed more characteristic for higher equilibrium conditions and slower growth kinetics.

Both the polar faces of the substrate produce graphene surfaces, where the small-range hexagonal network (seen as the fingerprint of individual graphene flakes) does not grow over the bunched steps which are therefore seen as separation lines for the graphene growth. A sketch can be found in Figure 6. A uniform and unlimited 2D growth over the whole surface remains, therefore, a challenge. The result from this interpretation would be that bands of graphene form between separating bunched steps, forming a structure that can be close to graphene nanoribbons. The width between these bunched steps seems to pin down the area, where graphene growth can occur in an unhindered way. The m-steps and small range steps are therefore considered as individual growth fronts. Monolayers would thus be limited to the higher end of a bunched step and be a function of the distance between the bunched steps. How temperature and duration affect the distance of the bunched steps and their height is difficult to state from this data set, as the experiments did not cover a big range of temperatures and growth times because the goal was to improve the general quality of the graphene and to study the interrelation of surface structures and electrical conductivity. However, it is suggested that a higher growth temperature and a lower growth duration led to a lower degree of medium and small range structures and less order on the surface of the sample with a higher probability of interconnection between ribbons of graphene between bunched steps. The resulting surface would be best described as weakly interconnected nanoribbons. The fact that these ribbons are connected in places, however, can be seen when studying the wrinkle structures in the AFM images. They do not always end at a bunched step but may run over it, indicating that there has to be a connection surpassing a bunched step and thus a closed surface of graphene.

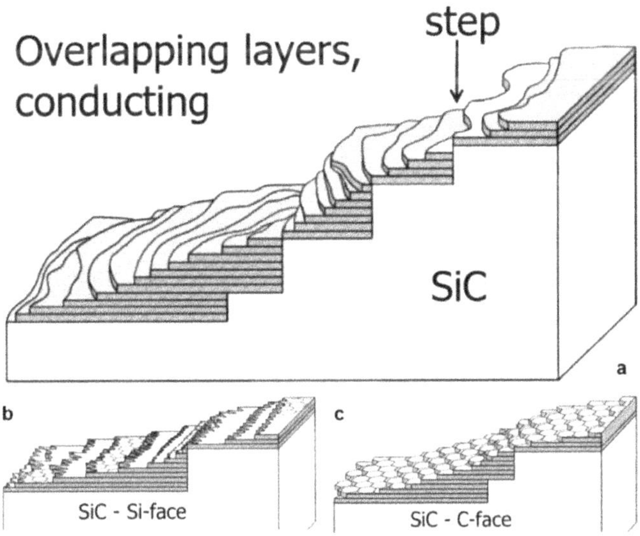

Figure 6. Schematic representation of the stock work of graphene layers on a SiC substrate. (**a**): An over-representation of atomic steps on the depicted substrate to be able to demonstrate the structure better. The substrate was an on-axis, industrially cut SiC wafer. Thus, some atomic steps could not be avoided, even though technically the substrate was cut in crystallographic plane orientation. Overlapping graphene layers allow for step-normal conductivity, while steps that are not completely overgrown account for a limited conductivity; (**b**,**c**): schematic figure, showing the difference in growth geometry of graphene on the two polar faces of SiC.

4.1. Four-Point Probe Measurements

The first important feature to notice for all the samples was that the electrical properties depend strongly on the direction of the measurement. Macroscopically, the layers look very uniform on the SiC surface, and hence their global electrical resistivity was measured using the four-point-probe technique. In each measurement, nearly a 3 mm area was measured, and the resistivity was calculated using Equation (3). Some samples showed directional differences in their electrical resistivity that could reach up to one order of magnitude. The reason for this will be subject to discussion in the section below. The resistivity values for graphene sheets of two, five and ten layers (these thickness values were being assumed based on various measurements) were calculated for each individual sample. Based on the observations depicted in Table 1 (IR, IS, width of 2D), the most likely average layer-thickness was chosen (indicated in green in Table 2).

Table 2. Results of the resistivity measurements for all samples. The values indicate the average of several measurements at different spots at the same orientation, using Equation (3) for an assumed graphene sheet of 2 layers (6.7 Å), 5 layers (16.75 Å) and 10 layers (33.5 Å). The samples StGr-1, StGr-2, StGr-6, Agr-12, Agr-13 were grown on Si-face. The other samples were grown on the C-face of SiC. Sample StGr-8 did not show a complete cover of graphene on the substrate and thus did not produce any reliable results. Numbers colored in green were the ones that were found to represent the whole samples' graphene thickness best, resulting from all collected data.

Samples	p [Ω.cm] as an Average of All Values in the Direction of the Lowest Values			p [Ω.cm] as an Average of All Values in the Direction of the Highest Values (Perpendicular to the Lowest Values)	SiC-polar Face
	Assumed 2 Layers	Assumed 5 Layers	Assumed 10 Layers		
StGr-1	4.60×10^{-7}	1.15×10^{-6}	2.30×10^{-6}	1.01×10^{-6}	Si
StGr-2	1.30×10^{-7}	3.24×10^{-7}	6.49×10^{-7}	2.82×10^{-7}	Si
StGr-3	8.37×10^{-7}	2.09×10^{-6}	4.19×10^{-6}	2.98×10^{-6}	C
StGr-4	2.56×10^{-8}	5.20×10^{-7}	1.04×10^{-6}	6.29×10^{-7}	C
StGr-5	3.43×10^{-8}	3.89×10^{-7}	7.79×10^{-7}	3.26×10^{-7}	C
StGr-6	3.56×10^{-7}	4.90×10^{-7}	1.78×10^{-6}	1.58×10^{-6}	Si
StGr-7	1.54×10^{-6}	3.84×10^{-6}	7.68×10^{-6}	3.09×10^{-6}	C
StGr-8	-	-	-	-	C
AGr-12	6.61×10^{-7}	1.65×10^{-6}	3.31×10^{-6}	1.47×10^{-5}	Si
AGr-13	4.49×10^{-7}	2.25×10^{-6}	4.49×10^{-6}	1.79×10^{-5}	Si

The resistivity values are ranged around 10^{-6} to 10^{-8} Ω.cm independent of the polarity of the substrate. This range occurs in the vicinity of the theoretical value of pristine graphene, which was computed to be around 1.04×10^{-6} Ω.cm [6]. Graphene that is bent, strained or has other structural modifications, which is assumed for graphene grown by the sublimation method, should show even higher resistivity values [32], but this was not the case for all the samples in this study. The strain in the graphene which was already observed in Raman measurements and by wrinkles in the AFM is related to thermal expansion differences between the substrate and the graphene layer on top as well as to the adaptation of the graphene to roughness of the substrate which can either be a relict or derived from the growth itself. The lowest values of strain were achieved for samples (StGr-2, 4, 5, 7 and AGr 12 and 13), indicated by the lowest measured 2D peak shift. StGr 2, 4 and 5, where this observation coincides with a relatively well established continuous few-layer graphene coverage, showed the lowest measured electrical resistivity values by far with the addition of StGr 1. Samples StGr 7 and AGr 12 and 13 showed quite thick graphene stacks in a few places (StGr 7) or developed thick graphitized surfaces that could almost be described as graphite, which is the reason why these samples have a higher resistivity (ca. 1 order of magnitude) compared with samples StGr 1,2, 4 and 5. Samples such as StGr-8 and 3 showed the highest resistivity values by far, which is easily explained because these samples had relatively low graphitization coverages compared with the others. Generally, there is no obvious difference between the values for the Si- and the C-face grown graphene. Samples that were regarded as best from the Raman and AFM investigation methods (e.g., sample StGr-1, 2, 4, 5) also displayed the lowest electrical resistivity values compared with the others.

The values that are calculated are subject to the interpretation of how many layers of actual graphene were grown. Additionally, structural differences such as homogeneity and graphene coverage in percent play a huge role. Most of the samples showed characteristic monolayer Raman peaks in some places, and yet non-graphitized spaces in others. Therefore, the thickness of these graphene sheets comes close to the ideal monolayer. However, as an average coverage of layers, it is assumed to lie most probably between 2 and 5 layers. For few samples (StGr-1, 2, 4, 5 and 7), a thickness of 2 layers is assumed, because all these samples comparatively often showed a strong substrate fingerprint in Raman and in the AFM measurements. The other samples are interpreted to have a graphene cover of at least 5 or 10 layers. This is assumed because of the high occurrence of graphite-typical peaks in the Raman measurements, as well as the thick and scaly appearance they showed in the AFM images.

Interpreted Results of Layer Thickness and Directional Dependence

By a closer examination, it becomes striking that most of the samples had a directional dependence (growth/step direction) on the determined electrical resistivity values. It ranged between a factor of 2 and about one order of magnitude between the samples. This observation is examined in detail by marking the orientation of the striations on the samples and noting the directional difference to the aforementioned parallel and diagonal four-point probe measurements. It was found that the highest resistivity always coincides to a direction being as close to perpendicular to the striations as possible. A sketch of this directional dependence can be seen in Figure 7. The striations were found to coincide with the direction and spacing of the bunched steps that were described in the AFM section. It is safe to assume that bunched steps act as obstacle for the electrical current to flow in the cases of all samples prepared in this study. This may be due to a high strain in the graphene lattice that produces a high scattering potential for electrons along the bunched steps, consequentially lowering conductivity (i.e., increased resistivity). Additionally, as was observed in the AFM images (Figure 3a), graphene films do usually not overgrow these bunched steps immediately so that the current may be limited to some specific spots at a bunched step where the step height is either low enough for graphene to grow over it or the stacking has become thick enough so that two graphene sheets on terraces can combine, as indicated by the wrinkles in the AFM images in some places.

Following this interpretation, the difference in growth of graphene on C versus Si polar face is considered important: As the C-face produced generally thicker stacks of graphene, the likelihood of a combination of two sheets over a bunched step seems higher on that polar face. Additionally, bunched steps on the C-face were approximately two or three times as high as on the Si-face. However, bunched steps had a higher tendency to split into two individual ones too. These split bunched steps rejoined to form a big step in other places again, which, created interspersed islands of possible connectivity. This favors the connection of graphene stacks over a bunched step on the C face too and might indicate that C-face sublimated graphene would be better in its electrical properties than the Si-face-grown graphene.

AFM images also showed that graphene crystals were generally smaller when the graphene was grown on the Si-face, which was also found by B. K. Daas [33]. This corresponds to a higher scattering of electrons on grain boundaries and a therefore lower net-conductivity which is the reason why the Si-face is not superior to the C-face as both effects affect each other in a reciprocal manner. For the case of the high-temperature-grown samples (AGr-12 and 13), it is assumed that the graphene stack became thick enough so that bunched steps no longer posed an obstacle. They were simply overgrown. Both samples were produced with a much higher growth temperature and a higher growth-duration accounting for the formation of thick graphene stacks that were confirmed by Raman measurements. This may rule out the effect of bunched steps acting as obstacles for current flow on these two samples completely. However, AGr-12 and 13 are still subject to the effect of scattering on grain boundaries. As stated, growth of graphene on the Si-face happens with the formation of small spikes that start growing from the vicinity of a bunched step (perpendicular to it) while the C-face produces crystals that are less anisotropic in shape. The long axis of the spikes of the Si face thus creates more crystal boundaries in a direction parallel to the bunched steps compared with perpendicular to it and thus produce more scattering in that direction compared with the C-face, where hexagonal, isotropic growth was observed. According to literature [1], C-face graphene should demonstrate better electrical properties. Observed resistivity values are expected to be significantly lower than those measured on the Si-face grown graphene. This was observed in our study as well. However, it also needs to be said that by taking the results of other measurement techniques into account, i.e., AFM, it was found that especially the C-face of the graphene showed big spots, where no growth had occurred.

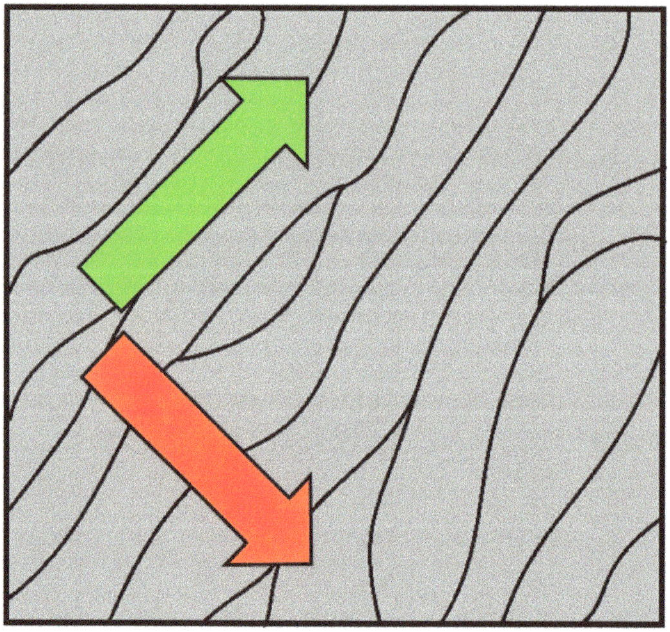

Figure 7. Schematic illustration of the directional dependence of the electrical resistivity on the structure of the layer. The box indicates a sample with steps and bunched steps that are indicated by the black lines. The green arrow indicates the direction that was found to yield lowest electrical resistivity (i.e., higher conductivity) values for thin graphene films. The red arrow indicates the direction of the highest resistivity values found on those samples. For thick graphene and graphite films, the directions are completely opposite.

5. Conclusions

The sublimation growth of epitaxial graphene on SiC was found to yield a few-layer graphene in most parts of the substrate up to 90% coverage. The substrate surfaces are covered by large-area flakes. The samples have at least one form of anisotropy in their morphology, which could be connected to direction-dependent behavior in the conductivity measurements. Differences in electrical resistivity range from 2-times up to one order of magnitude. The lowest resistivity values were found for those samples that showed the highest degree of monolayer coverage combined with the highest amount of graphene coverage on its surface. These resistivity values were in the range of 2×10^{-8} Ω.cm. Directly measured sheet resistance values using the described 4-point probe technique ranged around 5 to 10 Ω.

Contradictory to the literature, which usually describes the preparation of epitaxial graphene on the C-face being hard to control, we obtained the best results on this face. This may be due to the difference in growth mechanism of the forming graphene layers, as depicted in Figure 6b,c, under the investigated and optimized growth conditions. However, occurrence of bunched steps of more than 10 nm height as a form of structural defect was observed from the growth process. The combined observations of AFM, optical microscopy and resistivity measurements indicate that these bunched steps also act as a barrier for the charge flow. In extreme cases, this may even result in a cluster of sub-parallel nanoribbons-like structures that are separated from each other. This is valid for the both polar faces, but the Si-face characteristically produces lower step heights. Bunched steps that form on the SiC substrate during the sublimation phase can indirectly be influenced, in terms of height and of distance to each other, via the main growth parameters such as temperature and time. The samples that had bunched steps in the periodicity of about 30 μm were found to display the best combination of properties for a later application. Still, the growth temperatures and growth durations that ranged between 1900 and 2000 °C would have to be optimized further to produce the highest surface coverage while maintaining thin graphene sheets and a good bunched step periodicity. The ramping time to achieve this high growth temperature may act as a limiting factor, since a relatively fast ramping would be needed to prevent extreme graphene formation already happening during the ramp-up period.

Any structural defects such as step bunches or clusters will deteriorate the device performance, and anisotropy in any properties such as conductivity is also not beneficial from device perspective. Further detailed investigation on the reduction of defects and anisotropy removal is currently underway, including simulation studies and the results will be reported. Even though the formation of anisotropic bunched terraces in combination with the growth mechanism of graphene on SiC substrate is not particularly ideal for closed isotropic sheet formation, it may offer a novel possibility to obtain separated nanoribbon-like structures on SiC substrates. With more controllability and a clear separation between the steps, this observation may lead to new possible application of that method for devices that rely more on nanoribbons than on isotropic sheets of graphene.

Author Contributions: Conceptualization, R.R.S.; methodology, S.A.P. and R.R.S.; formal analysis, S.A.P.; investigation, S.A.P. and R.R.S.; resources, R.R.S.; data curation, S.A.P.; writing—original draft preparation, S.A.P.; writing—review and editing, R.R.S.; visualization, S.A.P.; supervision, R.R.S.; project administration, R.R.S.; funding acquisition, R.R.S. All authors have read and agreed to the published version of the manuscript.

Funding: The financial support given by the Bavarian Equal Opportunity Promotion (Bayerische Gleichstellungsförderung, BGF) is gratefully acknowledged.

Data Availability Statement: Not applicable.

Acknowledgments: The authors are very grateful to P. Gille for all his encouragements during this work and Guntram Jordan for his help in doing AFM measurements. This article is part of Stefan A. Pitsch's master thesis work.

Conflicts of Interest: The authors declare that they have no known competing financial interests or personal relationships that could have appeared to influence the work reported in this paper.

References

1. Ferrari, A.C.; Bonaccorso, F.; Fal'Ko, V.; Novoselov, K.S.; Roche, S.; Bøggild, P.; Borini, S.; Koppens, F.H.L.; Palermo, V.; Pugno, N.; et al. Science and technology roadmap for graphene, related two-dimensional crystals, and hybrid systems. *Nanoscale* **2015**, *7*, 4598–4810. [CrossRef]
2. Lebedev, A.; Davydov, S.; Eliseyev, I.; Roenkov, A.; Avdeev, O.; Lebedev, S.; Makarov, Y.; Puzyk, M.; Klotchenko, S.; Usikov, A. Graphene on SiC Substrate as Biosensor: Theoretical Background, Preparation, and Characterization. *Materials* **2021**, *14*, 590. [CrossRef] [PubMed]
3. Yazdi, G.R.; Iakimov, T.; Yakimova, R. Epitaxial Graphene on SiC: A Review of Growth and Characterization. *Crystals* **2016**, *6*, 53. [CrossRef]
4. Geim, A.K. Graphene: Status and Prospects. *Science* **2009**, *324*, 1530–1534. [CrossRef] [PubMed]
5. Neto, A.C.; Guinea, F.; Peres, N.M.; Novoselov, K.S.; Geim, A.K. The electronic properties of graphene. *Reviews of modern physics* **2009**, *81*, 109. [CrossRef]
6. Novoselov, S. *Scientific Background on the Nobel Prize in Physics: Graphene*; The Royal Swedish Academy of Sciences: Stockholm, Sweden, 2010.
7. Cao, X.; Qi, D.; Yin, S.; Bu, J.; Li, F.; Goh, C.F.; Zhang, S.; Chen, X. Ambient Fabrication of Large-Area Graphene Films via a Synchronous Reduction and Assembly Strategy. *Adv. Mater.* **2013**, *25*, 2957–2962. [CrossRef]
8. Virojanadara, C.; Syväjarvi, M.; Yakimova, R.; Johansson, L.; Zakharov, A.; Balasubramanian, T. Homogeneous large-area graphene layer growth on 6 H-SiC (0001). *Phys. Rev. B* **2008**, *78*, 245403. [CrossRef]
9. Chen, X.; Zhang, L.; Chen, S. Large area CVD growth of graphene. *Synth. Met.* **2015**, *210*, 95–108. [CrossRef]
10. Li, X.; Cai, W.; An, J.; Kim, S.; Nah, J.; Yang, D.; Piner, R.; Velamakanni, A.; Jung, I.; Tutuc, E.; et al. Large-Area Synthesis of High-Quality and Uniform Graphene Films on Copper Foils. *Science* **2009**, *324*, 1312–1314. [CrossRef]
11. Huang, M.; Ruoff, R.S. Growth of Single-Layer and Multilayer Graphene on Cu/Ni Alloy Substrates. *Acc. Chem. Res.* **2020**, *53*, 800–811. [CrossRef] [PubMed]
12. Longo, R.C.; Ueda, H.; Cho, K.; Ranjan, A.; Ventzek, P.L.G. Mechanisms for Graphene Growth on SiO_2 Using Plasma-Enhanced Chemical Vapor Deposition: A Density Functional Theory Study. *ACS Appl. Mater. Interfaces* **2022**, *14*, 9492–9503. [CrossRef] [PubMed]
13. Izhevskyi, V.; Genova, L.; Bressiani, J.; Bressiani, A. silicon carbide. Structure, properties and processing. *Cerâmica* **2000**, *46*, 4–13. [CrossRef]
14. van den Berg, J.J. Spintronics and Thermoelectrics in Exfoliated and Epitaxial Graphene: Rijksuniversiteit Groningen. Ph.D. Thesis, University of Groningen, Groningen, The Netherlands, 2016.
15. Norimatsu, W.; Kusunoki, M. Formation process of graphene on SiC (0 0 0 1). *Phys. E* **2010**, *42*, 691–694. [CrossRef]
16. Kim, M.; Hwang, J.; Shields, V.B.; Tiwari, S.; Spencer, M.G.; Lee, J.-W. SiC surface orientation and Si loss rate effects on epitaxial graphene. *Nanoscale Res. Lett.* **2012**, *7*, 186. [CrossRef]
17. De Heer, W.A.; Berger, C.; Ruan, M.; Sprinkle, M.; Li, X.; Hu, Y.; Baiqian, Z.; Hankinson, J.; Conrad, E. Large area and structured epitaxial graphene produced by confinement controlled sublimation of silicon carbide. *Proc. Natl. Acad. Sci. USA* **2011**, *108*, 16900–16905. [CrossRef]
18. Kobayashi, K.; Tanabe, S.; Tao, T.; Okumura, T.; Nakashima, T.; Aritsuki, T.; O, R.-S.; Nagase, M. Resistivity anisotropy measured using four probes in epitaxial graphene on silicon carbide. *Appl. Phys. Express* **2015**, *8*, 36602. [CrossRef]
19. Kakanakova-Georgieva, A.; Ivanov, I.G.; Suwannaharn, N.; Hsu, C.-W.; Cora, I.; Pécz, B.; Giannazzo, F.; Sangiovanni, D.G.; Gueorguiev, G.K. MOCVD of AlN on epitaxial graphene at extreme temperatures. *Crystengcomm* **2020**, *23*, 385–390. [CrossRef]
20. Chatterjee, A.; Kruskopf, M.; Wundrack, S.; Hinze, P.; Pierz, K.; Stosch, R.; Scherer, H. Impact of Polymer-Assisted Epitaxial Graphene Growth on Various Types of SiC Substrates. *ACS Appl. Electron. Mater.* **2022**, *4*, 5317–5325. [CrossRef]
21. Zebardastan, N.; Bradford, J.; Lipton-Duffin, J.; MacLeod, J.; Ostrikov, K.K.; Tomellini, M.; Tomellini, M.; Motta, N. High quality epitaxial graphene on 4H-SiC by face-to-face growth in ultra-high vacuum. *Nanotechnology* **2022**, *34*, 105601. [CrossRef]
22. Yakimova, R.; Virojanadara, C.; Gogova, D.; Syväjärvi, M.; Siche, D.; Larsson, K.; Johansson, L.I. Analysis of the Formation Conditions for Large Area Epitaxial Graphene on SiC Substrates. *Mater. Sci. Forum* **2010**, *645–648*, 565–568. [CrossRef]
23. Lee, J.-K.; Yamazaki, S.; Yun, H.; Park, J.; Kennedy, G.P.; Kim, G.-T.; Pietzsch, O.; Wiesendanger, R.; Lee, S.; Hong, S.; et al. Modification of Electrical Properties of Graphene by Substrate-Induced Nanomodulation. *Nano Lett.* **2013**, *13*, 3494–3500. [CrossRef] [PubMed]
24. Srivastava, N.; He, G.; Luxmi; Mende, P.C.; Feenstra, R.; Sun, Y. Graphene formed on SiC under various environments: Comparison of Si-face and C-face. *J. Phys. D Appl. Phys.* **2012**, *45*, 154001. [CrossRef]
25. Zhang, W.; Van Duin, A.C.T. Atomistic-Scale Simulations of the Graphene Growth on a Silicon Carbide Substrate Using Thermal Decomposition and Chemical Vapor Deposition. *Chem. Mater.* **2020**, *32*, 8306–8317. [CrossRef]
26. Bao, J.; Yasui, O.; Norimatsu, W.; Matsuda, K.; Kusunoki, M. Sequential control of step-bunching during graphene growth on SiC (0001). *Appl. Phys. Lett.* **2016**, *109*, 81602. [CrossRef]
27. Malard, L.M.; Pimenta, M.A.; Dresselhaus, G.; Dresselhaus, M.S. Raman spectroscopy in graphene. *Phys. Rep.* **2009**, *473*, 51–87. [CrossRef]
28. Pierson, H.O. *Handbook of Carbon, Graphite, Diamonds and Fullerenes: Processing, Properties and Applications*; William Andrew: Noyes, NJ, USA, 1994.

29. Ferralis, N.; Maboudian, R.; Carraro, C. Evidence of structural strain in epitaxial graphene layers on 6H-SiC (0001). *Phys. Rev. Lett.* **2008**, *101*, 156801. [CrossRef]
30. Musumeci, C. Advanced Scanning Probe Microscopy of Graphene and Other 2D Materials. *Crystals* **2017**, *7*, 216. [CrossRef]
31. Ferrari, A.C.; Meyer, J.C.; Scardaci, V.; Casiraghi, C.; Lazzeri, M.; Mauri, F.; Piscanec, S.; Jiang, D.; Novoselov, K.S.; Roth, S.; et al. Raman spectrum of graphene and graphene layers. *Phys. Rev. Lett.* **2006**, *97*, 187401. [CrossRef]
32. Wojtaszek, M. Graphene: A Two Type Charge Carrier System. Master's Thesis, Faculty of Mathematics and Natural Sciences, Groningen, The Netherlands, 2009.
33. Daas, B.; Omar, S.U.; Shetu, S.; Daniels, K.M.; Ma, S.; Sudarshan, T.; Chandrashekhar, M.V.S. Comparison of epitaxial graphene growth on polar and nonpolar 6H-SiC faces: On the growth of multilayer films. *Cryst. Growth Des.* **2012**, *12*, 3379–3387. [CrossRef]

crystals

MDPI

Article

Exploring Reduced Graphene Oxide Sheets Stabilized by Cu(II) and Cu(I) Cations in Ethanol

Aya Jezzini [1,2], Anne Davidson [1], Tayssir Hamieh [2,3,*] and Joumana Toufaily [2]

1 Laboratoire de Réactivité de Surface LRS, UMR 7197, Sorbonne Université, 75005 Paris, France; jizziniaya@hotmail.com (A.J.); anne.davidson@sorbonne-universite.fr (A.D.)
2 Laboratory of Materials, Catalysis, Environment and Analytical Methods (MCEMA), LEADDER, Faculty of Sciences, Lebanese University, Hadath P.O. Box 6573, Lebanon; joumana.toufaily@ul.edu.lb
3 Faculty of Science and Engineering, Maastricht University, P.O. Box 616, 6200 MD Maastricht, The Netherlands
* Correspondence: t.hamieh@maastrichtuniversity.nl

Abstract: In this study, ultrasound treatment was used to exfoliate commercially available graphite flakes into reduced graphene oxide (rGO) dispersed in ethanol. After centrifugation, solid copper chloride trihydrate was added, resulting in a green liquor containing Cu(II), Cu(I), and rGO. These liquors exhibited good and rapid photocatalytic activity in the degradation of eosin and bromophenol blue dyes (elimination in a few seconds) under visible-light irradiation. UV–visible spectroscopy confirmed the presence of rGO and Cu species. The size and morphology of the rGO sheets were investigated by several methods (SAXS, wide-angle XRD, SEM, and TEM). Negative UV peaks indicated light emission, which was independently verified by fluorescence. Intense plasmon peaks, with absorbances greater than 10, were observed after adding copper chloride salt. These plasmons were eliminated by a high dilution before the described catalytic tests were performed.

Keywords: photocatalysis; Cu^{2+} and Cu^{+}; graphite; ultrasound; plasmons; dye decomposition

Citation: Jezzini, A.; Davidson, A.; Hamieh, T.; Toufaily, J. Exploring Reduced Graphene Oxide Sheets Stabilized by Cu(II) and Cu(I) Cations in Ethanol. *Crystals* **2024**, *14*, 654. https://doi.org/10.3390/cryst14070654

Academic Editor: Ioannis Spanopoulos

Received: 4 June 2024
Revised: 8 July 2024
Accepted: 10 July 2024
Published: 16 July 2024

1. Introduction

A single or a few layers of 2D graphene sheets have excellent mechanical and optical properties. They comprise carbon atoms arranged to form a honeycomb structure, defined by carbon atoms in a sp^2 hybridization, as described in a general public magazine in 2008 [1]. The word graphene begins with graph—to recall the carbon species. Its ene end is associated with its polycyclic aromatic hydrocarbon nature [2]. The electrons inside the sigma bonds and between the carbon atoms are stable. Its extraordinary optical properties are a result of the electrons inside the atomic $2p_z$ orbitals of the carbon atoms. These atomic orbitals overlap and extended valence bonding π and conducting anti-bounding π^* molecular orbitals are formed. Intraband and interband electronic transitions in the spectral ranges of UV, visible, and far infrared are observed.

The introduction of graphene in photocatalysis was reviewed in 2016 [3]. Several methods have been developed, used, and discussed in the literature to achieve the exfoliation of graphite. Current exfoliation methods and techniques used to produce single-layer materials from graphite precursors were described by Minzhen et al. [4]. The UV–visible NIR absorption spectra of reduced graphene particles are complex, mainly because of electron interactions. The first detailed general description of electron interactions was published recently by Maier [5]. Transition metal ions can be coordinated on a reduced graphene surface [6–8]. With added copper chloride, reduced graphene oxide (rGO) particles, called plasmons, exhibit strong absorption [9]. If graphene particles are suspended in a solvent, they will have indirect semi-conducting properties, and a band-gap can be detected between their valence and conduction bands [10,11]. These bands' positions in energy are influenced by the graphene nanoparticles' sizes, shapes, and defects (O-containing and structural). This study addresses four questions:

1. Is the added copper chloride hydrated salt necessary to observe plasmons?
2. Can plasmon positions and intensities be adjusted by dilution?
3. Can plasmon coupling with C-C vibrations be evidenced?
4. Can plasmons be eliminated to study photocatalytic properties attributed to copper species and reduced graphene particles?

These questions are addressed using UV-visible–NIR spectra. Small negative peaks in the UV range suggest fluorescence, and this was independently verified for the graphene sheets before the addition of copper. Diluted and special cells (0.2 cm instead of 1 cm optical path) were compared to suppress plasmons. When suppressed, band-gap measurements of the C species were conducted using classical and new methods [10–15]. Optical characterizations and band-gap values were compared with published data [16,17]. Catalytic tests for eosin and bromophenol blue dye decomposition, under visible-light irradiation over copper/graphene, are also introduced.

Several authors have described nanoscrolls and their potential uses [18–21]. We have reported that ultrasound techniques can be used to exfoliate them, and 3D graphite GR particles can be used to obtain flat particles for use as supports for Zn–ferrite nanoparticles to eliminate antibiotic traces of the antibiotic amoxycillin [15]. In the most common preparation method using ultrasound in water, graphene oxide (GO) sheets are obtained using the Hummers method. GO is then reduced to give rGO, and three distinct reducers are used: hydrazine, ascorbic acid, and natural extracts of Amaranthus hybrids [18]. Here, ethanol was used as solvent and also as a reducer [15]. Some of the reported data concern samples prepared using ultrasound indirectly in thermostatic beakers, samples introduced in a water bath, or a horn introduced inside the solution that can be water [19], ethanol [20], or water with strong mineral acids. Solid catalysts containing doped oxide nanoparticles dispersed on porous nanocrystals of zeolite FAU or rGO, for instance, are interesting in oxidation reactions of oxygen and also in the degradation of pollutants [22–32]. They have also been used for biomedical applications [29]. The catalysts Cu/Zn on graphene have been formed by CVD [33]. To avoid graphene agglomeration, neutral surfactants and negatively charged species can be added. But a simple addition of organics affects the physical and chemical properties of graphene sheets. Here, similar solids are obtained by using a simple contact of suspended graphene nanoparticles with the mineral species $CuCl_2{\cdot}3H_2O$ after an ultrasound treatment aimed at eliminating the nanoscrolls of graphene and a centrifugation to remove graphite 3D particles. Solids characterizations are performed by XRD, SEM, and TEM micrographs, associated with selected area electron diffraction. We will focus on the description of the plasmon intensities and positions as a function of the Cu/rGO dilution in ethanol.

2. Materials and Methods

2.1. Materials

Commercial, natural graphite flakes, GR, 99% carbon, 100 mesh, were sourced from Sigma Aldrich n° 808091, CAS 7782-42-5, Paris, France; they were polluted in carbon nanoscrolls similar to the ones described by several authors [18–21]. We used absolute alcohol, CAS 64-17-5, and $CuCl_2{\cdot}3H_2O$, CAS 10125-13-0, from the same company. All these reagents lacked intrinsic toxicity. Here, we focused on working with the catalysts in a liquid state and on their dilution in ethanol. The graphene suspensions contained reduced graphene rGO, and they were easily recognized because of their black color. Ultrasound treatment was applied directly inside the centrifugation vials to warrant the absence of unwanted contamination by metallic species and to ascertain that the containers always had the same shape and volume. The vials had a basic conical shape and a height of 115 cm. They contained between 40 and 45 mL of ethanol. The effect of dispersion by organic solvents such as N-methyl-pyrrolidone was described by Politano and Chiarello [34]. The aromatic molecules were replaced by ethanol and an inorganic copper salt because they can alter the electronic properties of rGO.

2.2. Methods

Scanning electron microscopy (SEM) imaging was performed with a Hitachi SU-70 FESEM (Hitachi, Paris, France). Transmission electron micrographs were registered on a JEOL JEM 2011 UHR (Jeol, Croissy-sur-Seine, France) microscope, operating at 200 kV, and equipped with an ORIUS Gatan Camera (Paris, France). For the observations, the liquids were deposited on 3 mm copper grids coated with an amorphous carbon film and let dry.

Ultrasonic irradiations were generated by a Vibracell VCX 500 apparatus (Bioblock, Paris, France) (500 W power, used at 40%, with a frequency of 20 kHz). The horn of this apparatus, 13 mm in diameter, was made with an alloy of Ti-6Al-4V and warranted for an overall volume of solution of 50 mL. The probe penetrated 6 cm inside the used liquid inside the vial. We used constant irradiations for 5, 6, or 7 min or for 2 h with and without pulses of 15 s and stops of 15 s. Experimental conditions, the influence of the ultrasound power, several sonication times, and the initial graphite concentration were detailed by Navik et al. [23]. The used initial concentration of graphite was equal to 5 mg/mL. The sonication was realized in a sonication bath with a power of 1.08 kW. In a previous study [23], Navik et al. were interested in graphene stabilization by an organic molecule, curcumin, stable for a graphene concentration of 1.44 mg/mL; their study also contained complementary data concerning the detection of graphene sheets in ethanol with several initial GR concentrations. There is one main difference between these published results and ours: the used ultrasound horn penetrates deeply into the solution. In our liquors, to avoid solids precipitation, we used 6 min or 2 h of ultrasound treatment, and the initial graphite was introduced in concentrations lower than 0.5 mg/mL. After the ultrasonic treatment, a centrifugation was applied at 3000 and/or 5000 rotations by min, rpm, and 12 min. Bulk 3D graphite was recovered at the bottom of the flask. Its 3D character was demonstrated by SAXS measurements (Figure S1), and it was eliminated. The transparent black liquor was then used directly to dissolve the salt, copper chloride trihydrate. The solution was changing color spontaneously and became green in less than 2 min. We were surprised to observe that the green color of the solution was extremely stable, remaining constant even after several months of storage in air, whereas the species that we have detected that can be green are Cu(I) cations, and are known to be instable; therefore, only the redox potential of the couple Cu(II) Cu (metal) is given in the literature. To explain their stability, a positive role of ethanol to protect them is proposed.

The presence of the graphene sheets in suspension in ethanol was studied first by WAXS wide-angle Scattering on a Bruker-type D8 Advance (Paris, France) within the range of 2θ from 0.3° to 90° with copper 1.54186 Å, a mixture of $K\alpha_1$ and $K\alpha_2$, and a ratio of 1 for 2. A Bragg–Brentano set-up was used to obtain graphics in intensities versus 2θ. Simulations were created using the Fullprof program. With this program, precise values of peak positions and intensities (integrated units) were obtained. Within the range of 23–31°, three observed peaks were labelled (1), (2), and (3). Peak (1) is a small peak on the left part. Peaks (2) and (3) are more than 5 times more intense and are on the right part. Two additional peaks attributed to X-ray diffusion are presented in Figure 1 to take into account the shape of the diffraction. Peaks (2) and (3) were calculated based on the following parameters: H = FWHM, the Full-Width at Half Maximum, ETA, η, a linear combination of a Lorentzian L, and a Gaussian G line shape with the same FWHM. Expression 1 can be written for XRD:

$$I(2\theta) = I_{hkl}[\eta\ L(2\theta - 2\theta_0) + (1 - \eta)G(2\theta - 2\theta_0)] \quad (1)$$

The Gaussian peak is given by Equation (2):

$$G(2\theta) = C_1 exp[-\pi(2\theta - 2\theta_0)^2/\beta^2] \quad (2)$$

where C_1 is the peak intensity, $2\theta_0$ is the 2θ position of the peak maximum, and the integral breadth, β, is related to the FWHM peak width.

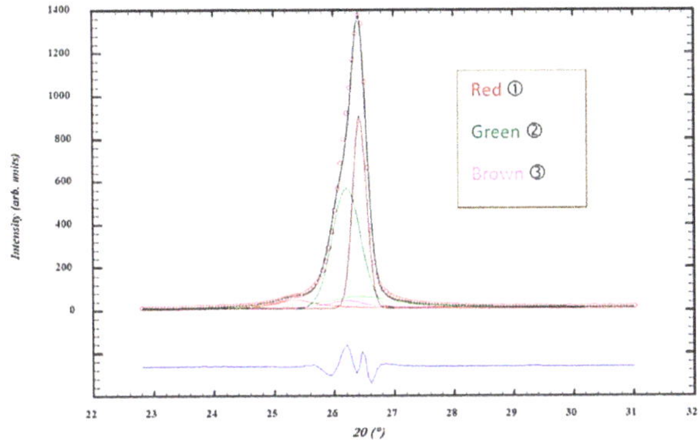

Figure 1. Commercial sample, GR. Possible decomposition of an XRD peak into five components: ①, ②, and ③ peaks. Both intense ② and ③ peaks are Gaussian in shape. I (observed) in black with red spots, I (calculated) for three peaks labelled (1) in red, (2) in green, and (3) in brown, and in blue is the difference between I (observed) and I (calculated).

A value of η close to zero indicates a Gaussian line shape. Intermediate values between 0 and 1 indicated a pseudo-Voigt shape with given % of G and L. Two very small diffraction peaks associated with X-ray diffusion are located at the same positions as peaks (2) and (3) and are labelled peaks (4) and (5).

UV–visible–NIR spectra were measured on a Varian 4000 spectrometer (Bimedics, Paris, France) within the range of 200–1400 nm. They were measured in the UV (200 to 400 nm), the visible (400 to 800 nm), and the NIR ranges inside two kinds of quartz cells, one with an optical path of 1 cm and a second with an optical path (internal) of 2 mm. Fluorescence measurements were carried out with a Horiba Jobin Yvon spectrometer Fluorolog®FL3-22 (Horiba, Paris, France). We tried two kinds of florescence measurements, the first one with diluted solutions in ethanol and the second one directly with the concentrated initial solutions obtained by applying ultrasound treatment for 6 min and without centrifugation.

Better results in fluorescence were obtained on the non-diluted solutions and for an angle between the excitation light and detector set at 90° to avoid detector saturation.

3. Results

3.1. Solid Graphite XRD to Study Mixture of Nanoscrolls and 3D Graphite

In XRD, enlargements between 23 and 31° were performed and analyzed with the program Fullprof. The positions of peaks (2) and (3), labels indicated in the experimental section, are equal to 26.208 and 26.427°, with respective FWHM values of 0.495 and 0.179°; their ETA values are both equal to zero. The two peaks are therefore Gaussian in shape. Peak (3) is the narrowest and can be assigned to 3D graphite (Figure 1).

Particles sizes can be estimated using the Scherer equation in the following form:

$$L_{hkl} = \frac{0.9\ \lambda}{\mathrm{FWHM}_{hkl}\ cos\ \theta}, \tag{3}$$

where FWHM = β is expressed in radians. Average particles sizes of 700 and 300 Å are obtained on peaks (2) and (3). These dimensions are too large for nanoparticles (Figure 1).

Peak (2) is attributed to graphene sheets, their varied sizes being associated with a large FWHM of this peak, significantly larger than the one of peak (3), attributed to graphite 3D particles. It is interesting to know that this decomposition of the X-ray diffraction peak is no longer detected when the GR commercial sample is submitted to an ultrasound treatment after its suspension in ethanol for 2 h. In that case, a single XRD peak at 26.54° with a Gaussian shape, which corresponds to the 002 diffraction of reconstructed 3D graphite domains, is observed.

The integrated intensities of peaks (1), (2), and (3) indicate that the percentage of 3D graphite is 60.52% and that of graphene stabilized in nanoscrolls is 36.33%.

3.2. *Observations Correlated to Graphene Particle Sizes*

3.2.1. Measurements Performed on SEM and TEM Images

The SEM image in Figure 2 is relative to graphene nanoscrolls. These nanoscrolls are decomposed after a simple manual grinding. Their diameter, as measured on adjacent nanoscrolls, is 235 nm. A correct evaluation of the two solids is difficult on SEM images only because only a given part of the sample is visible. But relative percentages of the two phases can be obtained using the XRD measurements.

Figure 2. SEM images of the nanoscrolls: other images already published [8]. Upper detector used for secondary electron detection. Magnification: ×25,000; scale bar: 2 μm.

After an ultrasound treatment of 6 min with pulses of 15 s and stops of 15 s, the nanoscrolls are no longer detected neither by microscopy nor by XRD. The fact that nanoscrolls can be transformed in flat domains was already known, with bubbles of cavitation being inserted within the main part of the scrolls and destroying them [6]. The present work complements previous studies by showing that the size of the flat nanoparticles is directly affected by the time of the applied ultrasound. Micrometer-sized particles observed after 6 min of ultrasound and drying are replaced by smaller particles of size lower than 100 nm after an ultrasound treatment of 2 h. An additional selected electron diffraction is presented in Figure 3 and was carried out on a graphene particle. This image is important, as it establishes that the studied specific sheet contains five distinct layers. The white spots are indeed grouped in thirds on the Bragg rings, but the medium spot is larger than the two other ones.

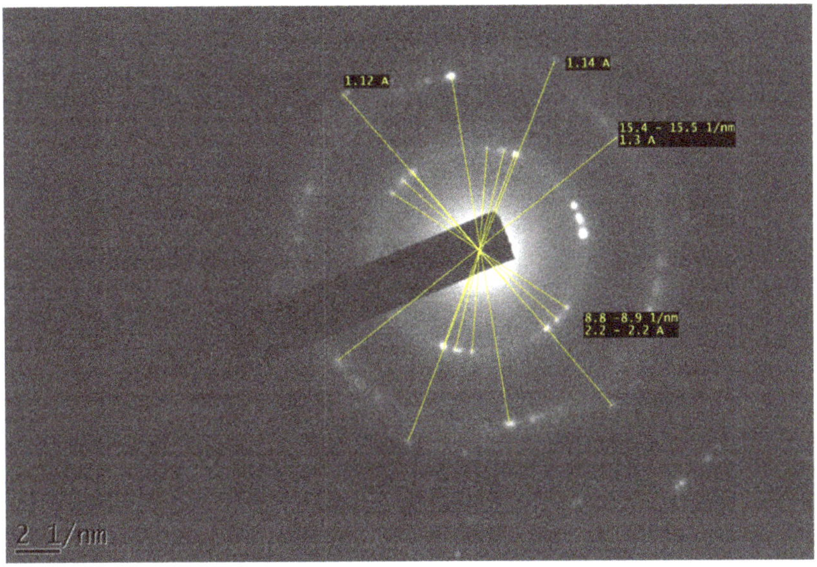

Figure 3. Selected area electron diffraction (SAED) on a graphene particle.

On the enlargement presented in Figure 4, several nanoparticles of copper oxide are seen aggregated and above the carbon surface, probably of a 3D particle of graphite of important size (more than 25 layers). A distance between fringes of 1.195 Å can be measured. This distance is confirmed by a Fourier mathematical transformation, and it is difficult to identify an inorganic species with one value only.

Figure 4. Details about copper oxide nanoparticles, external and covered by carbon layers. FFT on external oxide particles (×500,000; scale bar 10 nm).

The SEM and TEM images are important, as they evidence thin 2D graphene sheets (five layers in Figure 3) mixed with more thick ones and also with 3D graphite (Figure 4). The information about the copper species was difficult to analyze since dehydration was used. The dried powder was deposited on a carbon grid covered by a carbon film. The sample was then left to dry for a long time before being introduced inside the TEM apparatus.

3.2.2. UV–Visible–NIR Spectroscopy

Before Copper Introduction

The measured absorbance was used to estimate the concentration of graphene sheets in suspension. We used the Beer–Lambert relation, $A = \varepsilon\, c\, l$, and a cell with an optical path of $l = 1$ cm. The absorbance, A, has no units. C is the graphene concentration and ε the absorption coefficient in $L{\cdot}mol^{-1}{\cdot}cm^{-1}$.

Measurements were performed on two solutions, recovered after ultrasound treatments of 6 min. One solution is realized with pulses of 15 s for progressive dilutions in absolute ethanol and the other solution without. Four points corresponding to dilution by a factor of 0.10 to 0.35 were used, and the initial concentration of commercial graphite was 500 mg in 45 mL of absolute ethanol. Navik and Gai [23] have proposed for curcumin stabilized in water solutions of graphene that a preliminary value of (ε c) is 1650 mL/mg.

We present in Figure 5 the absorbance that can be measured after mixing several volumes, 1000, 750, and 500 µL, of the activated isolated solution immediately after ultrasound and centrifugation at 5000 rpm for 12 min with and without pulses. Dilution by absolute ethanol was carried out to obtain 4 mL of solution in each case (with 3 mL, 3.25 mL, and 3.5 mL, necessary to obtain an overall volume of 4 mL), and absorbance was measured on a quasi-flat straight line observed between 700 and 800 nm.

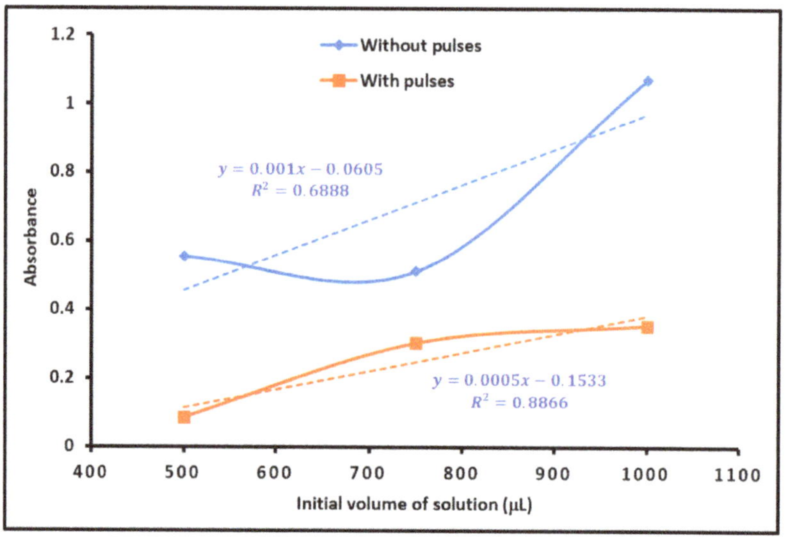

Figure 5. UV–visible results. Absorbance collected on dispersed graphene diluted in absolute ethanol: 1000, 750, and 500 µL diluted in 3.00, 3.25, and 3.50 mL of ethanol to obtain a total volume of 4 mL, studied in a 1 cm cell (in quartz) after ultrasound of 6 min, with and without pulses of 15 s and stops of 15 s, and centrifugation at 5000 rpm for 12 min.

Graphene was formed with pulses two more times. The corresponding overall weight of the C species in the solution was determined in an open Eppendorf tube and left to evaporate. No difference was measured with the two solutions made with 1000 and 750 µL with pulses: masses of 0.062 mg and 0.0605 mg of carbon were found. With the solution at

500 µL, the weight of carbon was equal to 0.0417 mg. Converted in µg by mL (with 4 mL total volume), this gives 15.5, 15.1, and 10.4 µg by ml of the diluted solution.

In practice, as detailed previously, dilution was necessary before collecting the UV–visible spectra. The necessary dilutions were performed taking an Eppendorf tube of 5 mL containing 250 µL of the selected "green" liquor, taken with a high-precision manual Eppendorf pipette BIOHT Proline 100–1000 µL, calibrated at 1% precision for 500 µL at room temperature (with a balance of precision of 0.001 g). The samples were dried in an autoclave at 60 °C for 2 h. The weight of the sample before and after drying was measured with a Mettler-Toledo scale (XPR204S maximum capacity 210 g, with a readability of 0.1 mg) after ethanol evaporation, and the final measured weight of the solid is due to the carbon contribution in the solution.

In Figure 6, a straight line is observed for the absorbance as a function of the detected C species in the flask containing the solution after ethanol evaporation. The linearity is better when the observed absorbance is plotted as a function of the concentration of carbon expressed in µg by ml. This line was used after to estimate the C amount in our ethanol solutions (without dilution). If we consider the point corresponding to 1000 µL of the solution prepared with 6 min of ultrasound and pulses of 15 s, it corresponds to 12.4 µg of C species by mL, and therefore there was 0.060 mg of suspended C species in the 5 mL of ethanol initially. Given the initial 500 mg of GR, this corresponds to a low percentage, less than 0.1% of the introduced C species. Compared with results published by Navik [22] that concern ultrasounds applied with a horn, a weight of C species in ethanol of 1.36 mg by ml, more than two times larger than ours, was obtained. The better obtained result was assigned to the use of large curcumin organic molecules that stabilize the graphene sheets. The power of the ultrasound treatment used was equal to 0.18, 0.54, 1.08, and 1.8 W in that publication, and we used about 40% of 500 W, which is more than 20 times more power but at 40% only. Despite the higher power that we used, our graphene suspension in ethanol was lower. More than the ultrasound power, this difference can be then assigned, according to Wallace and Shao [22], to the following parameters:

(1) The curcumin;
(2) The initial 3D graphite used;
(3) The duration of the ultrasound treatment applied (equal to 4 h).

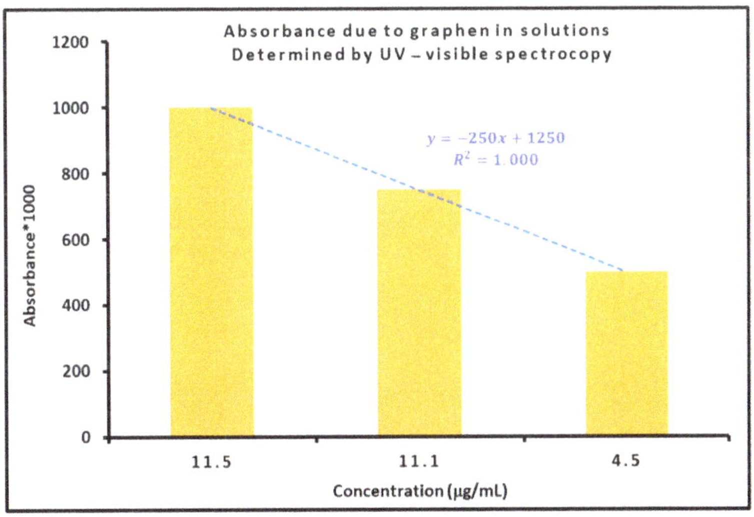

Figure 6. Absorbance measured as a function of the composition of the studied solution expressed in µg by mL.

A weight of 50 mg by mL was obtained by Wallace and Shao [22], and this concentration is equal to 500 mg in 45 mL for a duration time equal to 6 min in our work.

Our calculations showed an average absorption coefficient equal to 2500 $L \cdot g^{-1} \cdot m^{-1}$, which is very close to that obtained by Hernandez et al. [26], with an absorption coefficient equal to 2240 $L \cdot g^{-1} \cdot m^{-1}$. The difference between the two results does not exceed 10%.

After Copper Chloride Addition

After centrifugation at 3000 rpm for 12 min, the solution was recovered and the solid $CuCl_2 \cdot 2H_2O$ salt was added; then, the salt was fully dissolved in less than 2 min, and the liquor exhibited a "green" homogeneous color. This treatment was reproduced with centrifugation at 5000 rpm 12 min. Possible ligands around the Cu(II) cations, in our conditions, are chloride anions, water, and ethanol molecules. The possible exchange of water molecules by ethanol from the solvent has been studied by Larin et al. [35]. In this study, it was also indicated that links with chloride anions were implying strongly acidic HCl solutions. We = then neglected chloride ligands, and concluded that we are dealing with $[Cu(H_2O)_{6-n}$ (Ethanol)$_n$ with $n < 3$ complexes.

Figure 6 regroups the UV–visible spectra recorded with several graphene dispersions after the addition of $CuCl_2 \cdot 2H_2O$. Three distinct kinds of vibrations are observed: (1) in the UV range, vibrations of the C-C, C=C bonds of graphene and graphite; (2) still in the UV range between 300 and 400 nm, very intense vibrations due to collective excitation, in addition to plasmons of intensities greater than 10; (3) above 800 nm, broad vibrations of octahedral Cu(II) complexes (octahedral and tetrahedral).

On the right part of the visible spectrum, recorded with our more concentrated solution (Figure 6), are two broad peaks, one at 866 nm and a second one at 930 nm. It was observed that there was an exchange of ligands from water to ethanol, as detailed in Figure S2. Two distorted octahedral Cu(II) complexes are observed and assigned to trans and cis complexes of Cu(II), containing two ethanol and four water molecules. These species correspond to an axially distorted octahedral field in the visible spectrum that gives an averaged maximum at 866 nm. There is also a broad absorption observed at 930 nm on the spectrum of the most concentrated solution (Figures 7 and 8). Because of its position, this signal is attributed to Cu(I) species, possibly grafted on the graphene surface. They are stable in oxidation by oxygen molecules. Some protection by the solvent is necessary to explain their stability. Similar kinds of Cu species, grafted on C vacancies of carbon species on solid carbon diamonds, have been evidenced very recently [18]. They were investigated by demanding and expensive methods, EXAFS-XANES XANES [19], but also by less expensive operando IR measurements [18]. Here, we selected to use UV–visible–NIR spectroscopy.

Furthermore, on the left side of the spectrum, still with a concentrated solution, multiple and very intense peaks were detected between 309 and 371 nm (4.01 to 3.81 eV) with a second non-resolved broad peak between 262 and 371 nm. With progressive dilution, the narrow and intense peaks were eliminated, and the first peak was decomposed into at least three peaks, one at 230 nm and two above 250 nm. Further dilution shifted the broad peak to 265 nm, a position indicative of thin reduced graphene and attributed to the π-π^* electronic transition, confirming the absence of graphene oxide (GO) in our diluted samples. For oxidized samples, a peak near 230 nm was reported. The intense additional peaks were attributed to plasmons associated with light reflection by graphene suspended nanoparticles. We observed that a simple dilution can eliminate their very intense peaks. A very similar result was obtained by changing the thickness of the used cell, going from 1 cm to an overall path of 2 mm internal (and 4 mm external). As illustrated in Figure 7, the intense reflections of plasmons are eliminated.

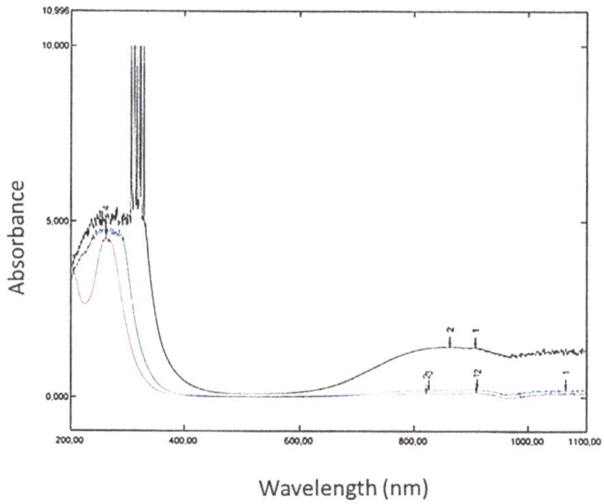

Figure 7. Influence of dilution by ethanol on UV–visible spectra of "green" liquors of Cu/graphene/ethanol. In black, preparation with 0.269 g commercial graphite: 45 mL absolute ethanol; ultrasound 7 min, and pulses of 15 s (in blue). Dilution "1": inside the UV cell, few drops inside 4 mL of absolute ethanol (in pink). Dilution "2": 1 mL of dilution "1" diluted with 3 mL ethanol (inside the visible cell).

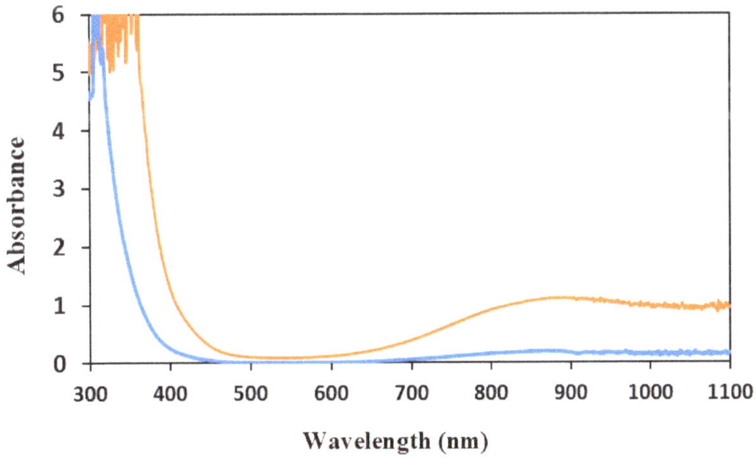

Figure 8. Influence of the optical path of UV–visible cells on plasmons coupling with the other vibrations: C-C vibrations on the left (from above to below 400 nm) and [Cu $Ethanol_2(H_2O)_4]^{2+}$ complex (864 nm instead of 800 nm for the hexa–aqua complex); orange in a cell of 1 cm and blue in a cell of 0.2 cm (internal).

It is incorrect to claim that graphene has a non-zero band-gap when it is dispersed inside a solvent dispersion. A band-gap, a difference in energy between its valence and its conduction band, is observed just after the plasmons in the UV spectral range, and its value is calculated here at the intersection between the most vertical line observed when going from 365 to 400 nm and the Ox axis. The spectra indicated a shift from 3.09 to 3.39 eV in its conversion of energy (eV) if the plasmons' contribution is removed. In the orange spectrum, coupling between them and the vibrations of graphene, its natural C-C, C=C bond is evidenced (Figure 8). Coupling between the added copper species and the plasmons is also demonstrated by the position of the broad peak due to the [Cu $(H_2O)_6]^{2+}$ complex, which is shifting from above 850 nm down to its usual value circa 800 nm for the hexa–aqua complex in water.

Reduced graphene is a semi-conductor, and its band-gap can be estimated after conversion of the Ox axis of the visible spectrum in energy ($E = (h\ c)/\lambda$, where h is the Planck constant, c is the celerity of light in vacuum, and λ is the light wavelength expressed in eV; the value is then converted into Joule and into eV, which gives $1238/\lambda$), with graphene highly dilutable in ethanol solutions. To obtain a Tauc plot, α, the absorption coefficient, is calculated, knowing it is equal to $(1/d)\ \ln(T)$, with d, the thickness of the sample, being 1 cm with our cell, and T, transmission, equal to 10-power, which is the absorbance. Alpha multiplied by the photons' energy (E α) power 0.5 is plotted versus the energy in eV, and straight lines are obtained. The corresponding band-gap for an indirect semi-conductor is obtained by calculation of the intercept of the lines with the vertical axis divided by the slope. An indirect value of 3.67 eV is obtained for the sample fabricated with 6 min of ultrasound. A value of 4.02 eV is obtained with the sample submitted to 2 h of ultrasound. The indirect value can be compared with published values: 4.7 eV (for a free-standing sheet of graphene) and 7.0 eV (graphite 3D) for a band attributed to a π-π^* interband transition, as proposed and calculated by Politano and Chiarello in 2014 [9].

For a given concentration, there is a small peak at 230 nm on Figure 7, observed only with the smallest thickness path cell. This peak is due to contamination by oxidized C species that can be OH, peroxide, or COOH groups. At the highest concentration, a superposition with a negative absorption is observed, and the negative peak is masking traces of oxidized species. When the negative peak is present, no information about the oxidized species is available. After the elimination of the strong plasmons, a maximum of absorption is only observed at 270 nm in Figure 7, and this can be attributed to the reduced graphene sheets, without coupling with the other vibrations. They are composed of C atoms hybridized with sp^2, forming closed domains, the sizes of which are clearly evidenced by the presented UV spectra.

Care is necessary before using this value. Indeed, in old publications, the graphene sheets were in general considered to have a zero band-gap. It is obviously not the case here since we are studying graphene sheets dispersed on a solvent. In ethanol and if plasmons are suppressed, we observe a very strong absorption in the UV spectrum, and the existence of a band-gap in energy will be then confirmed by fluorescence spectra, as investigated next. We emphasize the fact that the band-gap energy can be measured using the Tauc expression recommended for indirect semi-conductors. A value of 3.67 eV is obtained with the larger graphene particles obtained after 6 min of ultrasound (Figure 9). Small sheets obtained after 2 h of ultrasound give a larger energy value of 4.02 eV because of a smaller number of involved C atoms and a less efficient π delocalization.

Comparisons with published values for the band-gap of reduced graphene sheets are possible: values at 3.90, 3.60, and 3.20 eV have been published for nanoparticles obtained by the reduction of GO [16], with the decreasing value associated with differences in thicknesses. We assume that our two values of 4.02 and 3.67 eV are significantly different and rather associated with the dimension of graphene, from micron-size to smaller than 100 nm in the MEB and TEM images.

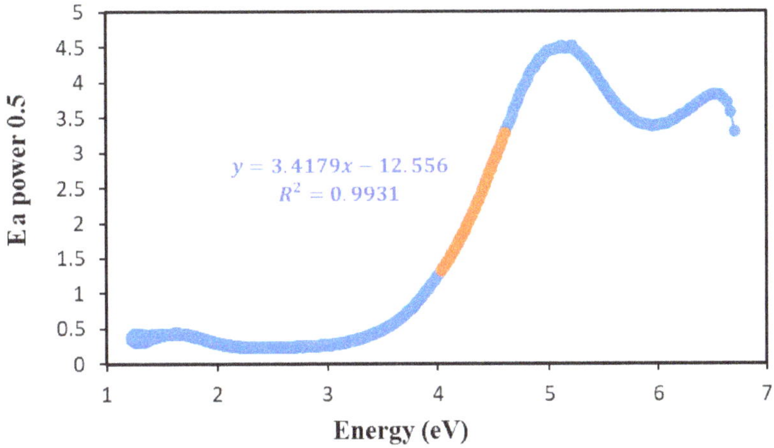

Figure 9. Associated TAUC plot for an indirect semi-conductor, obtained after 6 min of ultrasound treatment with pulses. The value of the band-gap is calculated by the extrapolation of the straight line in orange by dividing its origin (its common point with the Ox axis) by its slope. Here, a value of 3.67 eV is obtained.

3.2.3. Influence of Graphene Sizes Studied by Fluorescence (Emission) Spectra

The fluorescence of reduced graphene sheets was studied [30,31]. We tried two kinds of florescence measurements on our non-centrifugated liquors of graphene before the copper salt addition, the first one with ethanol and the second one directly with the concentrated initial black solutions, after a suspension of commercial graphite (GR) and ultrasound applied for 6 min and for 2 h. Better results were obtained without dilution and for an angle excitation light detector set at 90° to avoid detector saturation. The observed signals correspond to the superposition of signals attributed to graphite 3D and graphene 2D particles. The maximum shifts from 425 to 455 nm as the particle size increases from micron sizes to less than 100 nm in SEM and TEM images, involving fewer C atoms and lower energy π delocalization.

The fluorescence spectrum of reduced rGO dissolved in water, acetone, ethyl acetate, toluene, and benzene with an excitation pulse of 300 nm is already published [30,31]. It was demonstrated that the fluorescence indeed exists and that the fluorescence properties were strongly affected by the solvent polarity.

Ethanol's polarity is larger than acetone's, and a free OH group also offers the possibility of H-bonds; therefore, we were expecting to observe a fluorescence signal in ethanol, as suggested by negative peaks seen in the UV spectra.

The detected fluorescence signals are presented in Figure 10 (Figure 10a ethanol only and Figure 10b with 2 h of the ultrasound treatment without applied centrifugation), with two spectra recorded on the small particles of graphite 3D and graphene 2D obtained by a 2 h ultrasound treatment and with an ultrasound treatment of 6 min. The two measurements were performed with the same energy of excitation (358 nm), the same rate (1 nm/s), and the same slit widths (external 5 nm, internal 2 nm).

The signal keeps a detectible intensity between 400 and 625 nm. The fluorescence spectra of rGO dissolved in water, benzene, toluene, ethyl acetate, and acetone at the excitation pulse of 300 nm have already been reported. The signal detected here in ethanol is stronger with the larger graphene nanoparticles obtained with 6 min of ultrasound treatment than after 2 h of ultrasound treatment. Its maximum shifts from 425 to 455 nm if the graphene particle size increases from micron sizes to lower than 100 nm in SEM and TEM images with fewer involved C atoms and a lower energy π delocalization. The two measurements were performed with the same excitation of 358 nm, the same rate of 1 nm by s, and the same slit widths, external of 5 nm and internal of 2 nm.

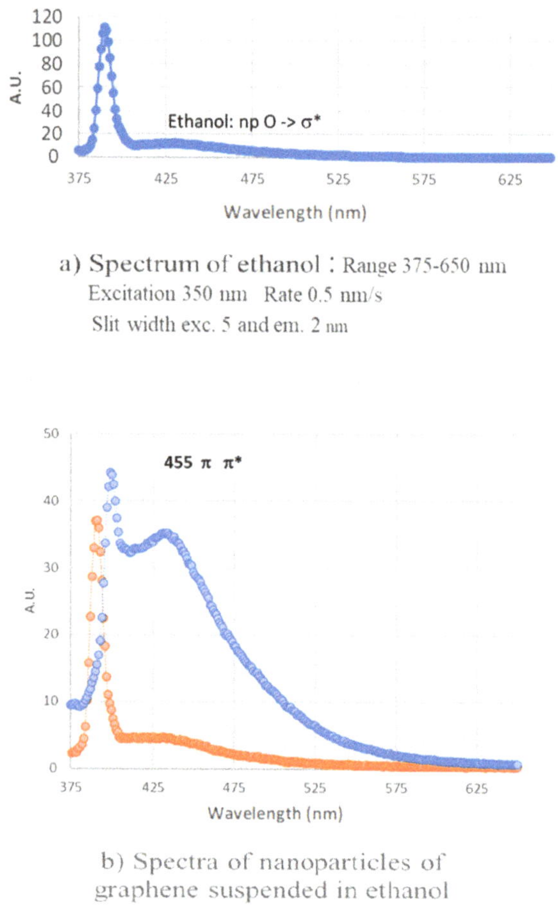

Figure 10. Fluorescence spectra of ethanol of two solutions of graphite 3D and graphene 2D suspended in ethanol prepared with 6 min (**a**) and with 2 h of the ultrasound treatment, respectively (excitation at 358 nm), without applied centrifugation (**b**). The existence of the 2D particles is demonstrated by SAXS measurements presented in the Supplementary Information—comparison of the SAXS results obtained on the powder after centrifugation and of the powder recovered by filtration on the solution on paper and drying in air.

3.2.4. Reactivity Test

UV–visible spectra were necessary to follow the oxidation of two organic dyes (eosin and bromophenol blue). With eosin, the initial concentration of the dye was equal to 44 mg in 100 mL of distilled water, while the blue bromophenol dye was in water. Dilutions were then prepared and were necessary to collect UV–visible–NIR spectra:

(1) An amount of 250 µL of the initial solution was diluted in 4 mL of distilled water, directly inside a 1 cm cell in quartz;
(2) Dilution 2 was prepared by the dilution of 250 µL of dilution 1 into 4 mL of distilled water;
(3) Dilution 3 was prepared with 250 µL of dilution 2 into 4 mL of water. To these dilutions, 250 µL of liquid was removed and replaced with 250 µL of the active liquor.

With the three solutions, the characteristic peaks of the eosin dye were located at 503 and 534 nm, with relative absorbances of 0.503 and 1.15 for dilution 3. The three peaks were eliminated in seconds with a few drops of the active green liquors that we tested both with small and with large rGO particles. A similar treatment was undertaken with a blue bromophenol dye in water. The treatments were also reproduced in ethanol.

It is important to know that no plasmons of rGO were present in the used dilutions. This was on purpose to study the material's photocatalytic activity in their absence.

4. Conclusions and Perspectives

Ultrasound treatments of ethanol dispersions of a commercial graphite (GR) were very efficient to obtain reduced and "flat" graphene sheets that we characterized by XRD, SEM, and TEM. UV–visible–NIR results were complemented by fluorescence spectroscopy. For ultrasound treatments at a power of 500 W (used at 40% and with a frequency of 20 MHz), a horn was directly introduced inside the solution, and the treatments were performed either with a constant method or using pulses of 15 s and stops of 15 s for overall times of 5–7 min and of 2 h.

Thanks to the systematic use of centrifugation vials having the same volume (45 mL) of solvent, the same amount of commercial C-based GR species, and similar centrifugation conditions, we can conclude that the graphene sheet dimension was directly affected by the time of the ultrasound treatment used. The main dimension of the graphene particles decreased from larger than a micron (ultrasound of 6 min) to smaller than 100 nm (after ultrasound of 2 h). Even after 2 h of ultrasound treatment, the C particles remained stacked, and more than 5 to 25 sheets were detected.

Interesting plasmons (collective excitation of electrons, considered as a gaseous plasma) were observed after centrifugation and the addition of $CuCl_2{\cdot}3H_2O$, and both their intensities (absorption that can be larger than 10) and their positions were affected by dilutions. The existence of a coupling between the plasmons and the vibrations of the C lattice was demonstrated. At a high dilution to remove the plasmons, the prepared Cl/Cu/graphene solutions were very strong oxidants, as demonstrated by catalytic tests performed with a dye (eosin and bromophenol blue in water and in ethanol). In our experimental conditions, the visible signal of the dye was indeed eliminated in a few seconds with both small and large rGO particles.

In the future, more concentrated solutions will be used to preserve plasmons. A more complete study including several solvents of varied polarity and using atomic force microscopy will be in progress. In addition, we will study the molecular interactions between copper, rGO, and added organic molecules, and we will propose a dynamic process.

Supplementary Materials: The following supporting information can be downloaded at https://www.mdpi.com/article/10.3390/cryst14070654/s1, Figure S1: Used apparatus. SAXS measurements were carried out with a XENOCS apparatus in the q range between 0.001 and 1.200 $Å^{-1}$, Figure S2: Evidence of intense plasmons in the visible and the NIR spectral range.

Author Contributions: Conceptualization, A.J., A.D. and J.T.; methodology, A.J., A.D. and J.T.; software, A.J., A.D. and J.T.; validation, A.J., A.D. and J.T.; formal analysis, A.J., A.D. and J.T.; investigation, A.J.; resources, A.D. and J.T.; data curation, A.J., A.D. and J.T.; writing—original draft preparation, A.J., A.D., T.H. and J.T.; writing—review and editing, A.J., A.D., T.H. and J.T.; visualization, A.J., A.D., T.H. and J.T.; supervision, A.D. and J.T.; project administration, A.D. and J.T.; funding acquisition, A.D., T.H. and J.T. All authors have read and agreed to the published version of the manuscript.

Funding: This research received no external funding.

Data Availability Statement: There are no additional data. The original contributions presented in the study are included in the article.

Acknowledgments: We would like to thank Mohamed Selmane, David Montero, and Dalil Brouri, working at Sorbonne University, for their help in the measurements of X-ray diffraction at small and large angles, as well as SEM and TEM measurements. The SEM and EDS were funded by Sorbonne Université, CNRS, and Région Ile de France. Thanks to the Fédération de Chimie et de Matériaux de Paris Centre (FCMat) for use of the X-ray diffusion and diffraction platform and the electronic microscopes. Vincent Vivier of the Laboratoire de Reactivité de Surface, Sorbonne Université, and all the technical staff of the Laboratory of Surface Reactivity in Sorbonne Université are also acknowledged.

Conflicts of Interest: The authors declare no conflicts of interest.

References

1. Fuchs, J.N.; Goerbig, M.O. Le graphène premier crystal bidimensionnel. *Pour La Sci. Phys.* **2008**, *367*, 36–43.
2. Mbayachi, B.; Ndayiragije, E.; Sammani, T.; Taj, S.; Mbuta, E.R.; Khan, A.U. Graphene synthesis, characterization and its applications: A review. *Results Chem.* **2021**, *3*, 100163. [CrossRef]
3. Li, X.; Yu, J.; Wageh, S.; Al-Ghamdi, A.A.; Xie, J. Graphene in Photocatalysis: A Review. *Small* **2016**, *12*, 6640–6696. [CrossRef]
4. Minzhen, C.; Thorpe, D.; Adamson, D.H.; Schniepp, H.C. Methods of Graphite Exfoliation. *J. Mater. Chem.* **2012**, *22*, 24992. [CrossRef]
5. Ikram, M.; Sarfraz, A.; Aqeel, M.; Ul-Hamid, A.; Imran, M.; Haider, J.; Haider, A.; Shahbaz, A.; Salamat, A. Visible-Light-Induced Dye Degradation over Copper-Modified Reduced Graphene Oxide. *J. Alloys Compd.* **2020**, *837*, 155588. [CrossRef]
6. Grasseschi, D.; Silva, W.C.; de Souza Paiva, R.; Starke, L.D.; do Nascimento, A.S. Surface coordination chemistry of graphene: Understanding the coordination of single transition metal atoms. *Coord. Chem. Rev.* **2020**, *422*, 21346. [CrossRef]
7. Huang, F.; Deng, Y.; Chen, Y.; Cai, X.; Peng, M.; Jia, Z.; Xie, J.; Xiao, D.; Wen, X.; Wang, N.; et al. Anchoring Cu_1 species over nanodiamond-graphene for semi-hydrogenation of acetylene. *Nat. Commun.* **2019**, *10*, 4431. [CrossRef]
8. Zhang, L.; Yang, X.; Yuan, Q.; Wei, Z.; Ding, J.; Chu, T.; Rong, C.; Zhang, Q.; Ye, Z.; Xuan, F.Z.; et al. Elucidating the structure-stability relationship of Cu single-atom catalysts using *operando* surface-enhanced infrared absorption spectroscopy. *Nat. Commun.* **2023**, *14*, 8311. [CrossRef]
9. Li, L.; Li, M.; Zhang, R.; Zhang, Q.; Geng, D. Liquid Cu–Zn catalyzed growth of graphene single-crystals. *New J. Chem.* **2023**, *47*, 20703–20707. [CrossRef]
10. Tabaja, N.; Brouri, D.; Casale, S.; Zein, S.; Jaafar, M.; Selmane, M.; Toufaily, J.; Davidson, A. Use of SBA-15 silica Grains for Engineering Mixtures of Co Fe and Zn Fe SBA-15 for Advanced Oxidation Reactions under UV and NIR. *Appl. Catal. B Environ.* **2019**, *253*, 369–378. [CrossRef]
11. Chettah, W.; Barama, S.; Medjram, M.-S.; Selmane, M.; Montero, D.; Davidson, A.; Védrine, J.C. Anatase titania activated by Cu(II) or Zn(II) nanoparticles for the photooxidation of methanol assisted by Rhodamine-B. *Mater. Chem. Phys.* **2021**, *257*, 123714. [CrossRef]
12. Ghazinejad, M.; Hosseini Bay, H.; Reiber Kyle, J.; Ozkan, M. Fluorescence quenching metrology of graphene. *SPIE Proc.* **2014**, *8994*, 169–175.
13. Jubu, P.R.; Yam, F.K.; Igba, V.M.; Beh, K.P. Tauc-plot scale and extrapolation effect on bandgap estimation from UV–vis–NIR data—A case study of β-Ga_2O_3. *J. Solid State Chem.* **2020**, *290*, 121576. [CrossRef]
14. Tabaja, N. Nanoparticules d'oxydes de fer et de Ferrites Obtenues par Nano-Réplication: Réactivité Chimique et Application en Dépollution des Eaux. Ph.D. Thesis, Université Pierre et Marie Curie-Paris VI, Paris, France, Université Libanaise, Beirut, Lebanon, 2015. Available online: https://theses.fr/070382972 (accessed on 10 May 2024).
15. Jezzini, A. $ZnFe_2O_4$ for Heterogeneous Photocatalysis in the Visible Spectrum. Ph.D. Thesis, Sorbonne Université, Paris, France, 2020.
16. Shenoy, R.U.K.; Rama, A.; Govindan, I.; Naha, A. The purview of doped nanoparticles: Insights into their biomedical applications. *OpenNano* **2022**, *8*, 100070. [CrossRef]
17. Larin, G.M.; Minin, V.V.; Levin, B.V.; Buslaev, Y.A. *Complexation in the $CuCl_2$_C_2H_5OH_H_2O System*; N. S. Translated from Izvestiya Akademii Nauk SSSR, Seriya Khimicheskaya; Kurnakov Institute of General and Inorganic Chemistry, Academy of Sciences of the USSR: Moscow, Russia, 1990; Volume 12, pp. 2725–2729.
18. Mak, K.F.; Ju, L.; Wang, F.; THeinz, F. Optical spectroscopy of graphene: From the far infrared to the ultraviolet. *Solid Stat. Commun.* **2012**, *152*, 1341–1349. [CrossRef]
19. Kumar, P. Bandgap Measurement of Reduced Graphene Oxide Monolayers through Scanning Tunneling Spectroscopy. *arXiv* **2019**, arXiv:1911.06230.
20. Li, H.; Papadakis, R.; Hassan, S.; Jafri, M.; Thersleff, T.; Michler, J.; Ottosson, H.; Leifer, K. Superior adhesion of graphene nanoscrolls. *Common. Phys.* **2018**, *1*, 44. [CrossRef]
21. Atri, P.; Tiwari, D.C.; Sharma, R. Synthesis of reduced graphene oxide nanoscrolls embedded in polypyrrole matrix for supercapacitor applications. *Synth. Met.* **2017**, *227*, 21–28. [CrossRef]
22. Wallace, J.; Shao, L. Defect-induced carbon nanoscroll formation. *Carbon* **2015**, *91*, 96. [CrossRef]

23. Jezzini, A.; Chen, Y.; Davidson, A.; Hamieh, T.; Toufailly, J. Photocatalytic Decomposition of Amoxicillin Using Zinc Ferrite Nanoparticles. *Crystals* **2024**, *14*, 291. [CrossRef]
24. Faniyi, O.; Fasakin, O.; Olofinjana, B.A.; Adekunle, S.; Oluwasusi, T.V.; Eleruja, M.A.; Ajayi, E.O.B. The comparative analyses of reduced graphene oxide (rGO) prepared via green, mild and chemical approaches. *SN Appl. Sci.* **2019**, *1*, 1181. [CrossRef]
25. Thiurnina, A.V. Ultrasonic exfoliation of graphene in water: A key parameter study. *Adv. Mater.* **2010**, *22*, 1039–1059. [CrossRef]
26. Navik, R.; Gai, Y.; Wang, W.; Zhao, Y. Curcumin-assisted ultrasound exfoliation of graphite to graphene in ethanol. *Ultrason. Sonochem.* **2018**, *48*, 96–102. [CrossRef]
27. Muthosamy, K.; Manickam, S. State of the art and recent advances in the ultrasound-assisted synthesis, exfoliation and functionalization of graphene derivatives. *Ultrason. Sonochem.* **2017**, *39*, 478–493. [CrossRef]
28. Simani, M.; Dehghani, H. The study of electrochemical hydrogen storage behavior of the UiO-66 framework on the metal/reduced graphene oxide substrate. *Fuel* **2023**, *341*, 127624. [CrossRef]
29. Kuterasiński, L.; Podobiński, J.; Madej, E.; Smoliło-Utrata, M.; Rutkowska-Zbik, D.; Datka, J. Reduction and Oxidation of Cu Species in Cu-Faujasites has been Studied by IR Spectroscopy. *Molecules* **2020**, *25*, 4765. [CrossRef]
30. Rosa Axet, M.; Durand, J.; Gouygou, M.; Serp, P. Surface coordination chemistry on graphene and two-dimensional carbon materials for well-defined single atom supported catalysts. *Adv. Organomet. Chem.* **2019**, *71*, 53–73. [CrossRef]
31. Yao, D.; Wang, Y.; Li, Y.; Li, A.; Zhen, Z.; Lv, J.; Sun, F.; Yang, R.; Luo, J.; Jiang, Z.; et al. Scalable synthesis of Cu clusters for remarkable selectivity control of intermediates in consecutive hydrogenation. *Nat. Commun.* **2023**, *14*, 1123. [CrossRef] [PubMed]
32. Zhao, Y.; Tao, L. Towards catalytic reactions of Cu single-atom catalysts: Recent progress and future perspective. *Chin. Chem. Lett.* **2024**, *35*, 108571. [CrossRef]
33. Bhaumik, A.; Narayan, J. Reduced Graphene Oxide-Nanostructured Silicon Photosensors with High Photoresponsivity at Room Temperature. *ACS Appl. Nano Mater.* **2019**, *2*, 2086–2098. [CrossRef]
34. Politano, A.; Chiarello, G. Plasmon modes in graphene: Status and prospect. *Nanoscale* **2014**, *6*, 10927–10940. [CrossRef] [PubMed]
35. Drosou, M.; Mitsopoulou, C.A.; Orio, M.; Pantazis, D.A. EPR Spectroscopy of Cu(II) Complexes: Prediction of *g*-Tensors Using Double-Hybrid Density Functional Theory. *Magnetochemistry* **2022**, *8*, 36. [CrossRef]

crystals

MDPI

Article

Ultraviolet Exposure Improves SERS Activity of Graphene-Coated Ag/ZrO_2 Substrates

Hanna Bandarenka [1], Aliaksandr Burko [1], Diana Laputsko [1], Lizaveta Dronina [1], Nikolai Kovalchuk [1], Alise Podelinska [2,3], Uladzislau Shapel [2], Anatoli I. Popov [2] and Dmitry Bocharov [2,*]

1 Micro- and Nanoelectronics Department, Belarusian State University of Informatics and Radioelectronics, 220013 Minsk, Belarus; h.bandarenka@bsuir.by (H.B.); a.burko@bsuir.by (A.B.); d.laputko@bsuir.by (D.L.); e.dronina@bsuir.by (L.D.); n.kovalchuk@bsuir.by (N.K.)
2 Institute of Solid State Physics, University of Latvia, LV-1063 Riga, Latvia; alisep@cfi.lu.lv (A.P.); uladzislau.shapel@cfi.lu.lv (U.S.); popov@latnet.lv (A.I.P.)
3 Institute of Physics, University of Tartu, 50411 Tartu, Estonia
* Correspondence: bocharov@cfi.lu.lv

Abstract: This study reveals a significant improvement in surface-enhanced Raman scattering (SERS) activity of Ag/ZrO_2 substrates covered with a few-layer graphene preliminary exposed to ultraviolet (UV) light. The SERS-active substrates are formed by the "silver mirror" deposition of Ag nanoparticles on annealed zirconia blocks. The film composed of ~3 graphene layers is grown on copper foil by a chemical vapor deposition and then wet-transferred to the SERS-active substrates. The graphene-free Ag/ZrO_2 samples are found to provide an enhancement of the Raman scattering from rhodamine 6G (R6G) at a micromolar concentration, which is associated with combined effects from the surface plasmon resonance in the Ag nanoparticles and a charge transfer facilitated by zirconium dioxide. It is revealed that the SERS signal from the analyte molecules can be suppressed by a UV exposure of the Ag/ZrO_2 samples due to photocatalytic activity of the wide band gap semiconductor. However, if the samples are covered with a few-layer graphene ($Gr/Ag/ZrO_2$) it prevents the dye molecule decomposition upon the UV treatment and improves SERS activity of the substrates. The 365 nm treatment leads to a 40% increase in the 10^{-6} M R6G SERS spectrum intensity, while the 254 nm irradiation causes it to rise by 47%, which is explained by different responses from the surface and bulk zirconia crystals to the short and long UV wavelengths. This enhancement is attributed to the distinct responses of surface and in-depth zirconia crystals to varied UV wavelengths and underscores the pivotal role of graphene as a protective and enhancing layer.

Keywords: a few-layer graphene; wide bandgap semiconductor; zirconia dioxide; silver nanoparticles; surface-enhanced Raman scattering; photocatalytic activity; charge transfer

Citation: Bandarenka, H.; Burko, A.; Laputsko, D.; Dronina, L.; Kovalchuk, N.; Podelinska, A.; Shapel, U.; Popov, A.I.; Bocharov, D. Ultraviolet Exposure Improves SERS Activity of Graphene-Coated Ag/ZrO_2 Substrates. *Crystals* **2023**, *13*, 1570. https://doi.org/10.3390/cryst13111570

Academic Editors: Justina Gaidukevic and Jurgis Barkauskas

Received: 7 October 2023
Revised: 25 October 2023
Accepted: 31 October 2023
Published: 3 November 2023

1. Introduction

In recent years, a trend towards the use of graphene-containing nanomaterials as tools for the surface enhancement of Raman scattering (SERS) has become strongly pronounced [1,2]. In general, a graphene story in SERS spectroscopy began a little over a decade ago from a formation of gold nanoparticles and dendrites on graphene oxide nanosheets [3,4]. Graphene has been found to quench fluorescence from analyte molecules [5] and to enhance Raman scattering via chemical mechanism [6]. In [7] the SERS-active substrate based on the paper strip modified with graphene and Ag nanoparticles was shown to facilitate detection of rhodamine 6G at 10^{-19} M concentration. A bit later, the few-layer graphene was found to be an efficient protective layer to prevent burning analyte molecules during SERS analysis, which provided the detection of lactoferrin protein adsorbed on the SERS-active surface from the 10^{-18} M solution [8]. Therefore, graphene has been proved to possess multifunctional properties that are extremely fruitful for SERS spectroscopy in terms of decreasing electromagnetic field influence on analyte molecules leading to the side effects in the SERS spectra [9,10].

Wide band gap (WBG) semiconductors are other rather "fresh" materials that can potentially lead to engineering the SERS-active substrates possessing athermal and self-cleaning properties [11–14]. The first feature allows SERS measurements to be free of the thermal-induced destruction of analyte molecules, thus improving analysis reliability [11]. At the same time, the UV-facilitated cleaning of the SERS-active substrates which have passed through the analysis cycle is of high demand in affordability terms [14]. Considering that SERS-active substrates are consumable tools of SERS spectroscopy, they should be cost-effective or reusable for wide implementation of such analytical techniques in clinical practice, forensic science, and pharmaceutics, to name a few.

Taking the above into account, the question arises: what kind of breakthroughs could the combination of graphene, plasmonic nanostructures, and WBG semiconductors bring to ultrasensitive molecular analysis by SERS spectroscopy? Very recently, gold-coated TiO_2 microparticles wrapped with reduced graphene oxide piqued the interest of researchers owing to their capability of amplifying SERS activity and providing photocatalytic degradation of organic pollutants [15]. In such cases, the Raman signal enhancement is attributed to a plethora of mechanisms inherent in these heterostructures. The presence of noble metal nanostructure in the substrate bolsters the electromagnetic field at the surface—a phenomenon attributed to its surface plasmonic resonance [16]. This enhancement in the electromagnetic field is instrumental in amplifying the Raman signal of molecules adsorbed on the surface. Graphene, characterized by its extraordinary electrical conductivity, augments the SERS effect. In conjunction with noble metal, graphene facilitates an increase in the charge transfer, contributing further to the electromagnetic field at the surface [17]. This synergy between graphene and metal emerges as a pivotal factor in the SERS activity improvement [18]. Moreover, graphene's pronounced absorption of UV light enriches the light–matter interaction at the hosting WBG semiconductor surface, offering the avenue for its prominent photocatalytic activity for the destruction of organic molecules after their detection [15]. The multifaceted role of graphene is not just limited to the charge transfer and UV absorption; it also acts as a heatsink and, thus, slows changing the analyte molecules structure during SERS analysis on the substrate surface [8]. In addition to the synergistic actions of graphene and plasmonic nanostructure, the WBG semiconductor plays a principal role. It avails additional sites for the adsorption of molecules and, in unison with Ag and graphene, is expected to augment Raman signals via charge transfer. However, it is imperative to note that the SERS and photocatalytic effects are contingent upon specific preparations and structure composition of the noble metal/WBG semiconductor and graphene coatings. Variations in these parameters can significantly influence the extent and efficiency of SERS activity upon visible light and photocatalytic features while the UV is irradiating.

Summarizing the above, the combination of graphene, plasmonic structures, and the WBG semiconductor opens new prospects in bringing SERS spectroscopy closer to real practical application. The goal of this work was to reveal the applicability of the sandwich structure based on a few-layer graphene, silver particles, and zirconia substrates for design and engineering the nanomaterial combining expressed SERS activity and photocatalytic properties. The silver was selected since it bears the most powerful surface plasmon efficiency among other metals [19]. Zirconia was used as a host semiconductor for the plasmonic nanomaterial because it is often characterized by band gap over 5 eV [20]. This is good for the photocatalytic activity at the far-UV light, which, in turn, can provide the most effectual destructive effect on the molecules to clean the SERS-active substrate from them after analysis. Furthermore, modification of ZrO_2-based structures with silver have already been reported elsewhere as a powerful approach to improve the zirconia surface reactivity for electrochemical determination of tinidazole antibiotic [21], photo electro chem ical degradation of methylene blue [22,23], and quantitative SERS detection of hexavalent chromium in the wastewater [24].

2. Materials and Methods

2.1. Materials

The Cercon ht 98-mm disk, manufactured by Dentsply International Inc., York, PA, USA, was used to prepare ZrO_2-based substrates. Rhodamine 6G (R6G), $AgNO_3$ (ACS reagent, ≥99.0%), NaOH (reagent grade, 97%), NH_3 (anhydrous, ≥99.98%), and $C_4H_4KNaO_6 \cdot 4H_2O$ (ACS reagent, ≥99.0%) were supplied by Sigma-Aldrich and used without additional purification. For the transfer of silver films from the ZrO_2 substrates, a poly mer support constructed from SU-8 2015 was used, sourced from MicroChem, Newton, MA, USA. Water was purified with Milli-Q system (Millipore, Bedford, MA, USA).

2.2. Fabrication of SERS-Active Substrates

The Cercon disk was meticulously cut into blocks measuring 8 × 6 × 14 mm, serving as the ZrO_2-based substrates. The blocks were annealed in air following the temperature regimes displayed in Table 1, which led to their approx. 20% shrinking. Then we formed a plasmonic coating composed of silver nanoparticles on the ZrO_2 blocks. Prior to this, each block was treated with a piranha solution, consisting of H_2SO_4:H_2O_2 at a 3:1 volume ratio, and maintained at a temperature between 70–80 °C for 5 min to slightly roughen its surface for further improvement of metal coating adhesive strength. Ag nanoparticles were grown by the "silver mirror" method in a mixture of basic solution (4 mL), reducing solution (1 mL), and ethanol (1 mL) [11,25]. Basic and reducing solutions were water solutions of 0.07 M $AgNO_3$, 0.2 M NaOH, 0.6 M NH_3, and 0.4 M $C_4H_4KNaO_6 \cdot 4H_2O$, respectively. The ZrO_2 blocks were immersed in the "silver mirroring" solution for 6 min. After the silver deposition, the samples were rinsed with deionized water and air-dried at 21 °C temperature.

Table 1. Regimes of the ZrO_2 block annealing.

Step	Starting Temperature, °C	Final Temperature, °C	Time, min	Rate, °C/min
1	50	–2	0	–
2	50	300	40	7.5
3	300	800	100	5.0
4	800	1300	200	2.5
5	1300	1530	46	5.0
6	1500	1530	120	Keeping 1530 °C
7	1530	20	–	Cooling

2.3. Few-Layer Graphene Formation and Transfer

The synthesis of few-layer graphene (Gr) was performed by chemical vapor deposition (CVD) from methane, a method thoroughly detailed in reference [8]. In this process, copper foil samples, electrochemically polished to perfection, were initially annealed at a temperature of 1050 °C for 1 h. This was conducted under the constant purge flows of argon and hydrogen, delivered at flow rates of 100 and 150 cm^3/min, respectively. Then, the H_2 flow rate was reduced to 15 cm^3/min and methane was bled in at the flow rate of 1.3 cm^3/min for 30 min. After the synthesis, the surface of the Cu foil facing the walls of the quartz reactor was worked up in an oxygen plasma to remove so-called unwanted graphene. The grown few-layer graphene was transferred to the ZrO_2-based blocks using a wet chemical technique. Firstly, the samples were immersed an aqueous solution of $FeCl_3$ until complete copper dissolution while the graphene film remained floating on the solution's surface. Before the graphene transfer, this solution was three times totally replaced with deionized water by pipette suction and filling. This step was introduced to minimize undesirable adsorption of the side reaction products on the surface of the SERS-active substrates. The transferring procedure was conducted by the ZrO_2-based blocks' immersion in the solution under the graphene film and its lifting up. As a result, the graphene film remained on the surface of the substrate.

2.4. Characterization of the Experimental Samples

The ZrO_2 and Ag/ZrO_2 substrates underwent a thorough structural characterization, facilitated by the advanced Hitachi S-4800 (Hitachi Limited, Tokyo, Japan) scanning electron microscope (SEM) stationed in Tokyo, Japan. Elemental composition of the same samples was defined with a Bruker QUANTAX 200 (Bruker Nano GmbH, Adlershof, Berlin, Germany) energy dispersive X-ray (EDX) spectrometer embedded onto the scanning electron microscope. This equipment provided the elemental distribution and composition, ensuring an exhaustive characterization.

Raman and SERS analysis was performed with a 3D confocal Raman microscope–spectrometer Confotec NR500 (SOL Instruments, Minsk, Belarus). The measurements were taken using 473 or 633 nm laser and a 100× objective (NA = 0.95). The power of both laser beams passed through the objective was 4.66 ± 0.005 mW.

The spectra of the few-layer graphene were collected using the 633 nm laser. The few-layer graphene was transferred to the cover glass and Ag/ZrO_2 sample for the Raman and SERS measurements to find out if the SERS activity affects the graphene structure.

The reproducibility of the SERS signal provided by the Ag/ZrO_2 and $Gr/Ag/ZrO_2$ substrates was studied with a testing analyte presented by rhodamine 6G (R6G) molecules dissolved in deionized water. Prior to the SERS measurements, the substrates were kept in the 10^{-6} M analyte solution for 2 h, rinsed with deionized water for 30 s to remove an excess of unabsorbed molecules, and air-dried. The arrays of spectra (100 in total) were collected by scanning the 100 × 100 μm spot on the SERS-active surface with a focused 473 nm laser beam.

2.5. Study of Photocatalytic Activity

Photocatalytic activity of the zirconia-based samples was assessed by comparing specific band intensity in the R6G SERS spectra before and after the UV exposure. Firstly, the dye molecules were deposited on the Ag/ZrO_2 and $Gr/Ag/ZrO_2$ samples from 10^{-6} M R6G solutions followed by collecting their SERS spectra. Then, each substrate was immersed in a Petri dish of 3 cm diameter filled with 3 mL of deionized water and subjected to the UV irradiation for different times. Parallel with this process, control samples, utilizing commercial SERS-active substrates referenced in [26], underwent identical UV treatment procedures to offer a comparative analysis of the photocatalytic activity. Testing the photocatalytic activity in a Petri dish filled with water was based on the rationale that it facilitates the desorption of decomposed dye molecules from the substrate surface into the liquid medium, resulting in an improved cleaning compared to cleaning solely in water under UV-free conditions. The UV exposure was provided by a VL-6. LC Vilber appliance (Vilber, Collegein, France) equipped with 254 and 365 nm wavelength lamps. This equipment ensured uniform and precise UV irradiation, which is essential for an accurate assessment of photocatalytic activity. After that, all the samples were rinsed with deionized water for 30 s, air-dried, and the SERS measurements were repeated.

3. Results and Discussion

3.1. Structure of the Experimental Samples

3.1.1. ZrO_2 and Ag/ZrO_2 Substrates

The ZrO_2 block surface is characterized by grainy morphology, as seen in Figure 1a. The grains, displaying a size variation from 225 nm to approximately 1.5 μm, predominantly fall within the 400–600 nm range. The black curve in the Raman spectrum of Figure 1b reveals the polycrystalline nature of the zirconia block (ZrO_2—1), evidenced by the distinct bands representative of tetragonal (~243, ~298, ~458, and ~631 cm^{-1}) and monoclinic (~200 and ~374 cm^{-1}) lattices [27,28]. The band at 458 cm^{-1} can be attributed to tetragonal zirconia stabilized with yttrium oxide [29], of which a small amount is present in the Cercon disk.

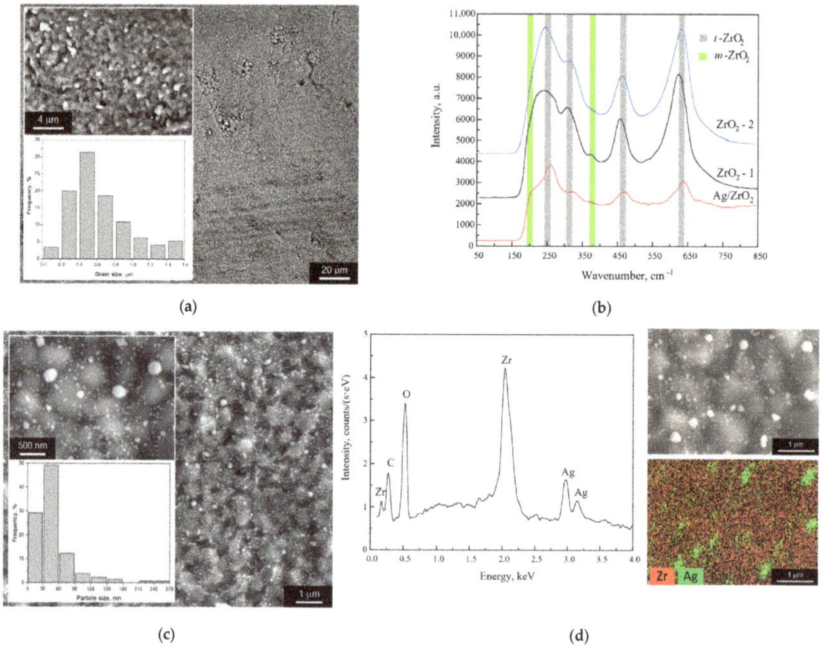

Figure 1. Structural characterization of the samples: (**a**) SEM top view images of the ZrO_2 substrate and the ZrO_2 grain size distribution histogram; (**b**) Raman spectra of the ZrO_2—1, Ag/ZrO_2, and ZrO_2—2 substrates; (**c**) SEM top view images of the Ag/ZrO_2 substrate and the Ag particle size distribution histogram; (**d**) EDX spectrum and EDX scan of the Ag/ZrO_2 substrate.

To infuse SERS activity into the zirconia substrate, it was submerged in the "silver mirror" solution, resulting in a uniform deposition of silver. This process is detailed through the following chemical reactions:

$$2AgNO_3 + 2NaOH \rightarrow Ag_2O + 2NaNO_3 \quad (1)$$

$$Ag_2O + 4NH_3{\cdot}H_2O = 2[Ag(NH_3)_2]OH + 3H_2O \quad (2)$$

$$6Ag^+ + C_4H_4O_6^{2-} + 8OH^- \rightarrow 6Ag + 2C_2O_4^{2-} + 6H_2O \quad (3)$$

The "silver mirror" chemical method was selected due to the WBG nature of zirconia [30,31]; other wet techniques, like corrosive and electrochemical deposition, are hardly applicable to growing silver particles on the ZrO_2 surface. Figure 1c depicts the SEM top view images of the zirconia block subjected to the silver deposition. The silver coating is constituted by nanoparticles, the majority of which have sizes of nanoscale range between a few nanometers and 90 nm. However, a small fraction, approximately 10%, of the Ag nanostructures, is characterized by larger sizes up to 270 nm as seen from the histogram in Figure 1c. The post-"silver mirroring" Raman spectrum of the zirconia block, depicted by the red curve in Figure 1b, is characterized by the suppressed Raman bands of ZrO_2 as a direct consequence of the silver coating. The silver deposition led to suppressing the 374 cm^{-1} band (responsible for the m-lattice) down to negligible intensity. This allows for the suggestion of the partial phase transformation of zirconia upon the Ag deposition. At the same time, the band at approx. 200 cm^{-1} appears more prominent after the silver deposition, which gives rise to controversial opinion. That is why we additionally studied

Raman spectrum of the zirconia sample subjected to removal of Ag coating with hydrochloric acid. This sample is marked as ZrO_2—2 in Figure 1b. Indeed, the new spectrum proves the Ag deposition can lead to the phase transformation since no band at approx. 375 cm^{-1} is observed and, also, the band at approx. 200 cm^{-1} looks more hidden by the t-ZrO_2 band at 250 cm^{-1}. This effect is explained by the surface zirconia lattice aligning to the Ag cubic lattice. Slight restructuring the zirconia surface is expected to improve an adhesion of the silver coating and provide faster charge transfer from the WBG substrate via the silver nanoparticles to the analyte molecules during the SERS measurements.

We also should note that no bands of Ag_2S typical for the Raman spectra of nanostructured silver [32,33] are observed. This means there is a rather good chemical stability of the Ag nanoparticles when taking electrons by (or oxidation with) sulfur ions from the environment, which is typical for silver surfaces [34]. The EDX spectrum and EDX scan in Figure 1d prove the Ag nanoparticles presence and uniform distribution on the sample surface.

3.1.2. Few-Layer Graphene

To study the Gr structure in more detail, it was characterized by Raman spectroscopy. As a rule, the graphene Raman spectrum contains G, 2D, and D prominent bands [35,36]. The first one (~1582 cm^{-1}) is attributed to a doubly degenerated phonon mode of E_{2g} symmetry from the center of the Brillouin zone. The 2D band (~2710 cm^{-1}) appears as a result of the resonant light scattering in which two equal-energy phonons are involved. However, the light is scattered in the opposite direction of the pulse, thus providing data on the graphene layers order. The last band (~1352 cm^{-1}) helps to reveal defects in the graphene layers. Therefore, the Raman mapping of graphene on copper was performed. A laser with excitation energy 2.62 eV (λ = 473 nm) was utilized. The size of the mapped area was 48 × 48 μm^2. The number of points on the map was 20 × 20. The intensities, positions, and half-widths of the Raman spectral bands were determined using the Lorentz approximation. The correlation between the position of the 2D band and the position of the G band was plotted based on data from Raman mapping (Figure 2). The correlation analysis of the G and 2D bands positions was based on that developed in the [37,38] approach, indicating a few-layer structure of graphene (~3 layers), which was confirmed using transmission optical spectroscopy (Figure 3).

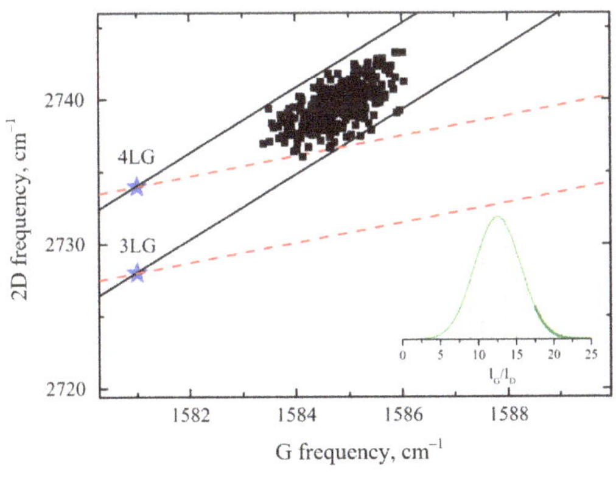

Figure 2. Experimental dependences of 2D band position on position of G band for graphene on copper. The bold solid line (with a slope of ~2.2) passing through the blue star marks are responsible for the biaxial strain in the ideal graphene. Dashed red lines (with a slope of ~0.7) are responsible for the charge carrier concentration with fixed biaxial strain. The blue star markers correspond to unstressed and undoped multilayer graphene [38] defined taking into account the excitation wavelength used in current work [39] Inset: histogram of the ratio of the intensities of the G band to the D band.

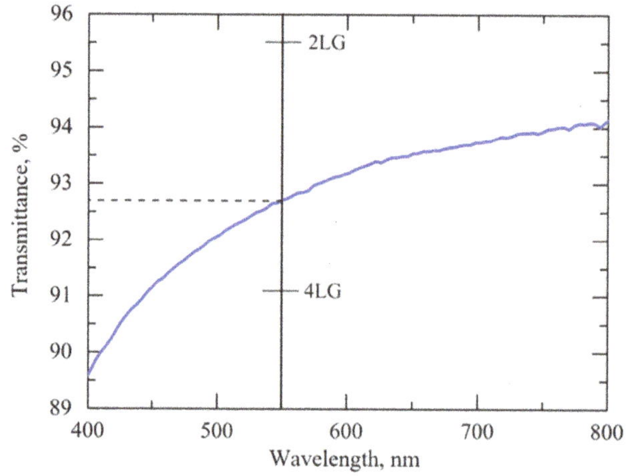

Figure 3. Transmission spectrum of graphene transferred onto the quartz glass.

The intensities, position of the G and 2D bands, and their full width at half-maximum were indicative of a few-layer structure of graphene (~3 layers), which was confirmed using transmission optical spectroscopy (Figure 3).

The transmittance at 550 nm is ~92.7%, which perfectly agrees with transmittance for 3 layered graphene (92.66%) calculated using the following formula [40]:

$$T = (1 + 1.13\pi\alpha N/2)^{-2}, \quad (4)$$

where N is the number of layers and α is the fine-structure constant [41].

The D ~1352 cm^{-1} band, which is responsible for deformation, disorder, and defects in the crystalline structure of graphene, has low intensity compared to the G band (inset Figure 2), so this proves the good quality of the graphene [35].

Subsequently, the few-layer graphene films formed by the CVD method were transferred from the copper foil onto the Ag/ZrO_2 substrates before further studying their structural properties. Figure 4 shows the SEM image of the silver-coated zirconium oxide surface after the Gr film transfer. It can be clearly seen that the surface of the Ag/ZrO_2 substrate has a coating with a small number of folds and holes, which should be a film containing graphene. This was also confirmed by the Raman spectroscopy analysis (Figure 5).

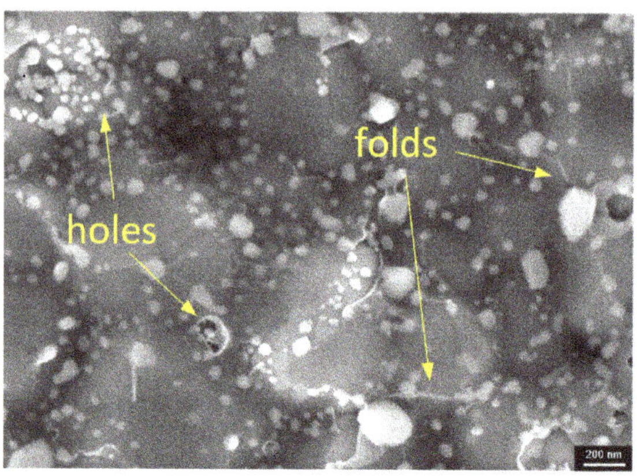

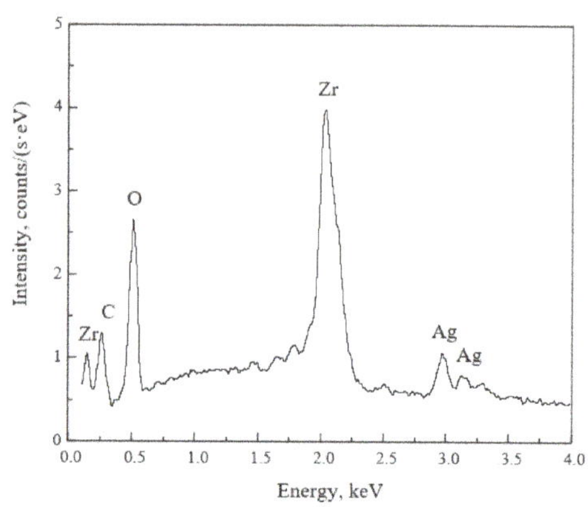

Figure 4. SEM top view image and EDX spectrum of the $Gr/Ag/ZrO_2$ sample.

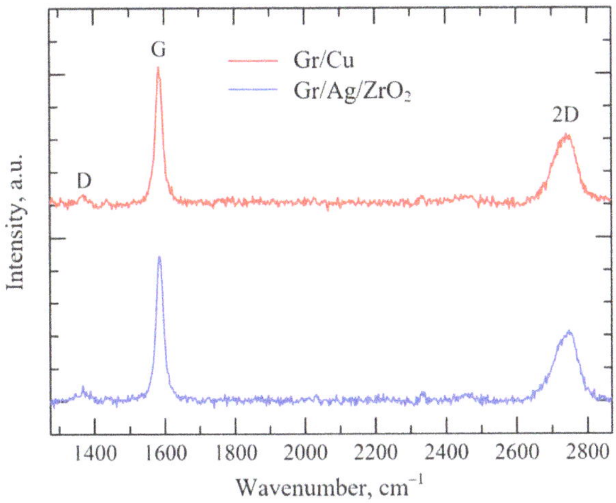

Figure 5. Typical Raman spectrum of graphene on copper and transferred onto an Ag/ZrO_2.

3.2. SERS Activity of the Experimental Samples

Before commencing the SERS measurements, the Ag/ZrO_2 and $Gr/Ag/ZrO_2$ substrates were immersed in 10^{-6} M R6G solution for 2 h for the analyte molecules adsorption. Then, the substrates were rinsed thoroughly with deionized water for 30 s for the removal of the unabsorbed molecules, before being dried. Such procedure was used to provide coating to the substrates with a monomolecular layer of the analyte molecules.

Both the substrates demonstrated SERS activity as seen in Figure 6a since all characteristic R6G Raman bands can be clearly distinguished in the SERS spectra [42,43]. Interestingly, the few-layer graphene provided a triple increase in the SERS spectrum intensity compared to that collected on the Gr-free substrates.

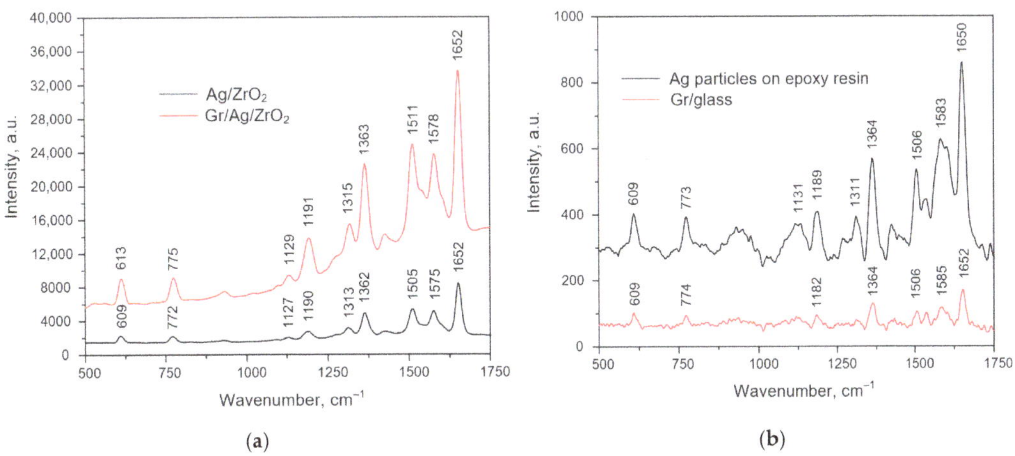

Figure 6. Mean SERS spectra of R6G molecules adsorbed from 10^{-6} M solution on the surface of the different substrates: (**a**) Ag/ZrO_2 and $Gr/Ag/ZrO_2$; and (**b**) Ag particles on epoxy resin and Gr/glass.

To further investigate the SERS activity intrinsic to the few-layer graphene, R6G molecules were adsorbed onto the Gr/glass sample. As seen in Figure 6b, the Gr film facilitates the weak enhancement of the analyte Raman spectrum, which is over two orders of magnitude lower compared with the SERS spectrum collected with the Gr/Ag/ZrO_2 substrate. Also, we tried to evaluate the contribution of the Ag nanoparticles to the Raman spectrum enhancement. The Ag layer was detached from the ZrO_2 block with the epoxy resin film, which was then coated with R6G molecules and studied using the Raman spectrometer. The black spectrum in Figure 6b proves the SERS activity of the Ag nanoparticles. However, even in combination with the few-layer graphene, it can hardly provide the multiplication of the Raman signal from the analyte molecules up to the SERS spectrum intensity achieved with the Gr/Ag/ZrO_2 substrate.

It should be noted that the ZrO_2 block was not found to enhance Raman signal from the analyte molecules at micromolar concentration. The collection of the Raman signal resulted in background free of characteristic analyte bands. Therefore, we suggest that the prominent SERS activity of the Gr-coated substrate is exclusively caused by the combination of the WBG semiconductor, plasmonic particles, and carbon nanostructure. The grainy ZrO_2 surface acts as a source of a charge transferred to the Gr film. Here, we would like to refer to a few papers that have already proved increasing the charge transfer effect in the defected WBG semiconductors [12]. Considering the submicron sizes of the ZrO_2 grains enriched with developed boundary network between them gives a rise to such an explanation. The defects enhance the charge transfer effect and significantly contribute to the electromagnetic enhancement provided by the silver coating.

3.3. *Photocatalytic Activity of the Ag/ZrO_2 Substrates*

Zirconia is known as a WBG semiconductor, which possesses photocatalytic activity [44]. In our study, we aimed to ascertain whether the granular ZrO_2 block imparts the self-cleaning property to the SERS-active substrate facilitated by the UV exposure and if the graphene coating enables an improvement of this property. Our findings revealed a discernible decline in the intensity of the SERS spectra of 10^{-6} M R6G post-exposure to both 365 and 254 nm lamp irradiation. The longer UV wavelength led to over 2–2.5 times (~40–50%) fewer intensive spectra (Figure 7a).

Noticeable was an increase in the 365 nm exposure time from 30 to 120 min which did not result in better analyte molecules decomposition. In contrast, the 254-nm UV exposure exhibited superior photocatalytic activity on the Ag/ZrO_2 sample, inducing approximately 72% spectral intensity reduction at 30 min and around 86% at 120 min of exposure, as illustrated in Figure 7b. The enhanced efficacy of the shorter wavelength UV light is attributable to its deeper penetration and interaction with the internal zirconia grains, while the near-UV primarily engages the surface layer [27]. The short-wavelength UV is hardly absorbed by zirconia [27] thus interacting just with the surface zirconia structures, while longer UV waves are well-absorbed thanks to the rather wide bandgap [44]. Therefore, 254 nm irradiation would interact with in-depth volume of the zirconia sample, which provides more zirconia crystals compared to the surface layer.

We considered that the UV exposure can cause organic dye decomposition, which can be expressed by the presence of d-metal like silver. To exclude such an effect, we made the SERS measurements with the control commercially available substrates based on the silvered silicon nanostructures [25]. Figure 7c shows the SERS spectra of 10^{-6} M R6G collected with such substrates before and after the UV exposure at 365 and 254 nm. Indeed, one can see a negligible decrease in the spectra intensity, which is below 30%. Thus, the R6G decomposition enabled by the ZrO_2-based substrates must be provided by the photocatalytic activity of the WBG semiconductor.

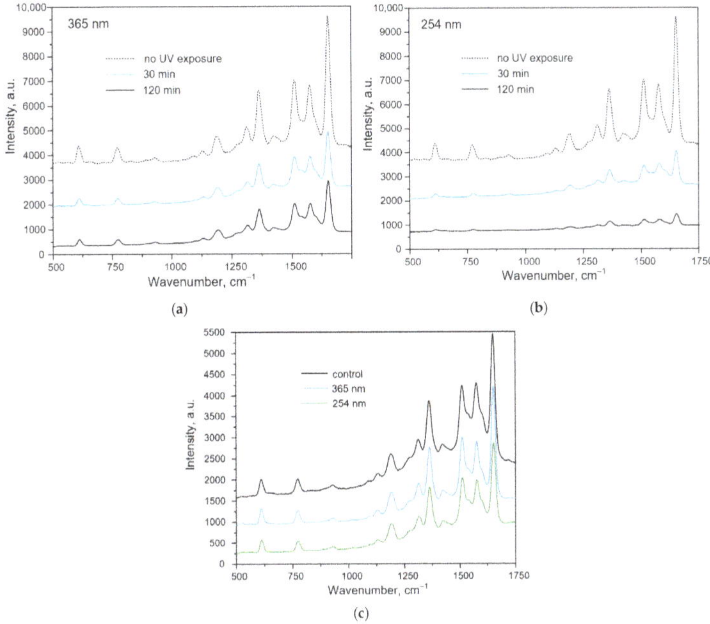

Figure 7. Mean SERS spectra of R6G molecules adsorbed from 10^{-6} M solution on the surface of (**a**,**b**) the Ag/ZrO_2 substrates, (**c**) the control SERS-active substrates, and exposed to 365 nm or 254 nm irradiation.

3.4. Effect of UV Exposure on SERS Activity of the $Gr/Ag/ZrO_2$ Substrates

The final step of the research was devoted to revealing the UV exposure effect on the $Gr/Ag/ZrO_2$ substrate properties. The expected improvement of the photocatalytic activity was not found. To the contrary, we observed an undoubted increase in SERS activity because the spectra intensity rose twice (~40%) after the 365 nm exposure and 2.3 time (~47%) in the case of 254 nm irradiation for 2 h (Figure 8), which is explained by the different response of surface and bulk zirconia crystals to the short and long UV wavelengths.

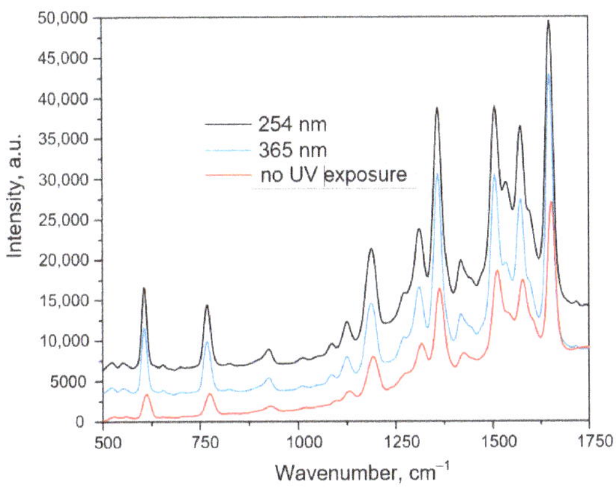

Figure 8. Mean SERS spectra of 10^{-6} M R6G on the $Gr/Ag/ZrO_2$ substrates before and after the UV exposure.

4. Conclusions

In this study, we investigated the synergistic effects of integrating few-layer graphene, silver particles, and zirconia substrates, unveiling a sophisticated platform for augmented SERS activity and photocatalytic properties.

The SERS-active substrates were formed by the chemical "silver mirror" technique to grow the Ag nanoparticles on the zirconia blocks cured in the furnace. The ZrO_2 surface was found to be composed of the polycrystalline grains with sizes varied from 225 nm to ~1.5 μm. In more detail, the zirconia blocks were shown to have tetragonal and monoclinic lattices as followed from the Raman spectra. The silver nanoparticles were characterized to have sizes of nanoscale range between a few nanometers and 90 nm. The film composed of the ~3 graphene layers was grown on copper foil by the CVD method and then wet-transferred to the SERS-active substrates. Graphene-free Ag/ZrO_2 samples provided the enhancement of the Raman scattering from 10^{-6} M R6G, which must be caused by the combined effects from the surface plasmon resonance in the Ag particles and a charge transfer facilitated by zirconium dioxide. However, the SERS signal from the R6G molecules was found to be suppressed by the UV exposure (365 and 254 nm) of the Ag/ZrO_2 samples due to their photocatalytic activity. At the same time, if the Ag/ZrO_2 samples were graphene coated, this prevented R6G decomposition upon the UV treatment and improved SERS activity of the substrates. The 365 nm exposure caused the 40% increase in the 10^{-6} M R6G SERS spectrum, while the 254 nm irradiation led to it rising by 47%. Such an effect is under discussion and can be associated with the different responses of surface and bulk zirconia crystals to the short and long UV wavelengths.

In closing, we argue the following key findings:

- Few-layer graphene on Ag-coated zirconia substrates boosts SERS-activity, attributed to enhanced charge transfer from zirconia and silver to analyte molecules via graphene;
- UV exposure, generally, suppresses SERS signals due to zirconia's photocatalytic activity, but graphene's integration counters this suppression, emphasizing its role in increasing the SERS signal;
- Our results highlight graphene's ability to enhance Raman scattering through its versatility in amplifying charge transfer and strengthening electromagnetic interactions.

From the above findings, future directions could involve refining photocatalytic applications by adjusting UV exposure intensity and duration to optimize both self-cleaning capabilities and the analytical performance of SERS-active substrates. Additionally, the synergy between graphene, silver nanoparticles, and zirconia should be further explored to enhance scalability, stability, and adaptability in real-world applications.

In essence, this research underscores a significant stride towards harnessing the combined potential of graphene, silver nanoparticles, and zirconia for advanced SERS applications.

Author Contributions: Conceptualization, H.B. and D.B.; methodology, A.B., L.D., A.P. and N.K.; software, A.B. and U.S.; validation, A.B., D.L., A.P. and H.B.; formal analysis, A.B.; investigation, L.D. and D.L.; resources, H.B., A.I.P. and D.B.; data curation, H.B. and U.S.; writing—original draft preparation, H.B.; writing—review and editing, H.B., A.P., A.I.P. and D.B.; visualization, H.B. and A.B.; supervision, H.B. and D.B.; project administration, H.B. and D.B.; funding acquisition, H.B. and D.B. All authors have read and agreed to the published version of the manuscript.

Funding: This research was funded by the Latvian Scientific Council grant LZP-2021/1-0464 and the State Program for Scientific Research of Belarus "Photonics and Electronics for Innovations" (subprogram "Photonics and its application", task 1.8; subprogram "Opto- and microwave electronics, task 2.2).

Data Availability Statement: Data are contained within the article.

Acknowledgments: The authors thank Dmitry Karpinsky very much for his kind support during the zirconia block annealing.

Conflicts of Interest: The authors declare no conflict of interest.

References

1. Cao, Y.; Cheng, Y.; Sun, M. Graphene-based SERS for sensor and catalysis. *Appl. Spectrosc. Rev.* **2023**, *58*, 1–38. [CrossRef]
2. Xu, W.; Mao, N.; Zhang, J. Graphene: A Platform for Surface-Enhanced Raman Spectroscopy. *Small* **2013**, *9*, 1206–1224. [CrossRef]
3. Goncalves, G.; Marques, P.A.A.P.; Granadeiro, C.M.; Nogueira, H.I.S.; Singh, M.K.; Grácio, J. Surface Modification of Graphene Nanosheets with Gold Nanoparticles: The Role of Oxygen Moieties at Graphene Surface on Gold Nucleation and Growth. *Chem. Mater.* **2009**, *21*, 4796–4802. [CrossRef]
4. Jasuja, K.; Berry, V. Implantation and Growth of Dendritic Gold Nanostructures on Graphene Derivatives: Electrical Property Tailoring and Raman Enhancement. *ACS Nano* **2009**, *3*, 2358–2366. [CrossRef]
5. Sil, S.; Kuhar, N.; Acharya, S.; Umapathy, S. Is Chemically Synthesized Graphene 'Really' a Unique Substrate for SERS and Fluorescence Quenching? *Sci. Rep.* **2013**, *3*, 3336. [CrossRef] [PubMed]
6. Yu, X.; Cai, H.; Zhang, W.; Li, X.; Pan, N.; Luo, Y.; Wang, X.; Hou, J.G. Tuning Chemical Enhancement of SERS by Controlling the Chemical Reduction of Graphene Oxide Nanosheets. *ACS Nano* **2011**, *5*, 952–958. [CrossRef] [PubMed]
7. Yang, C.; Xu, Y.; Wang, M.; Li, T.; Huo, Y.; Yang, C.; Man, B. Multifunctional paper strip based on GO-veiled Ag nanoparticles with highly SERS sensitive and deliverable properties for high-performance molecular detection. *Opt. Express* **2018**, *26*, 10023–10037. [CrossRef] [PubMed]
8. Zavatski, S.; Khinevich, N.; Girel, K.; Redko, S.; Kovalchuk, N.; Komissarov, I.; Lukashevich, V.; Semak, I.; Mamatkulov, K.; Vorobyeva, M.; et al. Surface Enhanced Raman Spectroscopy of Lactoferrin Adsorbed on Silvered Porous Silicon Covered with Graphene. *Biosensors* **2019**, *9*, 34. [CrossRef]
9. Yi, N.; Zhang, C.; Song, Q.; Xiao, S. A hybrid system with highly enhanced graphene SERS for rapid and tag-free tumor cells detection. *Sci. Rep.* **2016**, *6*, 25134. [CrossRef]
10. Jabłońska, A.; Jaworska, A.; Kasztelan, M.; Berbeć, S.; Pałys, B. Graphene and Graphene Oxide Applications for SERS Sensing and Imaging. *Curr. Med. Chem.* **2019**, *26*, 6878–6895. [CrossRef]
11. Girel, K.; Burko, A.; Zavatski, S.; Barysiuk, A.; Litvinova, K.; Eganova, E.; Tarasov, A.; Novikov, D.; Dubkov, S.; Bandarenka, H. Atomic layer deposition of hafnium oxide on porous silicon to form a template for athermal SERS-active substrates. *Appl. Phys. A* **2023**, *129*, 294. [CrossRef]
12. Song, G.; Cong, S.; Zhao, Z. Defect engineering in semiconductor-based SERS. *Chem. Sci.* **2022**, *13*, 1210–1224. [CrossRef]
13. He, R.; Lai, H.; Wang, S.; Chen, T.; Xie, F.; Chen, Q.; Liu, P.; Chen, J.; Xie, W. Few-layered vdW MoO_3 for sensitive, uniform and stable SERS applications. *Appl. Surf. Sci.* **2020**, *507*, 145116. [CrossRef]
14. Zhao, X.; Wang, W.; Liang, Y.; Fu, J.; Zhu, M.; Shi, H.; Lei, S.; Tao, C. Visible-light-driven charge transfer to significantly improve surface-enhanced Raman scattering (SERS) activity of self-cleaning TiO_2/Au nanowire arrays as highly sensitive and recyclable SERS sensor. *Sens. Actuators B Chem.* **2019**, *279*, 313–319. [CrossRef]
15. Zhao, B.; Liu, H.; Xia, L.; Wang, Z.; Zhang, C. Gold-Coated Flower-Like TiO_2 Microparticles Wrapped with Reduced Graphene Oxide for SERS Monitoring and Photocatalytic Degradation of Organic Pollutants. *ACS Appl. Nano Mater.* **2022**, *5*, 15341–15352. [CrossRef]
16. Panneerselvam, R.; Liu, G.K.; Wang, Y.H.; Liu, J.Y.; Ding, S.Y.; Li, J.F.; Wu, D.Y.; Tian, Z.Q. Surface-Enhanced Raman Spectroscopy: Bottlenecks and Future Directions. *Chem. Commun.* **2017**, *54*, 10–25. [CrossRef]
17. Ling, X.; Xie, L.; Fang, Y.; Xu, H.; Zhang, H.; Kong, J.; Dresselhaus, M.S.; Zhang, J.; Liu, Z. Can Graphene Be Used as a Substrate for Raman Enhancement? *Nano Lett.* **2010**, *10*, 553–561. [CrossRef] [PubMed]
18. Lu, G.; Li, H.; Liusman, C.; Yin, Z.; Wu, S.; Zhang, H. Surface Enhanced Raman Scattering of Ag or Au Nanoparticle-Decorated Reduced Graphene Oxide for Detection of Aromatic Molecules. *Chem. Sci.* **2011**, *2*, 1817–1821. [CrossRef]
19. McPeak, K.M.; Jayanti, S.V.; Kress, S.J.P.; Meyer, S.; Iotti, S.; Rossinelli, A.; Norris, D.J. Plasmonic Films Can Easily Be Better: Rules and Recipes. *ACS Photonics* **2015**, *2*, 326–333. [CrossRef]
20. Ciuparu, D.; Ensuque, A.; Shafeev, G.; Bozon-Verduraz, F. Synthesis and apparent bandgap of nanophase zirconia. *J. Mater. Sci. Lett.* **2000**, *19*, 931–933. [CrossRef]
21. Pandiyan, R.; Vinothkumar, V.; Chen, T.W.; Chen, S.M.; Abinaya, M.; Rwei, S.P.; Hsu, H.Y.; Huang, C.W.; Yu, M.C. Synthesis of Ag@ZrO_2 Nanoparticles: A Sensitive Electrochemical Sensor for Determination of Antibiotic Drug Tinidazole. *Int. J. Electrochem. Sci.* **2022**, *17*, 220414. [CrossRef]
22. Ama, O.M.; Wilson, A.W.; Ray, S.S. Photoelectrochemical Degradation of Methylene Blue Dye under Visible Light Irradiation Using EG/Ag-ZrO_2 Nanocomposite Electrodes. *Int. J. Electrochem. Sci.* **2019**, *14*, 9982–10001. [CrossRef]
23. Shinde, H.M.; Kite, S.V.; Shirke, B.S.; Garadkar, K.M. Eco-Friendly Synthesis of Ag-ZrO_2 Nanocomposites for Degradation of Methylene Blue. *J. Mater. Sci. Mater. Electron.* **2021**, *32*, 14235–14247. [CrossRef]
24. Zhou, L.; Yang, J.; Wang, X.; Song, G.; Lu, F.; You, L.; Li, J. Ag Nanoparticles Decorated Ag@ZrO_2 Composite Nanospheres as Highly Active SERS Substrates for Quantitative Detection of Hexavalent Chromium in Waste Water. *J. Mol. Liq.* **2020**, *319*, 114158. [CrossRef]
25. Alyea, H.N.; Dutton, F.B. *Tested Demonstrations in Chemistry*, 6th ed.; Division of Chemical Education of the American Chemical Society: Easton, PA, USA, 1965.
26. Plasmonics.by. Available online: http://plasmonics.by/en/ (accessed on 2 October 2023).
27. Basahel, S.N.; Ali, T.T.; Mokhtar, M.; Narasimharao, K. Influence of crystal structure of nanosized ZrO_2 on photocatalytic degradation of methyl orange. *Nanoscale Res. Lett.* **2015**, *10*, 73. [CrossRef] [PubMed]

28. Ciszak, C.; Mermoux, M.; Gutierrez, G.; Leprêtre, F.; Duriez, C.; Popa, I.; Fayette, L.; Chevalier, S. Raman spectra analysis of ZrO_2 thermally grown on Zircaloy substrates irradiated with heavy ion: Effects of oxygen isotopic substitution. *J. Raman. Spectrosc.* **2019**, *50*, 425–435. [CrossRef]
29. Li, C.; Li, M. UV Raman spectroscopic study on the phase transformation of ZrO_2, Y_2O_3–ZrO_2 and SO_4^{2-}/ZrO_2. *J. Raman. Spectrosc.* **2002**, *33*, 301–308. [CrossRef]
30. Králik, B.; Chang, E.K.; Louie, S.G. Structural properties and quasiparticle band structure of zirconia. *Phys. Rev. B* **1998**, *57*, 7027. [CrossRef]
31. Emeline, A.; Kataeva, G.V.; Litke, A.S.; Rudakova, A.V.; Ryabchuk, V.K.; Serpone, N. Spectroscopic and Photoluminescence Studies of a Wide Band Gap Insulating Material: Powdered and Colloidal ZrO_2 Sols. *Langmuir* **1998**, *14*, 5011–5022. [CrossRef]
32. Alekperov, O.; Jahangirli, Z.; Paucar, R. First-principles lattice dynamics and Raman scattering in ionic conductor β-Ag_2S: Lattice dynamics and Raman scattering in ionic conductor β-Ag_2S. *Phys. Status Solidi B* **2016**, *253*, 2049–2055. [CrossRef]
33. Xiu, Z.; Zhang, Q.; Puppala, H.L.; Colvin, V.L.; Alvarez, P.J.J. Negligible Particle-Specific Antibacterial Activity of Silver Nanoparticles. *Nano Lett.* **2012**, *12*, 4271–4275. [CrossRef]
34. Greenwood, N.N.; Earnshaw, A. *Chemistry of the Elements*, 2nd ed.; Elsevier: Amsterdam, The Netherlands, 1997; ISBN 9780750633659.
35. Das, A.; Chakraboty, B.; Sood, A. Raman spectroscopy of graphene on different substrates and influence of defects. *Bull. Mater. Sci.* **2008**, *31*, 579–584. [CrossRef]
36. Calizo, I.; Balandin, A.A.; Bao, W.; Miao, F.; Lau, C.N. Temperature Dependence of the Raman Spectra of Graphene and Graphene Multilayers. *Nano Lett.* **2007**, *7*, 2645–2649. [CrossRef]
37. Lee, J.E.; Ahn, G.; Shim, J.; Lee, Y.S.; Ryu, S. Optical separation of mechanical strain from charge doping in graphene. *Nat. Commun.* **2012**, *3*, 1024. [CrossRef]
38. Ke, Q.; Wang, J. Graphene-based materials for supercapacitor electrodes—A review. *J. Mater.* **2016**, *2*, 37–54. [CrossRef]
39. Ferrari, A.; Basko, D. Raman spectroscopy as a versatile tool for studying the properties of graphene. *Nat. Nanotech.* **2013**, *8*, 235–246. [CrossRef] [PubMed]
40. Zhu, S.-E.; Yuan, S.; Janssen, G.C.A.M. Optical transmittance of multilayer graphene. *Europhys. Lett.* **2014**, *108*, 17007. [CrossRef]
41. Nair, R.R.; Blake, P.; Grigorenko, A.N.; Novoselov, K.S.; Booth, T.J.; Stauber, T.; Peres, N.M.; Geim, A.K. Fine Structure Constant Defines Visual Transparency of Graphene. *Science* **2008**, *320*, 1308. [CrossRef]
42. Watanabe, H.; Hayazawa, N.; Inouye, Y.; Kawata, S. DFT Vibrational Calculations of Rhodamine 6G Adsorbed on Silver: Analysis of Tip-Enhanced Raman Spectroscopy. *J. Phys. Chem. B* **2005**, *109*, 5012–5020. [CrossRef] [PubMed]
43. Hildebrandt, P.; Stockburger, M. Surface-Enhanced Resonance Raman Spectroscopy of Rhodamine 6G Adsorbed on Colloidal Silver. *J. Phys. Chem.* **1984**, *88*, 5935–5944. [CrossRef]
44. Rani, V.; Sharma, A.; Kumar, A.; Singh, P.; Thakur, S.; Singh, A.; Le, Q.V.; Nguyen, V.H.; Raizada, P. ZrO_2-Based Photocatalysts for Wastewater Treatment: From Novel Modification Strategies to Mechanistic Insights. *Catalysts* **2022**, *12*, 1418. [CrossRef]

crystals

MDPI

Article

Graphocrown—A Novel, Two-Dimensional Oxocarbon: A Theoretical Study

Mikhail A. Kalinin [1], Maximilian Kriebel [2], Alexander S. Oshchepkov [1,3,*] and Dmitry I. Sharapa [4,*]

1 Organic Chemistry Department, Institute of Chemistry, Martin-Luther-University Halle-Wittenberg, Kurt-Mothes-Straße 2, 06120 Halle, Germany
2 Faculty of Physics, Astronomy, and Informatics, Institute of Physics, Nicolaus Copernicus University, ul. Grudziądzka 5, 87-100 Toruń, Poland
3 Department of Physics, Max Planck Institute for the Science of Light, Staudtstrasse 2, 91058 Erlangen, Germany
4 Institute of Catalysis Research and Technology, Karlsruhe Institute of Technology, Hermann-von-Helmholtz-Platz 1, 76344 Eggenstein-Leopoldshafen, Germany
* Correspondence: aleksandr.oshchepkov@mpl.mpg.de (A.S.O.); dmitry.sharapa@kit.edu (D.I.S.)

Abstract: An innovative 2D-material, graphocrown, was designed and studied. Our graphocrown computations revealed a higher stability compared to previous materials studied with the same generalized C_2O formula. The energetic benefit of the graphocrown formation from benzenehexol was also evaluated. The structure and properties of graphocrowns with various layer arrangements were analysed and compared. In addition, the formation of potassium complexes with the new material was studied. It was found that graphocrown binds potassium better than 18-crown-6, and the intercalation of graphocrown with potassium is more favourable, compared to graphite. Finally, the band structure, as well as the mobility of the charge carriers in the graphocrown, were investigated.

Keywords: 2D-materials; COF; oxocarbon; crowns; intercalation; DFT

Citation: Kalinin, M.A.; Kriebel, M.; Oshchepkov, A.S.; Sharapa, D.I. Graphocrown—A Novel, Two-Dimensional Oxocarbon: A Theoretical Study. *Crystals* **2023**, *13*, 909. https://doi.org/10.3390/cryst13060909

Academic Editor: Jan Macutkevic

Received: 8 May 2023
Revised: 28 May 2023
Accepted: 29 May 2023
Published: 2 June 2023

1. Introduction

The design of promising materials is key to the development of applied science and technology [1]. One such material is graphene, whose theoretical research was initiated by P.R. Wallace in 1947, long before any real samples of the material were available [2]. The unique properties of graphene, predicted and then derived from experiments such as unusual thermal conductivity [3], high electron mobility [4,5], and optical transparency [6] gave a big jump-start to the development of electronics [7]. The graphene family can now be implemented in a wide range of applications in photovoltaics [8] and photodetection [9], painting [10] and sensing [11], drug delivery [12] and medicine [13]. However, unmodified graphene is a semi-metal with a zero gap, and has an extremely low on/off current ratio, which prevents it from being used in semiconductor devices such as field-effect transistors and logic circuits [14]. One of the approaches for tuning the electronic properties would be the introducing of heteroatoms—the formation of heterographenes [15]. For the first period (where the atoms match well the size of carbon and each other), there are several observed or predicted regular 2D materials—g-C_3N_4, [16] CB [17] and h-BN [18]. At the same time, carbon oxygenates are of much less interest, except for graphene oxide [19]—a mysterious irregular material with a not-fully understood structure. However, strong oxidising agents lead to the destruction of the p-p conjugation of the sp^2 carbon network, and to a loss of conductivity in graphene oxides [20]. In addition, the inevitable heterogeneities and defects arising from the chemical modification of graphene can lead to the inevitable loss of the expected properties of the final materials [21]. The irregular distribution of functional groups and coordination centres can also have a negative impact on the properties of graphene oxide/metal composites [22].

Another approach to developing materials is to use relatively simple oxocarbon molecules as a starting structure [23]. The enormous variability in structures/properties of the oxocarbon molecules opens up excellent prospects for the development of materials for electronics engineering that will surpass their existing analogues [24–26]. One of the first such materials was graphether, a new two-dimensional oxocarbon that could be synthesised by assembling dimethyl ether molecules [27]. The graphether had superior thermodynamic stability, a direct bandgap of 2.39 eV and strong light absorption in the UV region [28]. Further investigating the graphether-based nanoribbon, it was found that symmetry plays a key role in the band structure, which can cause an indirect–direct bandgap transition, following parity-oddness in nanoribbons [29]. However, more complex symmetric oxocarbons are not known yet, nor are the coordination properties directly related to their structure.

Inspired by graphether, as well as the coordination properties of crown ethers [30,31], in particular integrated into the graphene flaxes [32,33], we decided to combine these concepts and developed a covalent organic framework (COF) that would consist of graphene-planar crown ethers—the graphocrown. One of the key features of the new material is its preorganisation and fixed pore size. This is particularly important in the fields of selective membrane engineering [34], extraction [35], and catalysis [36]. Graphocrown could theoretically be used more successfully in environmental and medical problems, replacing its pure graphene analogues [37].

Unlike the pristine graphene/graphite, hexagonal boron nitride or graphether, but like the carbon nitride (g-C_3N_4) [16], PSN [38] and CTF [38], graphocrown is a porous material. Interest in such materials has been growing in recent years [39]. The main difference in relation to the experimentally most well-known 2D porous material of a graphene-like structure—g-C_3N_4—is the planarity of graphocrown (discussed further), while g-C_3N_4 has a wave-distorted structure [40].

In this paper we have developed a 2D-material based on the new concept of the graphocrown, and investigated its properties. We believe that this research will create a stimulus for the development of materials science, creating a platform for the interaction of coordination chemistry, supramolecular chemistry, electrochemistry and physics.

2. Materials and Methods

Computational Details

Periodic DFT calculations were performed with PBE functional [41] in the VASP simulation package [42,43]. There was an energy cut-off of 600 eV and quite a dense k-point grid ($N_{kpoints} \times l_{vector} \approx 40$ Å). Dispersion correction of Grimme with zero dumping was applied [44]. More details can be found in Supplementary Materials. Gas phase molecules (benzenehexol, water) were calculated in large cubic cells (30 × 30 × 30 Å).

Cluster-model DFT calculations were performed with ORCA 5.0.4 [45,46] with PBE/def2-TZVP for the smaller systems and r2SCAN-3c [47] for the larger stripes (mentioned in the text).

The semi-empirical MD calculation was performed with the EMPIRE program [48,49] via PM3 Hamiltonian [50], for a better comparison with the discussed literature data.

All the following calculations, with the exception of charge carrier mobilities were performed based on the VASP outputs (wavefunctions, charge densities, etc.) [51–53]. Bader charges were evaluated using the script of the Henkelman group. Bader critical points and noncovalent interactions were generated by the Critic2 code [54,55]. Band structure and DOS were visualised with the help of p4vasp software. VBM and CBM wavefunctions were produced with the LOBSTER software [56,57] and visualised by the VMD [58] and wxDragon programs (wxDragon is copyright, 1994–2012, by Bernhard Eck). See details in Supplementary Materials (SM). The general procedure was previously described in [59] and further technical details are given in Supplementary Materials. The visualisation of the structures was performed with ChemCraft [60].

For calculations of charge carrier mobilities, the local molecular properties (Local Ionization Energy (IEL) [61] and Local Electron Affinity (EAL) [62,63]) were calculated from the molecular orbital densities and their respective energies. These calculations were performed with the EMPIRE program [48,49] and eh5cube code [64] via the semi empirical Hartree–Fock molecular orbital theory, using an AM1 [65] and PM6 [66] Hamiltonians. These local properties allow for the description of the interaction of a charge carrier of complicated quantum mechanical systems into simple scalar potentials embedded in 3-dimensional space. The effective potential for a hole is then given by the local ionization energy $V_{hole}(x) = IEL(x)$ and the effective potential for an electron is given by minus the local electron affinity $V_{electron}(x) = -EAL(x)$. It is possible to perform the propagation of an excess charge carrier using these energy landscapes, which then can reveal information about the conduction characteristics of quantum systems [67,68]. The interpretation of the effective potentials alone can already help to estimate some conductivity characteristics.

Although the proposed graphocrown material belongs to the hexagonal lattice system (the primitive cell shown on the left in Figure 1), for clarity, in this paper we focused on the simulation of the orthorhombic supercell with 2 layers and 72 atoms (Figure 1 on the right).

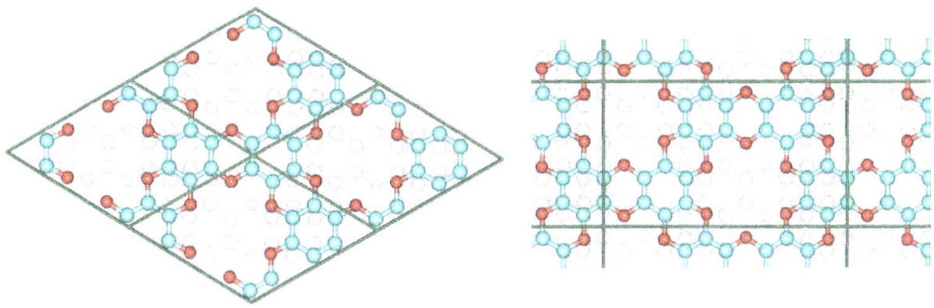

Figure 1. The hexagonal unit cell (**left**) and the orthorhombic supercell (**right**) of bilayer graphocrown. The bottom layer coloured to white for clarity.

3. Results

3.1. Single Layer

Although the single-layer graphocrown (free-hanging graphocrown) is not the main focus of this work, we would like to start discussion from there. While phonon calculations demonstrate the stability of this material, according to work [69] it is not a sufficient criterion. Criticising artificial stabilisation with periodic boundary conditions, the authors of [69] rely on terminated flakes (where edge effects can, and quite often actually do, cause the distortion, as is discussed below). While for chemists the planarity of these fragments should be obvious, we followed the approaches of that work and the requirements of the "topology conservation theorem", to demonstrate that periodic boundary conditions do not introduce any artificial stabilization of the unstable structure.

First of all, according to TCT, "the free-standing constituting fragments (unit cells) must perfectly fit planar low-dimensional space". The dibenzodioxin unit is well known for being perfectly planar, and the minimal crown cycle (Figure 2a) that can be compiled from such dioxins keeps the perfect D_{6h} point group (PBE/def2-TZVP, ORCA). Thus, there is no internal stress perpendicular to the plane to be compensated for.

As the next step, we cut two different stripes from the single-layer graphocrown (in perpendicular directions), terminated them with hydrogens, and tried to bend them (Figure 2b,c). The relaxed scan in respect to the bending/torsional angle was performed in ORCA 5.0.4 with the r^{2}SCAN-3c approach, which we consider as more reliable than the PM3 Hamiltonian intensively used in some work [69]. For consistency, we also cut graphene stripes of same shape. Figure 3 demonstrates the fact that graphocrown stripes can be bent (actually much more easily than graphene stripes), but still have minimal energy, being planar. In fact, due to hydrogen repulsion on the armchair edge, the graphene flakes actually have a destabilizing distortion. This also brings us to the controversy of the approach suggested in work [69]—while we agree that the improper usage of PBC might artificially stabilize nonoptimal conformations, the edge effects of an improperly terminated cluster approach might artificially destabilise planar structures, and the principal role here (for the single layers) is the lability or robustness of the "inner part" of the flake.

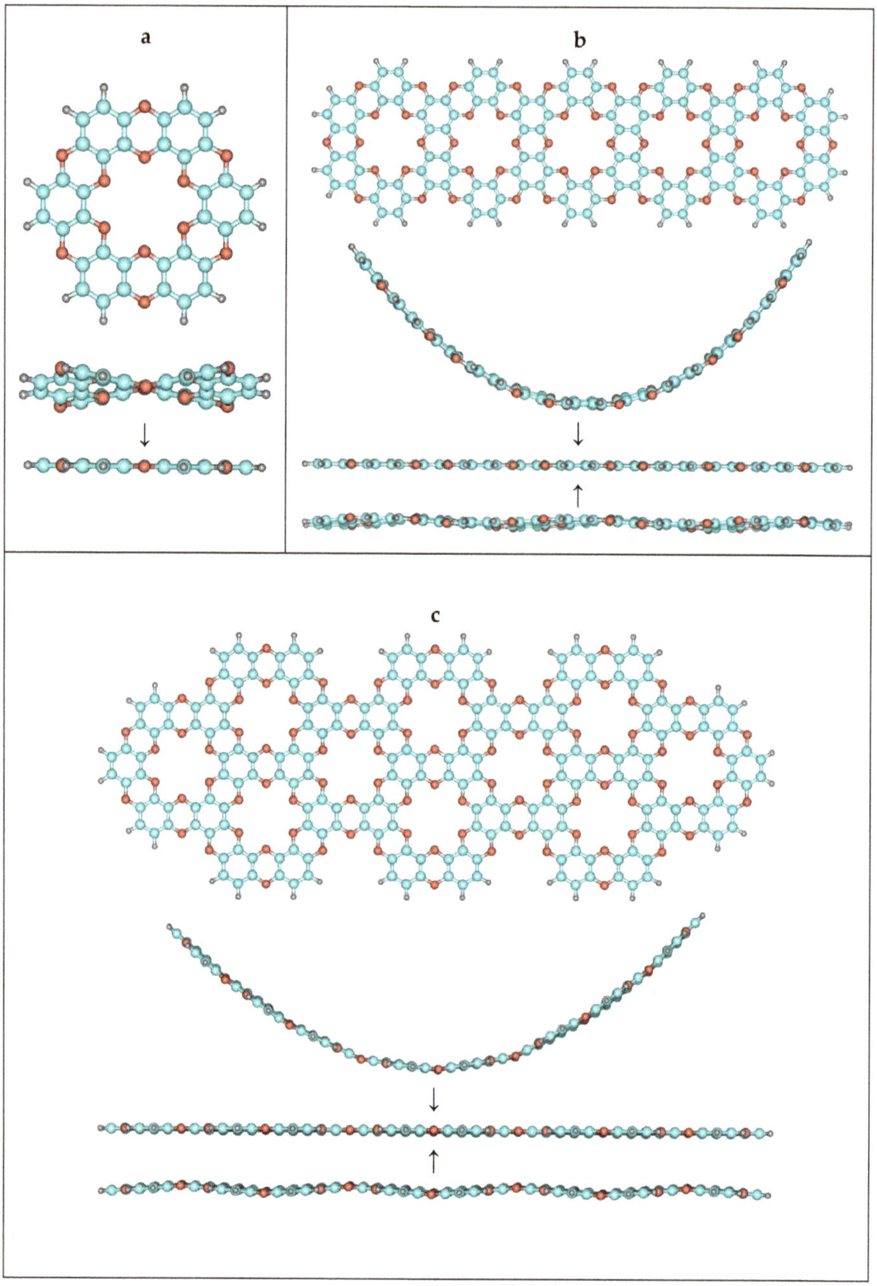

Figure 2. Investigation of the stability of bent and wave clusters. (**a**) primitive unit – "single ring" (**b**) acene-like cluster (**c**) teropyrene-like cluster.

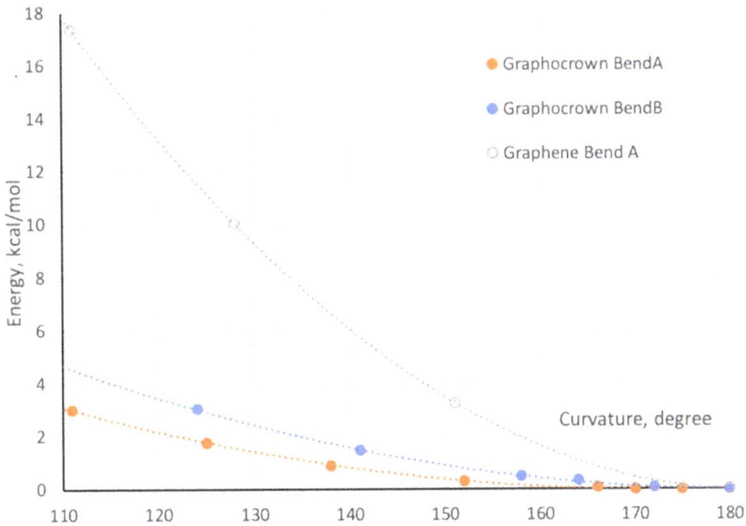

Figure 3. Bending energy of graphocrown and graphene. BendA corresponds to the bending shown in Figure 2b, and BendB corresponds to the bending shown in Figure 2c. For the case of graphene we built a cluster based on structure 2b, all coordinates can be found in Supplementary Materials.

We also generated a number of differently distorted flakes, and performed their optimization. It is important to mention that for some "wave-style" geometries the optimization (smoothing of waves) goes through a very narrow PES, and very tight optimization criteria should be applied to reach the true minima (planar).

Next, following work [69] we performed a PM3 optimization of a large graphocrown flake ($C_{528}H_{52}O_{238}$), which kept the flake planar as well as a 5 ns molecular dynamics simulation with a step of 0.5 fs with the NVT ensemble (details can be found in Supplementary Materials). Quite expectedly, the flake left the planarity (also increasing the total energy), and random concaves and convexities wandered over it (several snapshots can be found in Supplementary Materials). Under no condition should such behaviour be interpreted as material instability. The disruption of the pristine order is a natural feature of MD, and on "free-hanging" epitaxial graphene or boron nitride [70] similar waves or other fluctuating distortions were observed or simulated—the corresponding crystals are perfectly stable, as are the layers within them—and were perfectly planar.

Although the transformation between planar and nonplanar form could not be observed in PBC MD (as is specified in [69]), we built a waved single layer in PBC and performed its optimization in the cell that was able to change shape but keep the volume (ISYF = 4 in VASP). Once again, increase of numerical accuracy results in smoothening the waves.

3.2. Structure and Stability

Each layer of the pristine graphocrown is completely planar, almost independently of the stacking.

The investigated cell contains two layers of graphocrown, and thus several possible laying patterns were tested: "pore-to-pore", "pore-to-ring" and "displaced" (Table 1). As one can see, "displaced" stacking is the most preferable, with quite a difference when compared to the alternatives. This means that layer shift is not a low-barrier process. The exfoliation energy of this material was found to be 2.78 eV per simulation cell (72 atoms) or 116 meV per C_2O unit or 11.9 meV per Å^2 (for a single layer). For comparison, the exfoliation energy calculated for the graphene supercell with two layers and the same number of atoms was 3.49 eV per cell, 48 meV per C atom or 18.4 meV per Å^2 (for a single layer).

Table 1. Comparison of the different stacking possibilities of graphocrown.

Structure	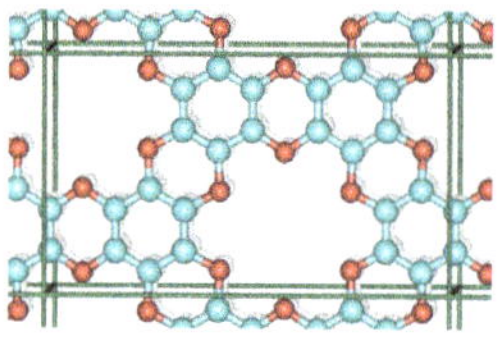 **"Pore-to-Pore"** **2a**	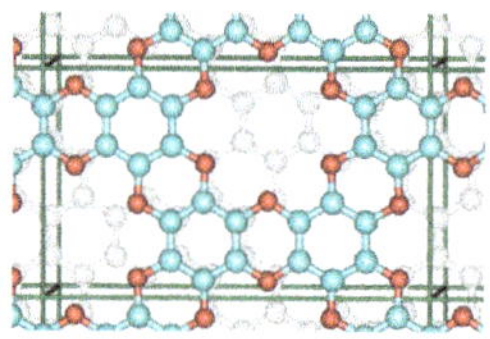 **"Pore-to-Ring"** **2b**	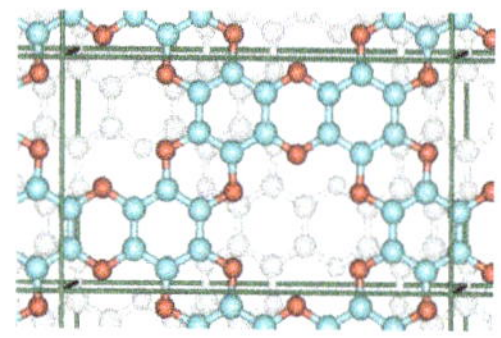 **"Displaced"** **2c**
a (Å)	14.25	14.24	14.24
b (Å)	8.22	8.22	8.22
c (Å)	7.10	6.71	6.50
Relative energy			
per supercell (eV)	0.65	0.3	0.0
per C_2O unit (meV)	26.9	12.4	0.0

The reason for the unfavourable nature of the "pore-to-pore" stacking is the repulsive interaction of the oxygen lone pairs which face each other as the structure remains planar. This can be nicely demonstrated by non-covalent interaction (NCI) plots (Figure 4). While the "displaced" stacking on the interlayer distance of 6.5Å shows just the "unshaped" vdW interactions, the "pore-to-pore" demonstrates the very "well-defined" interaction regions, in particular for the interaction between the oxygens.

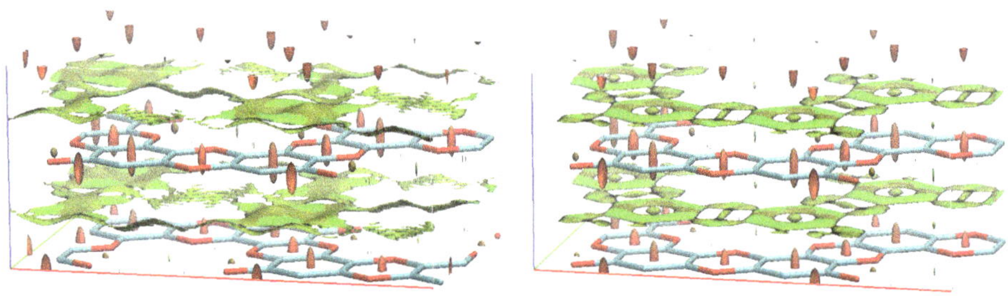

Figure 4. NCI in "displaced" and "pore-to-pore" stackings of graphocrown. The red spindles in the centres of the 6-member rings correspond to known intramolecular repulsive interactions.

Next, we decided to evaluate the stability of our material. Therefore, we compared our graphocrown with the graphether of Liu [27] (which has the same generalized formula, C_2O). It was found that graphocrown is significantly (over 1 eV per C_2O unit) more stable than graphether. Keeping in mind the demonstrated stability of graphether and the lack of obvious pathways for the decomposition of graphocrown, we rely on the reasonable stability of the suggested system.

Furthermore, we compared the stability of graphocrown with respect to the simplest educt—hexahydroxybenzene (benzenehexol)—from which graphocrown could be obtained by straightforward polycondensation:

$$8 \times C_6(OH)_6 = C_{48}O_{24} + 24 \times H_2O$$

where $C_{48}O_{24}$ corresponds to the bulk graphocrown, while benzenehexol and water were calculated in the gas phase. The enthalpy of this reaction is 183 meV per C_2O unit, and thus the reaction is endothermic. To the best of our knowledge, there are no detailed studies of products of the polycondensation of benzenehexol. On the other hand, a conductive two-dimensional MOF was experimentally obtained from hexahydroxybenzene and copper [71].

3.3. Electronic Structure

The band structure and DOS of graphocrown are presented in Figure 5. Please note that there are no discontinuities, but that the last three blocks correspond to edges that could not be included in the main sequence.

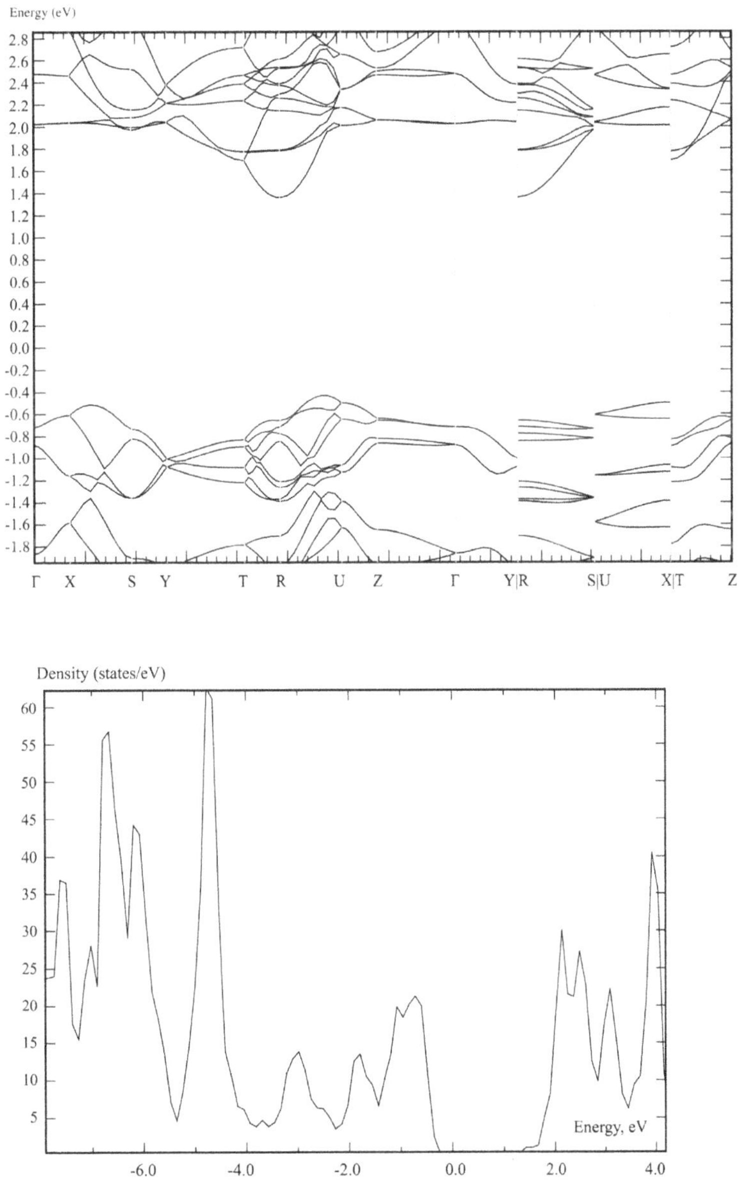

Figure 5. The band structure and DOS of graphocrown calculated on a PBE level.

The PBE-predicted bandgap of graphocrown (1.79 eV) is larger than that of grapheter (0.81 eV), GaSe (1.63 eV) or MoS_2 (1.75 eV), at the same level of theory [72].

For a more detailed insight into the near-band-edge states, the isosurfaces of the electron density of the valence band maximum (VBM) and conduction band minimum (CBM) are shown in Figure 6.

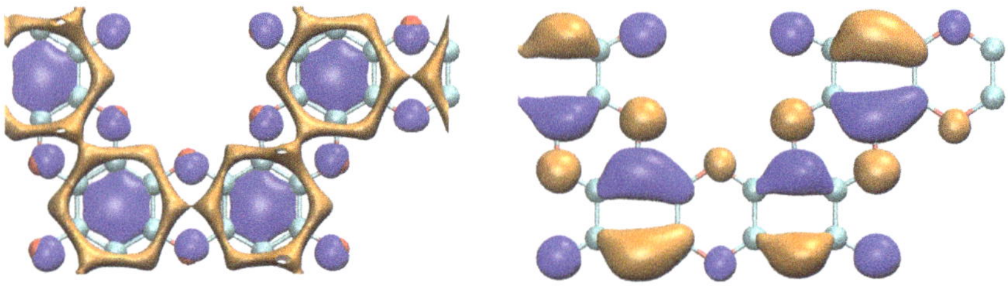

Figure 6. CBM (**left**) and VBM (**right**) at the Γ point of graphocrown. The calculations were performed for the "displaced" bulk, but only one layer is shown, for clarity. Different colours correspond to different phases of the wavefunction; more details can be found in the Supplementary Materials. CBM and VBM calculated at point R are provided in the Supplementary Materials.

One can recognize continuous "channels" in the lowest conduction band, while the isosurface of the highest valence band is "well-isolated". Obtained data indicate a higher electron than hole conductivity.

3.4. Carrier Mobility

The isosurfaces of the local ionisation potential (IEL) and local electron affinity (EAL) are shown in Figure 7.

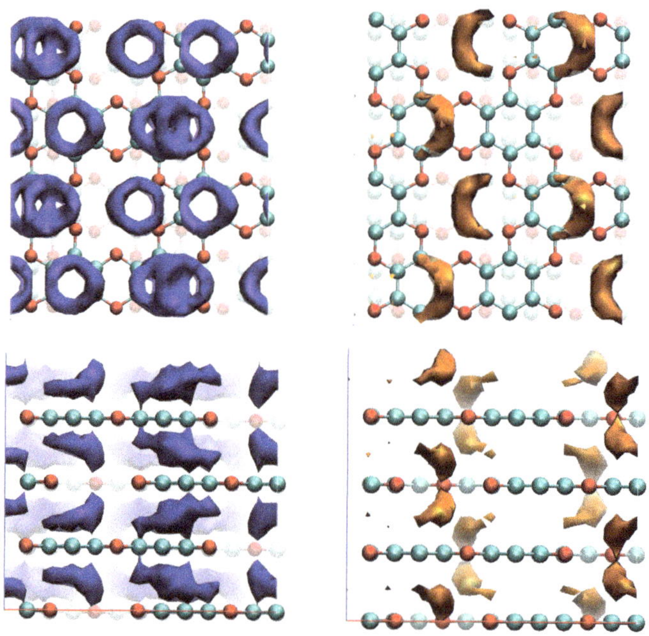

Figure 7. EAL (blue) and IEL (orange) isosurfaces for bulk graphocrown. The strong cueing was applied to differentiate flakes in the front line. Isosurface cutoffs—20 and 390 kcal/mol, respectively.

In the case of the EAL of graphocrown, one can observe a kind of Moire pattern: the benzene rings that have no direct neighbours in layers above or below form "single rings", while those with such neighbours form a more prominent "double ring, which provides a more efficient (and less coherent) conductive channel. For holes, on the other hand, we have an interesting feature: the most attractive sites seem to be not on atoms, but in between the oxygen rings.

No significant difference was observed between the results of the PM6 and AM1 calculations. The performed propagation simulations indicate that, perpendicular to the planes, the hole mobility is lower (by a factor of approximately 0.95) than the electron mobility.

3.5. Intercalation Complexes

"Normal" crowns, such as 18-Crown-6, coordinate the metal cation by rotating the carbon linkers so that the single oxygen pairs are directed toward it, making the complex planar. As was mentioned before, the oxygen's lone pairs in the graphocrown are not pointing to the middle of the pore. Indeed, by placing the potassium cations at different positions along the pore of the pore–pore graphocrown stack, we found that in the 2D-layer plane there was not a minimum, but a transition, state, with one of the possible minima (such coordinated diffusion through the pore would have a barrier just above 1 eV). In fact, the deepest energy minimum for the potassium cations was unexpectedly found in the "pore-to-ring" structure.

While the binding energy of the potassium cation in 18-Crown-6 was found to be about 2.5 eV (dipole correction has to be applied, and the check-up calculations with nonperiodic software (ORCA [45,46], PBE/def2-TZVP) resulted in 2.41 eV), the binding in graphocrown is much stronger (about 4.65 eV per potassium cation in the "pore-to-ring" geometry with monopole correction), see Figure 8. These results are in a good agreement with the stronger binding 1.4 eV in the 18-crown-6 incorporated into the graphene, in comparison to the binding into the "normal" 18-Crown-6 [33].

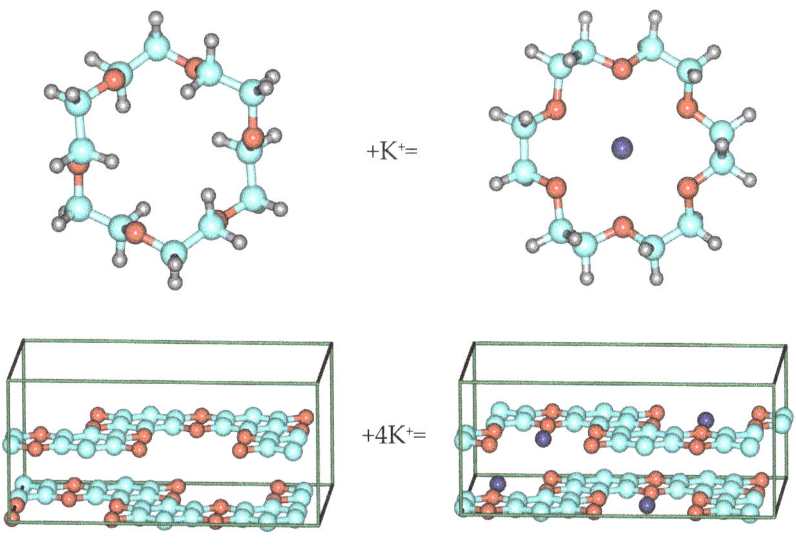

Figure 8. Complexes of 18-Crown-6 and graphocrown with potassium cation.

One can expect that the coordination preference would depend on the size and charge of the cations. Thus, large metal atoms (which can also transform into polycations) are preferred for "pore-to-pore" coordination. Indeed, we observed that, for example, cerium cations strongly contract the layers of the graphocrown, forming almost covalent bonds. It is also possible to speculate about the prospects of using this material for the extraction of heavy-atom cations, and to screen the periodic table for patterns and the most notable cases, but we would like to omit this task and concentrate on another class of material.

Since graphocrown does not form complexes similar to "normal" crowns, and there is no place in a bulk structure for the counterions (and calculation of the polycationic bulk is complicated from a physical, and questionable from a chemical, point of view), we directed our focus to the intercalation compounds similar to KC_8, where (initially neutral) metal atoms are located between the layers of the 2D-material.

We found out that in the neutral cell of $C_{48}O_{24}K_4$, potassium occupies the space between the centers of the pores in the "pore-to-pore" configuration (Figure 9).

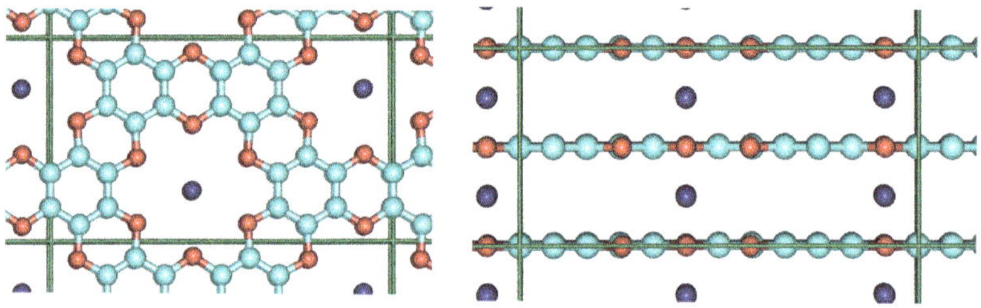

Figure 9. Top and side view of the K-graphocrown intercalation complex.

The Bader charge [53] analysis shows that in this system potassium atoms lose over 90% of their valence electron to the oxocarbon layers (similar to KC_8). While Bader charges (as well as many other charges) are often controversial, we also performed a cluster simulation with a sandwich of a potassium atom between two "single rings" (shown in Figure 2a). We would like to mention the fact that ionisation calculated in this way should likely be overestimated, in comparison to the PBC bulk. For this system on PBE/def2-TZVP we obtained a Mulliken charge on K—+0.71, and a Löwdin charge of +0.18. (See charge distribution in Supplementary Materials). Such a difference should not be confusing, as pure agreement between different charge models is well known [73]. One can also try to make some evaluation relative to other materials, i.e., comparing the results with LiF (controversial, yet sometimes used as example, as Li is definitely a worse electron donor than K, and there are also stronger acceptors than F). From such a comparison, one can see that the K in the K-graphocrown is 10 to 20% (Bader and Mulliken, respectively), less ionic than the Li in LiF. As thousands of different references can be suggested here, and as the results still remain speculative and controversial, no further comparisons were performed here.

A critical point search (QTAIM analysis) showed for each potassium atom twelve bond points between it and the surrounding oxygens (a positive Laplacian and "field value" of 0.006, which should correspond to an extremely weak hydrogen bond), and two bond points towards the surrounding potassium atoms (a "field value" of 0.003), Figure 10. Typical "field values" for C-C bonds in the system are about 0.2–0.3.

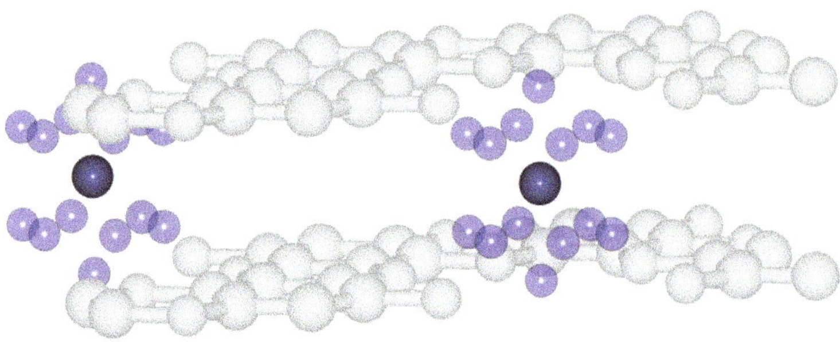

Figure 10. Critical bond points (QTAIM) in potassium intercalated graphocrown.

Finally, we performed a comparison of K-graphite (KC_8) and our K-graphocrown-intercalate in terms of their relative stability. For this purpose, we have constructed the following equation:

$$\text{K-graphite} + \text{graphocrown} \leftrightarrow \text{graphite} + \text{K-graphocrown}$$

All cell energies were scaled by one potassium atom (i.e., a K-graphite cell includes eight potassium atoms). Surprisingly, this migration (from KC_8 to K-graphocrown) appears to be energetically favorable (0.6 eV per each K-atom). While KC_8 is quite stable under certain conditions (being, however, a very active reducing agent), we assume that K-graphocrown (and other metal-intercalates) could be the observable compounds.

An interesting point is the question of the existence of a charge-carrier transfer channel through the metal atoms of the structure—an electrical wire 1 atom thick.

Contrary to the consistent results for pristine graphocrown, in the case of the potassium intercalate, the PM6 and AM1 Hamiltonians showed different results for EAL (Figure 11).

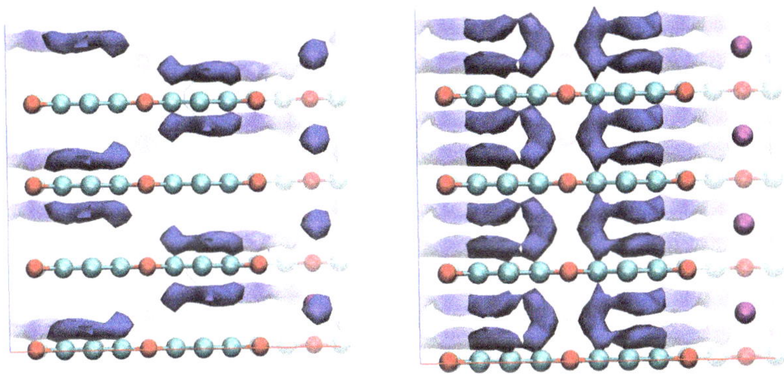

Figure 11. EAL with PM6 and AM1 Hamiltonians. Visualisation cutoffs: for PM6—45 kcal/mol, for AM1—30 kcal/mol.

While AM1 shows no energy minima on the potassium atoms, and thus no electron conduction channel through them, PM6 demonstrates a "coherent channel" going through this metal rod.

To the best of our knowledge, this is the first example where a change in the Hamiltonian has had a qualitative effect on EAL or IEL.

As a final remark regarding intercalation systems, we would like to mention the fact that such systems typically have a high number of low frequencies (especially for intercalated atoms), and to speculate on a significant role for anharmonicity. Since there is no convenient black-box method for accurate evaluation of anharmonicity, we prefer to refrain from making any ill-defined speculation on the entropy-related part.

3.6. Related Materials

Before coming to the end of this manuscript on the graphocrown crystal, we would like to spend a few more paragraphs on a discussion of other materials with similar or related structures.

The possibility of bending the graphocrown similarly to graphene opens the way to graphocrown nanotubes (Figure 12a). The formal polycondensation reaction of 1,3,5-triformylbenzene should end with a graphocrown analogue with oxygen atoms shifted to the methylene (CH_2) groups—a framework hydrocarbon with a primitive unit ($>C\text{-}CH_2\text{-})_n$ (Figure 12b). Further oxidation (dehydrogenation) of this structure could result in a graphocrown analogue with CH groups at the graphocrown site (Figure 12c). This structure could also be classified as a hydrogen-terminated extremely porous graphene. An interesting feature of this structure is that, although the structure can be nicely split into independent 18-membered aromatic rings that go around the pores, a similar attempt to split into Clar's sextets (of benzene rings) would end up in a polyradical system. Apart from graphether and graphene epoxide, one more C_2O polymorph is possible, at least theoretically: polyether of polyene-polyol (Figure 12d). Although this material has an extremely simple primitive cell and fully replicates graphocrown in types of bonds (each oxygen connected with two sp^2 carbons, each carbon connected with one oxygen and two of the same carbons), we have significant doubts about the stability of such a material. One can also try to build some other 2D carbon oxide structures with pores of a similar or larger size, for example using "carbon-rich" per-hydroxy-naphthalene and "oxygen-rich" "tetrahydroxy-ethylene" units (Figure 12e–h). We would like to underline the fact that we are not stating the planarity of any of the structures (12b–h), neither in terms of their stability or of any other property. In Supplementary Materials we provide the PBE-optimised (VASP) geometry of the 12 × 12 nanotube and polyene-polyol, as it might take some time for the engaged reader to build these structures. The remaining structures can be easily constructed from the graphene layer.

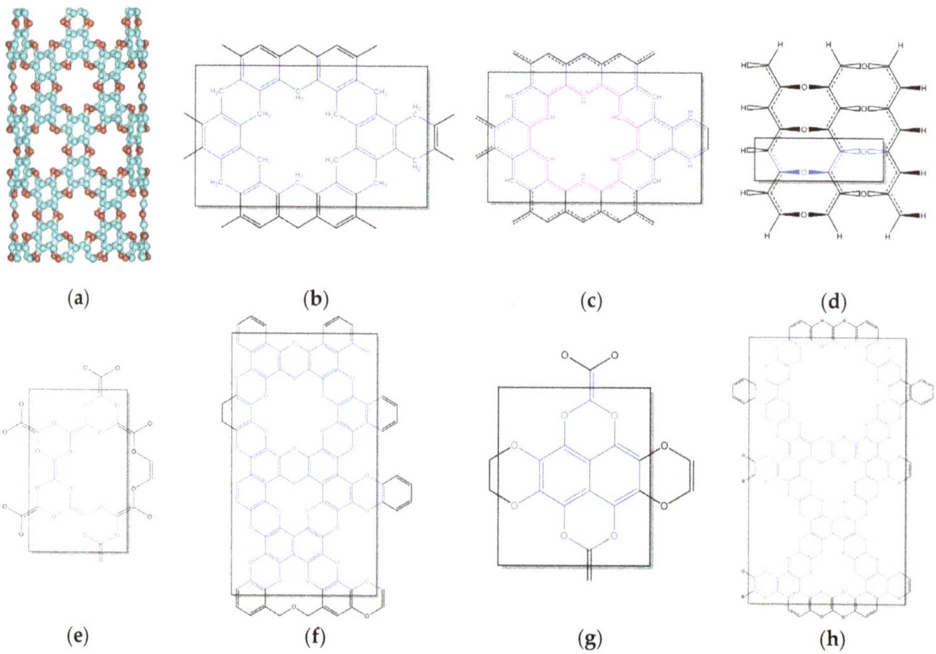

Figure 12. C_2O polymorphs and derivatives. (**a**) 12 × 12 nanotube, (**b**) 1,3,5-triformylbenzene polycondensate, (**c**) porous graphene, (**d**) polyether of polyene-polyol, (**e–h**) different polycondensates of "tetrahydroxy-ethylene" with benzenehexol and perhydroxy-naphtalene.

4. Conclusions

In this work we designed a new 2D-material—graphocrown—and studied its properties in comparison to known or previously proposed materials. Graphocrown was found to be more stable than graphether, despite the fact that its formation from benzenehexol was assumed to be an endothermic process. We found that the most favourable (displaced) stacking of monolayers was in the bulk material, and explained it through the non-covalent interactions. It was demonstrated that graphocrown binds potassium cations much more strongly than 18-crown-6 itself, and even more strongly than "crown-in-graphene". Furthermore, we showed that, similarly to graphene, graphocrown can form intercalation compounds with potassium (and potentially many other alkaline and non-alkaline metals), forming one-dimensional metal rods in the pores. In this process (according to the QTAIM approach), potassium atoms predominantly lose their electrons. The band structure as well as the charge-carrier mobility of graphocrown were studied. We hope that the results of this study will stimulate the development of a next generation of promising materials.

Supplementary Materials: The following supporting information can be downloaded at https://www.mdpi.com/article/10.3390/cryst13060909/s1. Additional details of performed calculations and postprocessing as well as coordinates of discussed systems can be found in Supplementary Materials.

Author Contributions: The research was conceptualized and proposed by D.I.S.; the methodology was discussed by D.I.S., M.K., M.A.K. and A.S.O.; DFT calculations were performed by D.I.S. and M.A.K.; calculations of charge-carrier mobilities were performed by M.K.; single-and multi-reference simulations were performed by D.I.S.; the supervision and project administration was by A.S.O. and D.I.S.; results were visualized by D.I.S. and M.K.; the original draft was written by D.I.S. and A.S.O.; review and editing were carried out by D.I.S., M.A.K. and A.S.O. All authors have read and agreed to the published version of the manuscript.

Funding: This research received no external funding.

Informed Consent Statement: Not applicable.

Data Availability Statement: Not applicable.

Acknowledgments: D.I.S acknowledge the support of Baden-Württemberg through bwHPC (bwUniCluster) and the German Research Foundation (DFG) through grant no INST 40/575-1 FUGG (JUSTUS 2 cluster).

Conflicts of Interest: The authors declare no conflict of interest.

References

1. Ares, P.; Novoselov, K.S. Recent Advances in Graphene and Other 2D Materials. *Nano Mater. Sci.* **2022**, *4*, 3–9. [CrossRef]
2. Wallace, P.R. The Band Theory of Graphite. *Phys. Rev.* **1947**, *71*, 622–634. [CrossRef]
3. Balandin, A.A.; Ghosh, S.; Bao, W.; Calizo, I.; Teweldebrhan, D.; Miao, F.; Lau, C.N. Superior Thermal Conductivity of Single-Layer Graphene. *Nano Lett.* **2008**, *8*, 902–907. [CrossRef] [PubMed]
4. Blase, X.; Benedict, L.X.; Shirley, E.L.; Louie, S.G. Hybridization Effects and Metallicity in Small Radius Carbon Nanotubes. *Phys. Rev. Lett.* **1994**, *72*, 1878–1881. [CrossRef]
5. Zhang, Y.; Tang, T.-T.; Girit, C.; Hao, Z.; Martin, M.C.; Zettl, A.; Crommie, M.F.; Shen, Y.R.; Wang, F. Direct Observation of a Widely Tunable Bandgap in Bilayer Graphene. *Nature* **2009**, *459*, 820–823. [CrossRef]
6. Blake, P.; Brimicombe, P.D.; Nair, R.R.; Booth, T.J.; Jiang, D.; Schedin, F.; Ponomarenko, L.A.; Morozov, S.V.; Gleeson, H.F.; Hill, E.W.; et al. Graphene-Based Liquid Crystal Device. *Nano Lett.* **2008**, *8*, 1704–1708. [CrossRef]
7. Urade, A.R.; Lahiri, I.; Suresh, K.S. Graphene Properties, Synthesis and Applications: A Review. *JOM* **2023**, *75*, 614–630. [CrossRef]
8. Taheri, B.; Yaghoobi Nia, N.; Agresti, A.; Pescetelli, S.; Ciceroni, C.; Del Rio Castillo, A.E.; Cinà, L.; Bellani, S.; Bonaccorso, F.; Di Carlo, A. Graphene-Engineered Automated Sprayed Mesoscopic Structure for Perovskite Device Scaling-Up. *2D Mater.* **2018**, *5*, 045034. [CrossRef]
9. Iqbal, M.A.; Anwar, N.; Malik, M.; Al-Bahrani, M.; Islam, M.R.; Choi, J.R.; Pham, P.V.; Liu, X. Nanostructures/Graphene/Silicon Junction-Based High-Performance Photodetection Systems: Progress, Challenges, and Future Trends. *Adv. Mater. Interfaces* **2023**, *10*, 2202208. [CrossRef]
10. Fu, P.; Teri, G.-L.; Chao, X.-L.; Li, J.; Li, Y.-H.; Yang, H. Modified Graphene-FEVE Composite Coatings: Application in the Repair of Ancient Architectural Color Paintings. *Coatings* **2020**, *10*, 1162. [CrossRef]
11. Pang, J.; Peng, S.; Hou, C.; Zhao, H.; Fan, Y.; Ye, C.; Zhang, N.; Wang, T.; Cao, Y.; Zhou, W.; et al. Applications of Graphene in Five Senses, Nervous System, and Artificial Muscles. *ACS Sensors* **2023**, *8*, 482–514. [CrossRef]
12. Ghamkhari, A.; Abbaspour-Ravasjani, S.; Talebi, M.; Hamishehkar, H.; Hamblin, M.R. Development of a Graphene Oxide-Poly Lactide Nanocomposite as a Smart Drug Delivery System. *Int. J. Biol. Macromol.* **2021**, *169*, 521–531. [CrossRef] [PubMed]
13. Nasker, S.S.; Ajayan, P.M.; Nayak, S. Emerging Trends and Future Direction of Graphene Family of Materials as Potential Antimicrobials: A Critical Review. *ACS Mater. Lett.* **2023**, *5*, 673–693. [CrossRef]
14. Li, X.; Wang, X.; Zhang, L.; Lee, S.; Dai, H. Chemically Derived, Ultrasmooth Graphene Nanoribbon Semiconductors. *Science* **2008**, *319*, 1229–1232. [CrossRef]
15. Zhao, W.; Papp, C.; Steinrück, H.-P. Heterographenes. In *Encyclopedia of Polymeric Nanomaterials*; Springer: Berlin/Heidelberg, Germany, 2014; pp. 1–15.
16. Thomas, A.; Fischer, A.; Goettmann, F.; Antonietti, M.; Müller, J.-O.; Schlögl, R.; Carlsson, J.M. Graphitic Carbon Nitride Materials: Variation of Structure and Morphology and Their Use as Metal-Free Catalysts. *J. Mater. Chem.* **2008**, *18*, 4893. [CrossRef]

17. Tian, X.; Xuan, X.; Yu, M.; Mu, Y.; Lu, H.-G.; Zhang, Z.; Li, S.-D. Predicting Two-Dimensional Semiconducting Boron Carbides. *Nanoscale* **2019**, *11*, 11099–11106. [CrossRef]
18. Watanabe, K.; Taniguchi, T.; Kanda, H. Direct-Bandgap Properties and Evidence for Ultraviolet Lasing of Hexagonal Boron Nitride Single Crystal. *Nat. Mater.* **2004**, *3*, 404–409. [CrossRef]
19. Compton, O.C.; Nguyen, S.T. Graphene Oxide, Highly Reduced Graphene Oxide, and Graphene: Versatile Building Blocks for Carbon-Based Materials. *Small* **2010**, *6*, 711–723. [CrossRef]
20. Kim, F.; Cote, L.J.; Huang, J. Graphene Oxide: Surface Activity and Two-Dimensional Assembly. *Adv. Mater.* **2010**, *22*, 1954–1958. [CrossRef]
21. Gutiérrez-Cruz, A.; Ruiz-Hernández, A.R.; Vega-Clemente, J.F.; Luna-Gazcón, D.G.; Campos-Delgado, J. A Review of Top-down and Bottom-up Synthesis Methods for the Production of Graphene, Graphene Oxide and Reduced Graphene Oxide. *J. Mater. Sci.* **2022**, *57*, 14543–14578. [CrossRef]
22. Bharatiya, D.; Parhi, B.; Sahu, H.; Swain, S.K. Factors Influencing the Dielectric Properties of GO/MO Nanocomposites: Review. *J. Mater. Sci. Mater. Electron.* **2023**, *34*, 452. [CrossRef]
23. Cai, Y.; Pan, Y.; Liu, L.; Zhang, T.; Liang, C.; Mou, X.; Ye, X.; Wang, W.; Dong, X. Succinct Croconic Acid-Based near-Infrared Functional Materials for Biomedical Applications. *Coord. Chem. Rev.* **2023**, *474*, 214865. [CrossRef]
24. Zhang, Z.; Ma, J.; Zhang, R.; Rajagopalan, R.; Tang, Y.; Ren, Y.; Wang, H. Oxocarbons Electrode Materials for Alkali Ion Batteries: Challenges, Strategies and Development. *Batter. Supercaps* **2021**, *4*, 1791–1802. [CrossRef]
25. Colmenero, F. Anomalous Mechanical Behavior of the Deltic, Squaric and Croconic Cyclic Oxocarbon Acids. *Mater. Res. Express* **2019**, *6*, 045610. [CrossRef]
26. Liu, L.; Cheng, B.; Yang, Z.; Wang, H.; Yue, C.; Hu, F. Oxocarbon Organic Conjugated Compounds for Lithium-Ion Batteries and Solar Cells: Progress and Perspectives. *Curr. Org. Chem.* **2020**, *24*, 200–215. [CrossRef]
27. Zhu, G.-L.; Ye, X.-J.; Liu, C.-S. Graphether: A Two-Dimensional Oxocarbon as a Direct Wide-Band-Gap Semiconductor with High Mechanical and Electrical Performances. *Nanoscale* **2019**, *11*, 22482–22492. [CrossRef] [PubMed]
28. Yang, X.; Han, D.; Fan, H.; Wang, M.; Du, M.; Wang, X. First-Principles Calculations of Phonon Behaviors in Graphether: A Comparative Study with Graphene. *Phys. Chem. Chem. Phys.* **2021**, *23*, 123–130. [CrossRef]
29. Jiang, Y.; Guo, Y.; Zeng, H.; Lin, L.; Yan, X. Symmetry-Dependent Electronic Structure Transition in Graphether Nanoribbons. *AIP Adv.* **2022**, *12*, 015118. [CrossRef]
30. Liu, Z.; Nalluri, S.K.M.; Stoddart, J.F. Surveying Macrocyclic Chemistry: From Flexible Crown Ethers to Rigid Cyclophanes. *Chem. Soc. Rev.* **2017**, *46*, 2459–2478. [CrossRef]
31. Oshchepkov, A.; Oshchepkov, M.; Arkhipova, A.; Panchenko, P.; Fedorova, O. Synthesis of 4-Nitro-N-Phenyl-1,8-Naphthalimide Annulated to Thia- and Azacrown Ether Moieties. *Synthesis* **2017**, *49*, 2231–2240. [CrossRef]
32. Fang, A.; Kroenlein, K.; Riccardi, D.; Smolyanitsky, A. Highly Mechanosensitive Ion Channels from Graphene-Embedded Crown Ethers. *Nat. Mater.* **2019**, *18*, 76–81. [CrossRef]
33. Guo, J.; Lee, J.; Contescu, C.I.; Gallego, N.C.; Pantelides, S.T.; Pennycook, S.J.; Moyer, B.A.; Chisholm, M.F. Crown Ethers in Graphene. *Nat. Commun.* **2014**, *5*, 5389. [CrossRef]
34. Pandey, R.P.; Kallem, P.; Hegab, H.M.; Rasheed, P.A.; Banat, F.; Hasan, S.W. Cross-Linked Laminar Graphene Oxide Membranes for Wastewater Treatment and Desalination: A Review. *J. Environ. Manag.* **2022**, *317*, 115367. [CrossRef]
35. Egorova, B.V.; Oshchepkov, M.S.; Fedorov, Y.V.; Fedorova, O.A.; Budylin, G.S.; Shirshin, E.A.; Kalmykov, S.N. Complexation of Bi^{3+}, Ac^{3+}, Y^{3+}, Lu^{3+}, La^{3+} and Eu^{3+} with Benzo-Diaza-Crown Ether with Carboxylic Pendant Arms. *Radiochim. Acta* **2016**, *104*, 555–565. [CrossRef]
36. Pemberton, B.C.; Raghunathan, R.; Volla, S.; Sivaguru, J. From Containers to Catalysts: Supramolecular Catalysis within Cucurbiturils. *Chem. A Eur. J.* **2012**, *18*, 12178–12190. [CrossRef] [PubMed]
37. An, Y.-C.; Gao, X.-X.; Jiang, W.-L.; Han, J.-L.; Ye, Y.; Chen, T.-M.; Ren, R.-Y.; Zhang, J.-H.; Liang, B.; Li, Z.-L.; et al. A Critical Review on Graphene Oxide Membrane for Industrial Wastewater Treatment. *Environ. Res.* **2023**, *223*, 115409. [CrossRef] [PubMed]
38. Lim, J.S.; Kim, G. First-Principles Modeling of Water Permeation through Periodically Porous Graphene Derivatives. *J. Colloid Interface Sci.* **2019**, *538*, 367–376. [CrossRef]
39. Liu, T.; Ding, J.; Su, Z.; Wei, G. Porous Two-Dimensional Materials for Energy Applications: Innovations and Challenges. *Mater. Today Energy* **2017**, *6*, 79–95. [CrossRef]
40. Gracia, J.; Kroll, P. Corrugated Layered Heptazine-Based Carbon Nitride: The Lowest Energy Modifications of C3N4 Ground State. *J. Mater. Chem.* **2009**, *19*, 3013. [CrossRef]
41. Perdew, J.P.; Burke, K.; Ernzerhof, M. Generalized Gradient Approximation Made Simple. *Phys. Rev. Lett.* **1996**, *77*, 3865–3868. [CrossRef]
42. Kresse, G.; Furthmüller, J. Efficiency of Ab-Initio Total Energy Calculations for Metals and Semiconductors Using a Plane-Wave Basis Set. *Comput. Mater. Sci.* **1996**, *6*, 15–50. [CrossRef]
43. Kresse, G.; Furthmüller, J. Efficient Iterative Schemes for Ab Initio Total-Energy Calculations Using a Plane-Wave Basis Set. *Phys. Rev. B* **1996**, *54*, 11169–11186. [CrossRef] [PubMed]
44. Grimme, S.; Antony, J.; Ehrlich, S.; Krieg, H. A Consistent and Accurate Ab Initio Parametrization of Density Functional Dispersion Correction (DFT-D) for the 94 Elements H-Pu. *J. Chem. Phys.* **2010**, *132*, 154104. [CrossRef]
45. Neese, F. The ORCA Program System. *WIREs Comput. Mol. Sci.* **2012**, *2*, 73–78. [CrossRef]

46. Neese, F. Software Update: The ORCA Program System, Version 4.0. *WIREs Comput. Mol. Sci.* **2018**, *8*, e1327. [CrossRef]
47. Grimme, S.; Hansen, A.; Ehlert, S.; Mewes, J.-M. R 2 SCAN-3c: A "Swiss Army Knife" Composite Electronic-Structure Method. *J. Chem. Phys.* **2021**, *154*, 064103. [CrossRef]
48. Hennemann, M.; Clark, T. EMPIRE: A Highly Parallel Semiempirical Molecular Orbital Program: 1: Self-Consistent Field Calculations. *J. Mol. Model.* **2014**, *20*, 2331. [CrossRef]
49. Margraf, J.T.; Hennemann, M.; Meyer, B.; Clark, T. EMPIRE: A Highly Parallel Semiempirical Molecular Orbital Program: 2: Periodic Boundary Conditions. *J. Mol. Model.* **2015**, *21*, 144. [CrossRef]
50. Stewart, J.J.P. Optimization of Parameters for Semiempirical Methods IV: Extension of MNDO, AM1, and PM3 to More Main Group Elements. *J. Mol. Model.* **2004**, *10*, 155–164. [CrossRef]
51. Tang, W.; Sanville, E.; Henkelman, G. A Grid-Based Bader Analysis Algorithm without Lattice Bias. *J. Phys. Condens. Matter* **2009**, *21*, 084204. [CrossRef]
52. Sanville, E.; Kenny, S.D.; Smith, R.; Henkelman, G. Improved Grid-Based Algorithm for Bader Charge Allocation. *J. Comput. Chem.* **2007**, *28*, 899–908. [CrossRef] [PubMed]
53. Henkelman, G.; Arnaldsson, A.; Jónsson, H. A Fast and Robust Algorithm for Bader Decomposition of Charge Density. *Comput. Mater. Sci.* **2006**, *36*, 354–360. [CrossRef]
54. Otero-de-la-Roza, A.; Blanco, M.A.; Pendás, A.M.; Luaña, V. Critic: A New Program for the Topological Analysis of Solid-State Electron Densities. *Comput. Phys. Commun.* **2009**, *180*, 157–166. [CrossRef]
55. Otero-de-la-Roza, A.; Johnson, E.R.; Luaña, V. Critic2: A Program for Real-Space Analysis of Quantum Chemical Interactions in Solids. *Comput. Phys. Commun.* **2014**, *185*, 1007–1018. [CrossRef]
56. Maintz, S.; Deringer, V.L.; Tchougréeff, A.L.; Dronskowski, R. LOBSTER: A Tool to Extract Chemical Bonding from Plane-Wave Based DFT. *J. Comput. Chem.* **2016**, *37*, 1030–1035. [CrossRef]
57. Nelson, R.; Ertural, C.; George, J.; Deringer, V.L.; Hautier, G.; Dronskowski, R. LOBSTER: Local Orbital Projections, Atomic Charges, and Chemical-bonding Analysis from Projector-augmented-wave-based Density-functional Theory. *J. Comput. Chem.* **2020**, *41*, 1931–1940. [CrossRef]
58. Humphrey, W.; Dalke, A.; Schulten, K. VMD: Visual Molecular Dynamics. *J. Mol. Graph.* **1996**, *14*, 33–38. [CrossRef]
59. Akhmetov, V.; Feofanov, M.; Sharapa, D.I.; Amsharov, K. Alumina-Mediated π-Activation of Alkynes. *J. Am. Chem. Soc.* **2021**, *143*, 15420–15426. [CrossRef]
60. Chemcraft-Graphical Software for Visualization of Quantum Chemistry Computations. Available online: https://www.chemcraftprog.com (accessed on 30 May 2023).
61. Sjoberg, P.; Murray, J.S.; Brinck, T.; Politzer, P.A. Average Local Ionization Energies on the Molecular Surfaces of Aromatic Systems as Guides to Chemical Reactivity. *Can. J. Chem.* **1990**, *68*, 1440–1443. [CrossRef]
62. Ehresmann, B.; Martin, B.; Horn, A.H.C.; Clark, T. Local Molecular Properties and Their Use in Predicting Reactivity. *J. Mol. Model.* **2003**, *9*, 342–347. [CrossRef] [PubMed]
63. Clark, T. The Local Electron Affinity for Non-Minimal Basis Sets. *J. Mol. Model.* **2010**, *16*, 1231–1238. [CrossRef]
64. Hennemann, M. EH5cube. 2019. Available online: https://www.ceposinsilico.de/products/empire.htm (accessed on 30 May 2023).
65. Dewar, M.J.S.; Zoebisch, E.G.; Healy, E.F.; Stewart, J.J.P. Development and Use of Quantum Mechanical Molecular Models. 76. AM1: A New General Purpose Quantum Mechanical Molecular Model. *J. Am. Chem. Soc.* **1985**, *107*, 3902–3909. [CrossRef]
66. Stewart, J.J.P. Optimization of Parameters for Semiempirical Methods V: Modification of NDDO Approximations and Application to 70 Elements. *J. Mol. Model.* **2007**, *13*, 1173–1213. [CrossRef] [PubMed]
67. Kriebel, M.; Sharapa, D.; Clark, T. Charge Transport in Organic Materials: Norm-Conserving Imaginary Time Propagation with Local Ionization Energy as the External Potential. *J. Chem. Theory Comput.* **2017**, *13*, 6308–6316. [CrossRef] [PubMed]
68. Kriebel, M.; Hennemann, M.; Beierlein, F.R.; Medina, D.D.; Bein, T.; Clark, T. Propagation of Holes and Electrons in Metal–Organic Frameworks. *J. Chem. Inf. Model.* **2019**, *59*, 5057–5064. [CrossRef] [PubMed]
69. Avramov, P.V.; Kuklin, A.V. Topological and Quantum Stability of Low-Dimensional Crystalline Lattices with Multiple Nonequivalent Sublattices. *New J. Phys.* **2022**, *24*, 103015. [CrossRef]
70. Talla, J.A.; Almahmoud, E.A.; Abu-Farsakh, H. Rippling Effect on the Electrical Properties of Boron Nitride Monolayer: Density Functional Theory. *Semiconductors* **2021**, *55*, 696–703. [CrossRef]
71. Park, J.; Hinckley, A.C.; Huang, Z.; Feng, D.; Yakovenko, A.A.; Lee, M.; Chen, S.; Zou, X.; Bao, Z. Synthetic Routes for a 2D Semiconductive Copper Hexahydroxybenzene Metal–Organic Framework. *J. Am. Chem. Soc.* **2018**, *140*, 14533–14537. [CrossRef]
72. Zhou, J.; Shen, L.; Costa, M.D.; Persson, K.A.; Ong, S.P.; Huck, P.; Lu, Y.; Ma, X.; Chen, Y.; Tang, H.; et al. 2DMatPedia, an Open Computational Database of Two-Dimensional Materials from Top-down and Bottom-up Approaches. *Sci. Data* **2019**, *6*, 86. [CrossRef]
73. Frank, J. *Introduction to Computational Chemistry*, 2nd ed.; John Wiley & Sons Ltd.: Hoboken, NJ, USA, 2007.

crystals

MDPI

Article

Adsorption and Sensing Properties of Formaldehyde on Chemically Modified Graphene Surfaces

Lunwei Yang [1,2,3,4], Wei Xiao [2,3,4,*], Jianwei Wang [2,3,4,*], Xiaowu Li [1] and Ligen Wang [2,3,4,*]

1 Key Laboratory for Anisotropy and Texture of Materials, Department of Materials Physics and Chemistry, School of Materials Science and Engineering, Ministry of Education, Northeastern University, Shenyang 110819, China; waterloortoy@stumail.neu.edu.cn (L.Y.); xwli@mail.neu.edu.cn (X.L.)
2 State Key Laboratory of Nonferrous Metals and Processes, GRIMN Group Co., Ltd., Beijing 100088, China
3 GRIMAT Engineering Institute Co., Ltd., Beijing 101407, China
4 General Research Institute for Nonferrous Metals, Beijing 100088, China
* Correspondence: xiaowei@grinm.com (W.X.); wangjianwei@grinm.com (J.W.); lgwang@grinm.com (L.W.)

Abstract: Chemically modifying graphene (such as chemical doping) is a commonly used method to improve its formaldehyde sensing properties, but the microscopic mechanisms of heteroatoms in the adsorption and sensing process are still unclear. In this paper, the adsorption and sensing properties of formaldehyde on graphene surfaces modified by X doping (X = B, N, O, P, S, Mg and Al) were systematically investigated by first-principles calculations. The adsorption geometries, adsorption energies, charge transfers, and electronic structures were obtained and analyzed. The adsorption strengths of HCHO molecule on the Mg- and Al-doped graphene surfaces were stronger than those of non-metal (B, N, O, P and S)-doped cases. These results showed that the Mg- or Al-doped graphene was better for HCHO detecting than the non-metal-doped graphene systems. The sensing properties were simulated by theNEGF method for the two-probe nano-sensors constructed from Al- and Mg-doped graphene. The maximum sensing responses of nano-sensors based on Al- and Mg-doped graphene were obtained to be 107% and 60%, respectively. The present study supplies a theoretical basis for designing superior graphene-based HCHO gas sensors.

Keywords: formaldehyde; graphene; first-principles calculation; sensing performance

Citation: Yang, L.; Xiao, W.; Wang, J.; Li, X.; Wang, L. Adsorption and Sensing Properties of Formaldehyde on Chemically Modified Graphene Surfaces. *Crystals* **2022**, *12*, 553. https://doi.org/10.3390/cryst12040553

Academic Editors: Chiashain Chuang, Dung-sheng Tsai, Sheng-Kuei Chiu, Pei Lay Yap and John Parthenios

Received: 8 March 2022
Accepted: 12 April 2022
Published: 15 April 2022

1. Introduction

As the simplest aldehyde molecule, formaldehyde (HCHO) is extensively utilized by organic chemists for producing resins due to its excellent solvent and adhesion properties [1]. With highly toxic, hazardous, and carcinogenic characteristics, long-term exposure to formaldehyde would severely impair human health [2]. Concerning human health, formaldehyde has been identified as a major contributor to sick building syndrome (SBS) [3–5], such as throat irritations, dizziness, dyspnea, and asthma-related symptoms. Criteria for the limitation and regulation of formaldehyde emissions from housing have been established in many countries [6]. For example, in China, the criterion is 0.10 mg/m^3 (GB/T 18883-2002) [7]. With increasing anthropogenic emissions, formaldehyde has slowly transformed from a household hazard to a life-threatening pollutant. As a priority indoor pollutant, there is an increasing interest in fast and simple formaldehyde monitoring methods, and this has stimulated the development of formaldehyde gas-sensing technologies.

Due to its novel and unique structure and physical and chemical properties, graphene has attracted tremendous attention. However, it is confirmed by theoretical and experimental studies that the adsorption of target gas molecules on pristine graphene is very weak and consequently, the change in conductivity by gas molecule adsorption is difficult to measure [8–11]. Following the explosion of interest in graphene used in molecular identifications, the idea of graphene doping emerged [12–16]. Dai et al. [12] performed a systematic theoretical study of the ability of graphene doped with B, N, Al, and S to chemically bind various common or polluting gases, and suggested that B and S-doped graphene could be a promising candidate for NO and NO_2 monitoring. Zhang et al. [13] found that defective and B-doped graphene showed higher adsorption strength to NH_3, NO, and NO_2, and demonstrated that the sensitivity of B-doped graphene-based NO_2 sensors is also greatly enhanced. Liu et al. [14] revealed that the sensitivity of graphene-based SO_2 sensors could be greatly improved by introducing Al doping.

In this paper, the adsorption and sensing performances of X-doped graphene (XG with X = B, N, O, P, S, Mg, and Al) for HCHO gas sensing were investigated by first-principal calculations. The adsorption structures, adsorption energies, charge transfers, and electric structures for HCHO adsorption on XG were obtained and analyzed. The two-probe HCHO nano-sensors were established and the current-voltage (I-V) curves were simulated to evaluate the sensing performance of heteroatom-doped graphene-based sensors for HCHO detection.

2. Computation and Method Details

Density functional theory (DFT) calculations are implemented in the Vienna ab initio simulation package [17–19], using the generalized gradient approximation (GGA) exchange-correlation functional method [20]. Projector-augmented wave (PAW) pseudopotentials with the valence electron configurations of 2s2p for B, C, N, and O, 1s for H, 3s3p for Mg, Al, P, and S, respectively, were employed with a plane-wave energy cutoff of 500 eV. A semiempirical DFT-D2 method [21] was applied to correct the van der Waals interaction between the gas molecule and the substrate. An energy convergence of 10^{-5} eV was achieved and the atomic forces were converged down to 0.01 eV/Å for atomic relaxations. The adsorption energy was calculated by

$$\Delta E_{ads} = E_{tot} - E_{sub} - E_{mol} + \Delta E_{ZPE} \quad (1)$$

where E_{tot}, E_{sub}, and E_{mol} are the total energies of the doped graphene substrate with gas molecule adsorption, the doped graphene, and the isolated gas molecule, respectively. ΔE_{ZPE} is the zero-point energy difference after and before adsorption of the gas molecule. A negative ΔE_{ads} represents that the adsorption makes the system more stable. Furthermore, the Bader charge analyses were carried out to determine the charge transfers between the gas and the substrate [22,23].

The sensing properties caused by adsorption molecules were investigated within a DFT-combined non-equilibrium Green's function (NEGF) method performed by NanoDcal code [24,25]. The electron-electron interaction was treated with PBE-96 functional [20]. A double-ζ plus polarization (DZP) basis set was adopted to describe the valence electron in this work. The cutoff energy of 100 Hartree was chosen for the plane-wave basis. The electric current I through the nano-sensor was defined using the Landauer-Büttiker formula [26]:

$$I = G_0 \int_{\mu_R}^{\mu_L} T(E, V_b)[f(E - \mu_L) - f(E - \mu_R)]dE \quad (2)$$

in which G_0 is the unit of the quantum conductance, $T(E, V_b)$ is the transmission coefficient of electrons incident at energy E under a bias voltage V_b and $f(E - \mu_{L/R})$ is the distribution function of the electrons in the left or right electrode. $\mu_{L/R}$ is the chemical potential of the left or right electrode. In the calculation of the electron transport properties, the bias voltage is the electrochemical potential difference between the two electrodes as $V_b = (\mu_L - \mu_R)/e$.

3. Results and Discussion

A (4 × 4 × 1) graphene supercell with 32 carbon atoms was modelled in our study, and the optimized lattice constant was a = b = 9.87 Å, which is in agreement with previous study results [27]. One carbon atom in the graphene supercell was replaced by a heteroatom X (X = B, N, O, P, S, Mg, and Al) and the X-doped graphene was fully relaxed. A typical XG structure is shown in Figure 1a. Because the radii of B, N, and O atoms are close to that of the

C atom, their XG structure remained as the planar form, similar to perfect graphene. Due to their large atomic radii of the dopants (P, S, Mg, and Al), these dopants protruded from the graphene plane, and carbocycle around the heteroatom were significantly deformed. The local deformations of the graphene surface induced by Mg and Al were generally greater than those of non-metal dopants (B, N, O, P and S). The above results are consistent with previous studies [12,28,29]. The local atomic structure and electronic structure changes induced by heteroatom doping are able to increase the chemical reactivity.

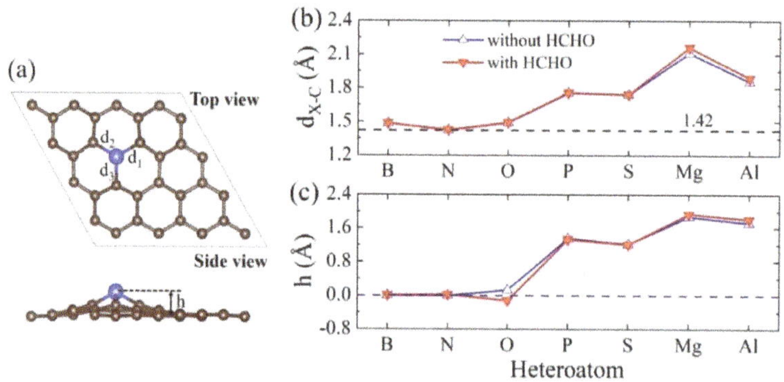

Figure 1. (**a**) Typical atomic geometry of heteroatom-doped graphene sheet. The blue ball indicates the heteroatom and the brown balls are carbon atoms, (**b**) average bond lengths ($d_{X\text{–}C} = (d_1 + d_2 + d_3)/3$) between the heteroatom and adjacent carbon atoms, and (**c**) heights of the dopant protrusion from the graphene with or without HCHO adsorption.

For a doping case, we placed the HCHO molecule around the heteroatom in various orientations and distances to produce several initial adsorption geometries and relaxed them to find the most stable atomic configuration with the lowest adsorption energy. For each case, greater-than-seven initial atomic geometries were fully relaxed. The most stable structures are collected in Figure 2, where the top and side views are the adsorption geometries viewed directly above and from the side of the graphene plane, respectively. The nearest distances between the heteroatom and oxygen atom in the adsorbed HCHO molecule, ZPE corrections, Bader charge results, adsorption energies, and C–O bond lengths of the adsorbed HCHO molecule are listed in Table 1 and also presented in Figure 3. The ZPE correction of gas molecule was calculated to improve the accuracy of the adsorption energies. For all heteroatom doping systems, the ZPE correction of the HCHO molecule was in a range from 0.020 to 0.042 eV. For B, N, O, S, and P doping systems, the adsorption energy of HCHO molecule was in the range from −0.173 to −0.252 eV, the nearest distances between the heteroatom and HCHO molecule were from 2.946 Å to 3.311 Å, and the charge gains of the adsorbed HCHO molecule were in the range of 0.022–0.074 e. The weak adsorption strength together with a large adsorption distance and small charge transfer implies that the HCHO molecule is physically adsorbed on the XG (X = B, N, O, S, and P) surface. Interestingly, the slightly protruded O dopant is pushed below the graphene plane after HCHO adsorption. For the Mg- or Al-doped system, it has a strong adsorption strength ($\Delta E_{ads} < -1.39$ eV) of the HCHO molecule with a short adsorption distance (d < 2 Å) and large charge transfer ($\Delta q > 0.36$ e). These results indicated the strong chemical X–O (X = Mg or Al) bonds formed between the XG substrate and HCHO molecule. However, the strong HCHO adsorption on the Mg- or Al-doped graphene implies a long recovery time [30,31]. The sensors based on the sensing material can detect a lower concentration of formaldehyde with higher sensitivity. In addition, the X–C bond lengths and the heteroatom heights shown in Figure 1b,c further support the strong adsorption of HCHO molecule on the XG (X = Mg or Al) substrate. For example, the N–C bond length and N dopant height did not change much after HCHO adsorption, while the Mg–C bond and Mg atom height became larger upon HCHO adsorption.

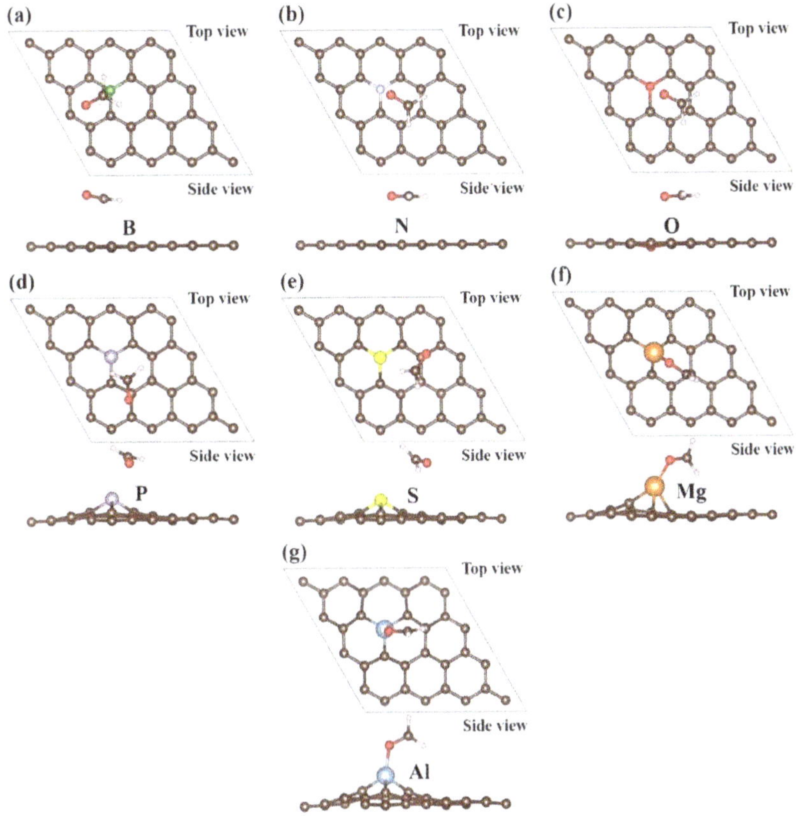

Figure 2. Top and side views of the lowest energy adsorption configurations for HCHO molecules on (**a–g**) B-, N-, O-, P-, S-, Mg-, and Al-doped graphene substrates. The red and white ball represents O and H atoms, respectively. Heterochromatic atoms embedded in graphene represent the heteroatom dopants.

Table 1. For the HCHO molecule adsorbed on XG (X = B, N, O, P, S, Mg, and Al) surfaces, the nearest distances between heteroatom X and the HCHO molecule, charge gains of the adsorbed HCHO molecule, zero-point energy corrections of the HCHO molecule, and adsorption energies and C–O bond length of the adsorbed molecule are presented.

Heteroatom	B	N	O	P	S	Mg	Al
d/Å	2.946	3.011	3.125	3.029	3.311	1.944	1.838
Δq/e	0.044	0.022	0.052	0.074	0.029	0.366	0.401
ΔE_{ZPE}/eV	0.025	0.032	0.031	0.030	0.020	0.035	0.042
ΔE_{ads}/eV	−0.233	−0.245	−0.252	−0.184	−0.173	−1.396	−1.413
d_{C-O}/Å	1.217	1.216	1.220	1.221	1.217	1.276	1.287

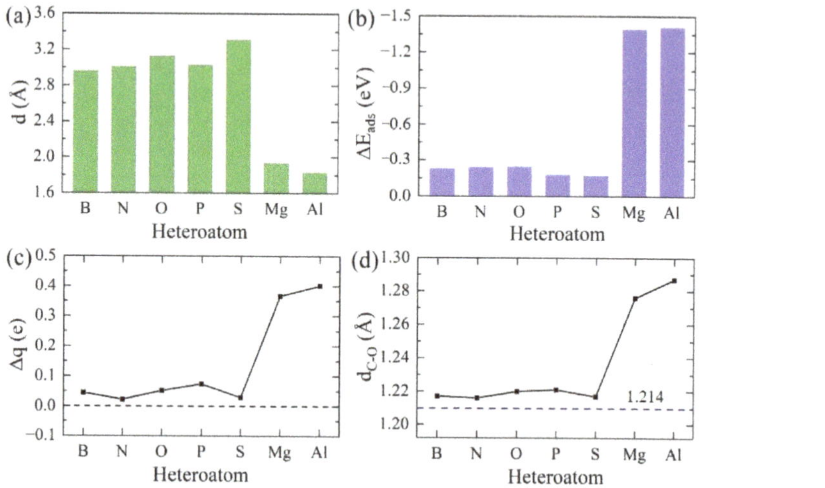

Figure 3. (**a**) The nearest adsorption distance between the heteroatom and HCHO molecule, (**b**) the adsorption energy of HCHO molecules on the XG substrate, (**c**) Bader charge gains Δq of HCHO molecule from the XG substrate, and (**d**) C–O bond length of HCHO molecule. The black dashed line in panel (**c**) indicates zero charge transfer, and the blue one in panel (**d**) denotes the C–O bond length of the free HCHO molecule.

As shown in Figure 3a,b, there is a clear connection between the adsorption energy and the adsorption distance, charge transfer, and C–O bond length. For a heteroatom doping system, if its adsorption energy is small, it usually has a larger adsorption distance, smaller charge transfer, and shorter C–O bond in general. For instance, for the S doping case the adsorption strength was the weakest, while the adsorption distance was the longest and the charge transfer and the C–O bond of adsorbed HCHO were the smallest. The Al doping case had the strongest adsorption, while the adsorption distance was short and the charge transfer and the C–O bond of HCHO were large. The charge transfer and the C–O bond length of the HCHO molecule in different heteroatom doping cases are plotted in Figure 3c,d. Compared with the free HCHO molecule ($d_{C\text{-}O}$ = 1.214 Å), the C–O bond length expanded to a different extent. The charge transfer results suggested that the HCHO molecule behaves as an acceptor obtaining electrons from the doped graphene substrate, which is in agreement with previous studies [30,32,33]. For B, N, O, S, or P doped systems, the adsorption strength was weak, which is consistent with the result that HCHO electron gain is less than 0.08 e and the C–O bond is stretched only 0.01 Å. The Al-doped graphene case had the maximum C-O bond expansion of over 0.07 Å with respect to that of the isolated HCHO molecule. The corresponding charge transfer was as much as 0.40 e. A large expansion of the C–O bond and charge gain implies that the HCHO molecule is very reactive and can form a strong chemical bonding with Al-doped graphene.

The densities of states for N-, P-, Mg- and Al-doped graphene with HCHO adsorption are shown in Figure 4. The TDOS of NG and PG systems had non-zero states at the Fermi level ("Fermi level" is the term used to describe the top of the collection of electron energy levels at absolute zero temperature), rendering them a metallic property. However, as shown in Figure 4a,b, the adsorbed HCHO did not contribute to the DOS at the Fermi level, so the NG and PG systems had little sensing response to HCHO molecule. The AlG and MgG systems are also metallic. Upon HCHO adsorption, the DOS around the Fermi level were substantially changed and the adsorbed HCHO created a DOS peak at the Fermi level. The strong interaction between the HCHO molecule and the Mg- or Al-doped graphene substrate may also be correlated to the HCHO density of states at the Fermi level. Besides, their DOSs become discontinuous (MgG-HCHO system around −0.5 eV,

AlG-HCHO system from 0.7 to 0.9 eV) after HCHO adsorption, which might decrease the conductivity of the MgG and AlG systems.

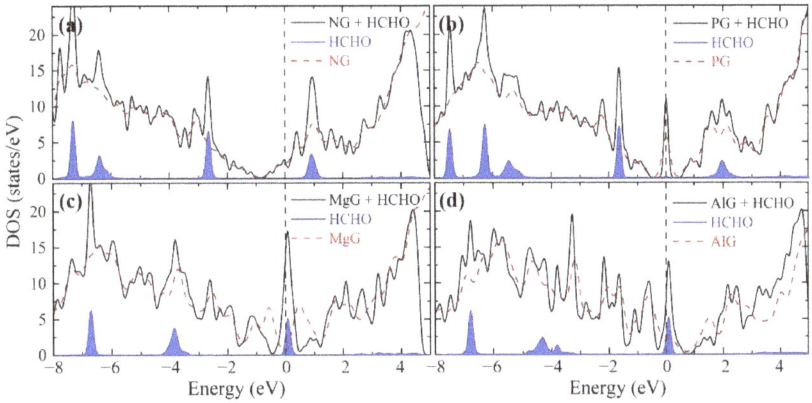

Figure 4. Total and projected densities of states for the (**a**) N-doped graphene (NG), (**b**) P-doped graphene (PG), (**c**) Mg-doped graphene (MgG), and (**d**) Al-doped graphene (AlG) systems with HCHO adsorption. The black and red dashed lines indicate total DOS of X-doped graphene with and without HCHO adsorption. The blue region represents the projected DOS of the adsorption HCHO molecule. The Fermi level is shifted to zero and represented by vertical dashed lines.

To evaluate the sensing performance of AlG and MgG for HCHO gas detecting, we analyzed the quantum transport properties with and without HCHO molecule adsorption employing the NEGF-DFT method with the NanoDcal code. Because of the anisotropy of graphene, two kinds of transport models were built for the AlG or MgG substrate with HCHO adsorption (Figure 2f,g), i.e., the current flows along the armchair and zigzag directions of the graphene lattice. For the two transport directions of the AlG nano-sensor for HCHO detection are shown in Figure 5, which consist of three parts, i.e., the central AlG substrate with HCHO adsorption scattering region and the two semi-infinite electrodes.

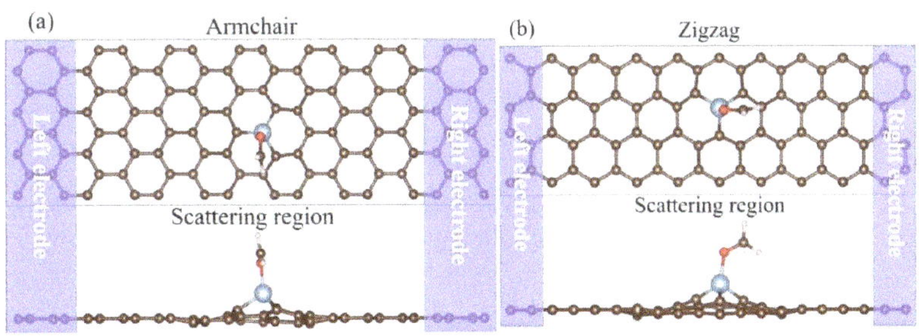

Figure 5. Top and side views of two-probe AlG nano-sensor for HCHO molecule detection along (**a**) armchair and (**b**) zigzag directions. The purple shadow regions represent the semi-infinite intrinsic left and right graphene electrodes, which are contacted with the central AlG substrate adsorbed by HCOH in the molecule scattering.

The current-voltage curves and sensing response (defined by $S = [(R - R_0)/R_0] = [(I_0 - I)/I] \times 100\%$, where $R_0(I_0)$ and R (R_0) is the resistances (currents) of the sensor before and after adsorption of HCHO under bias voltage, respectively) are presented in Figure 6. The currents are reduced upon HCHO adsorption. This phenomenon may be attributed to losing charge carriers of the AlG or MgG substrate after HCHO adsorption. We also observed the same phenomenon for the transition metal-doped graphene systems [30].

The sensing responses along the armchair and zigzag directions show the anisotropy of graphene's transport properties. The maximum sensing responses of AlG and MgG nano-sensors to HCHO molecule adsorption were 107% and 60%, respectively. For the Al- and Mg-doped systems, the average response was larger than 30%, especially for the armchair-orientated AlG nano-sensor, in which the average response reached 67%. The high sensitivity indicated that AlG and MgG can be applied as the sensing material for HCHO gas detecting.

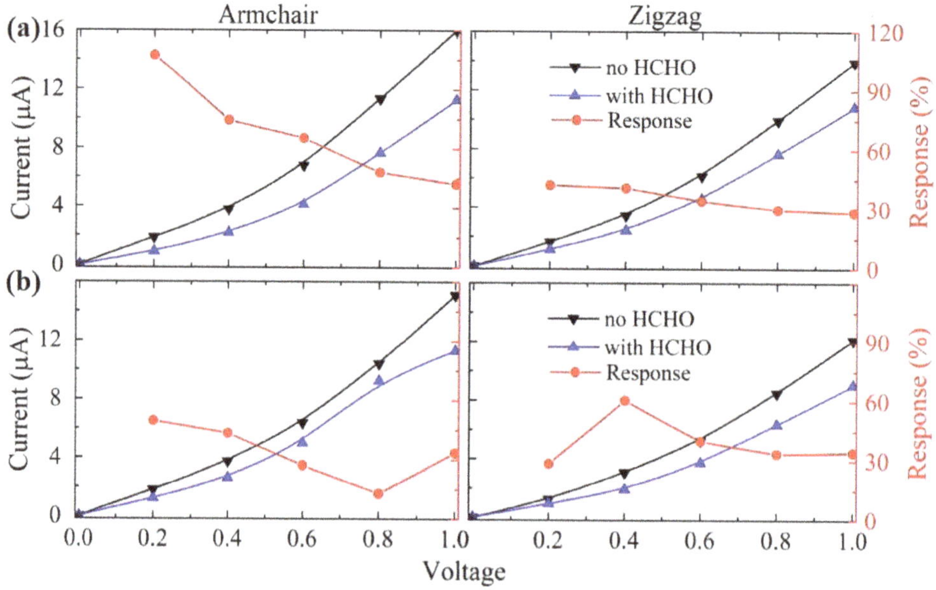

Figure 6. Current-voltage curves and sensing responses of (**a**) Al-doped graphene (AlG) and (**b**) Mg-doped graphene (MgG) nano-sensors without and with HCHO adsorbed. The current flows in the armchair and zigzag direction of the graphene lattice.

The transmission spectra of AlG and MgG nano-sensors before and after CHOH molecule adsorption under the bias from 0 to 0.8 V are plotted in Figures 7 and 8 to explore the mechanisms for HCHO detecting. The transmission coefficients of AlG and MgG nano-sensors with or without HCHO adsorption along two transport directions were almost zero at the bias window. As the bias increased, the area of transmission spectra in the bias windows became larger. These characteristics are similar to those of graphene nanoribbon and doped graphene [30,34]. According to Equation (2), the current flowing through the device is equal to the integral of the transmission coefficient in the bias windows. Figures 7 and 8 show that the transmission spectrum area of AlG and MgG nano-sensors in the bias window after HCHO adsorption was obviously smaller than before. Therefore, we can conclude that HCHO adsorption might reduce the current. It was also found that under zero bias, the transmission coefficient of MgG and AlG without or with HCHO adsorption were asymmetric with respect to the Fermi level. We can expect that the bias augment could not lead a linear increase of the area of transmission spectra, which could render a non-linear change in current.

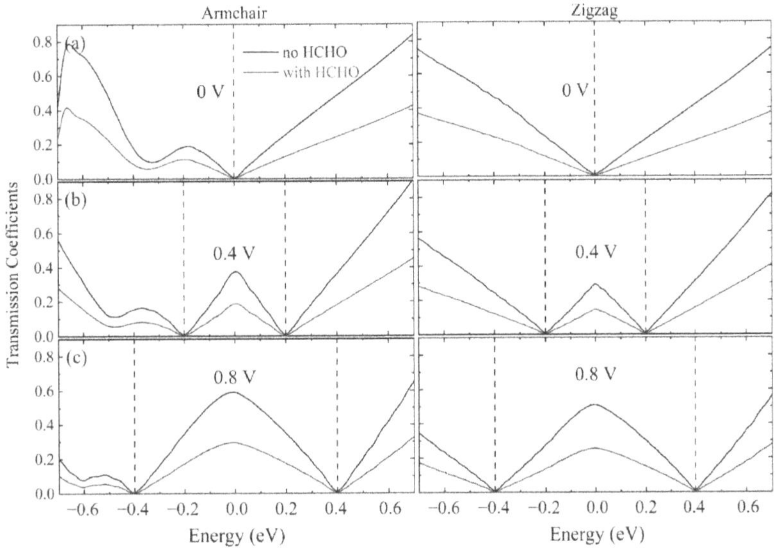

Figure 7. The transmission spectra of the AlG nano-sensor for HCHO detection along armchair and zigzag directions under the bias voltage of (**a**–**c**) 0, 0.4, and 0.8 V. The black dashed lines represent the bias windows.

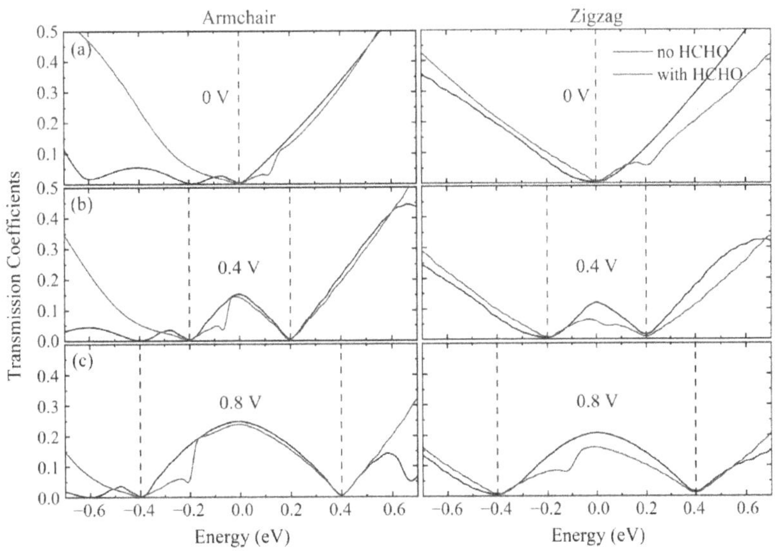

Figure 8. The transmission spectra of MgG nano-sensor for HCHO detection along armchair and zigzag direction under the bias voltage of (**a**–**c**) 0, 0.4, and 0.8 V. The black dashed lines represent the bias windows.

4. Conclusions

In summary, we investigated the adsorption and sensing properties of formaldehyde on graphene doped by various heteroatoms (B, N, O, P, S, Mg, and Al). The most stable atomic configurations of heteroatom-doped graphene adsorbed with HCHO, adsorption energies, Bader charge transfers, and electronic structures were picked up. We found that the adsorption strengths of the HCHO molecule on the Mg- and Al-doped graphene surfaces were stronger than those of non-metal (B, N, O, P and S)-doped cases. The results indicated that the Mg- or Al-doped graphene is better for HCHO detecting. The two-probe HCHO nano-sensors were built for the Mg- and Al-doped graphene along the armchair and zigzag directions and the I-V curves were simulated to evaluate the sensing performance by the NEGF-DFT method. The maximum sensing responses of Al- and Mg-doped graphene nano-sensors were 107% and 60%, respectively, while the average responses for both cases were larger than 30%. Based on our theoretical results, we predict that the Al- and Mg-doped graphene can be potentially used as sensing materials for highly sensitive HCHO detection.

Author Contributions: Funding acquisition, J.W.; investigation, L.Y.; methodology, W.X., X.L., and L.W.; project administration, J.W.; resources, J.W.; supervision, W.X., X.L., and L.W.; writing—original draft, L.W.; writing—review and editing, W.X. and L.W. All authors have read and agreed to the published version of the manuscript.

Funding: This work is financially supported by Beijing Natural Science Foundation (No. 2222084).

Institutional Review Board Statement: Not applicable.

Informed Consent Statement: Not applicable.

Data Availability Statement: Not applicable.

Acknowledgments: Not applicable.

Conflicts of Interest: The authors declare no academic misconduct or competing for financial interests.

References

1. Cockcroft, D.W.; Hoeppner, V.H.; Dolovtch, J. Occupational asthma caused by cedar urea formaldehyde particle board. *Chest* **1982**, *82*, 49–53. [CrossRef]
2. Tang, X.; Bai, Y.; Duong, A.; Smith, M.T.; Li, L.; Zhang, L. Formaldehyde in China: Production, consumption, exposure levels, and health effects. *Environ. Int.* **2009**, *35*, 1210–1224. [CrossRef]
3. Korpan, Y.I.; Gonchar, M.V.; Sibirny, A.A.; Martelet, C.; El'skaya, A.V.; Gibson, T.D.; Soldatkin, A.P. Development of highly selective and stable potentiometric sensors for formaldehyde determination. *Biosens. Bioelectron.* **2000**, *15*, 77–83. [CrossRef]
4. Grammer, L.C.; Harris, K.E.; Cugell, D.W.; Patterson, R. Evaluation of a worker with possible formaldehyde-induced asthma. *J. Allergy Clin. Immunol.* **1993**, *92*, 29–33. [CrossRef]
5. Dou, K.; Chen, G.; Yu, F.; Liu, Y.; Chen, L.; Cao, Z.; Chen, T.; Li, Y.; You, J. Bright and sensitive ratiometric fluorescent probe enabling endogenous FA imaging and mechanistic exploration of indirect oxidative damage due to FA in various living systems. *Chem. Sci.* **2017**, *8*, 7851–7861. [CrossRef]
6. Salthammer, T.; Mentese, S.; Marutzky, R. Formaldehyde in the indoor environment. *Chem. Rev.* **2010**, *110*, 2536–2572. [CrossRef]
7. Indoor Air Quality Standars. Available online: http://www.gb688.cn/bzgk/gb/newGbInfo?hcno=59392CE8FD4422912D71145 9014EA4A7 (accessed on 18 March 2022).
8. Giovanni, M.; Poh, H.L.; Ambrosi, A.; Zhao, G.; Sofer, Z.; Sanek, F.; Khezri, B.; Webster, R.D.; Pumera, M. Noble metal (Pd, Ru, Rh, Pt, Au, Ag) doped graphene hybrids for electrocatalysis. *Nanoscale* **2012**, *4*, 5002–5008. [CrossRef]
9. Hu, X.; Meng, F. Structure and gap opening of graphene with Fe doped bridged trivacancy. *Comput. Mater. Sci.* **2016**, *117*, 65–70. [CrossRef]
10. Krasheninnikov, A.V.; Lehtinen, P.O.; Foster, A.S.; Pyykko, P.; Nieminen, R.M. Embedding transition-metal atoms in graphene: Structure, bonding, and magnetism. *Phys. Rev. Lett.* **2009**, *102*, 126807. [CrossRef]
11. Zhou, M.; Lu, Y.H.; Cai, Y.Q.; Zhang, C.; Feng, Y.P. Adsorption of gas molecules on transition metal embedded graphene: A search for high-performance graphene-based catalysts and gas sensors. *Nanotechnology* **2011**, *22*, 385502. [CrossRef]
12. Dai, J.; Yuan, J.; Giannozzi, P. Gas adsorption on graphene doped with B, N, Al, and S: A theoretical study. *Appl. Phys. Lett.* **2009**, *95*, 232105. [CrossRef]
13. Zhang, Y.H.; Chen, Y.B.; Zhou, K.G.; Liu, C.H.; Zeng, J.; Zhang, H.L.; Peng, Y. Improving gas sensing properties of graphene by introducing dopants and defects: A first-principles study. *Nanotechnology* **2009**, *20*, 185504. [CrossRef]
14. Liu, X.Y.; Zhang, J.M.; Xu, K.W.; Ji, V. Improving SO_2 gas sensing properties of graphene by introducing dopant and defect: A first-principles study. *Appl. Surf. Sci.* **2014**, *313*, 405–410. [CrossRef]
15. Liu, X.Y.; Zhang, J.M. Formaldehyde molecule adsorbed on doped graphene: A first-principles study. *Appl. Surf. Sci.* **2014**, *293*, 216–219. [CrossRef]
16. Lv, R.; Dos Santos, M.C.; Antonelli, C.; Feng, S.; Fujisawa, K.; Berkdemir, A.; Cru-Silva, R.; Elias, A.L.; Perea-Lopez, N.; Lopez-Urias, F.; et al. Large-area Si-doped graphene: Controllable synthesis and enhanced molecular sensing. *Adv. Mater.* **2014**, *26*, 7593–7599. [CrossRef]

17. Kresse, G.; Furthmüller, J. Efficient iterative schemes for ab initio total-energy calculations using a plane-wave basis set. *Phys. Rev. B* **1996**, *54*, 11169. [CrossRef]
18. Kresse, G.; Furthmüller, J. Efficiency of ab-initio total energy calculations for metals and semiconductors using a plane-wave basis set. *Comput. Mater. Sci.* **1996**, *6*, 15–50. [CrossRef]
19. Kresse, G.; Joubert, D. From ultrasoft pseudopotentials to the projector augmented-wave method. *Phys. Rev. B* **1999**, *59*, 1758–1775. [CrossRef]
20. Perdew, J.P.; Burke, K.; Ernzerhof, M. Generalized gradient approximation made simple. *Phys. Rev. Lett.* **1996**, *77*, 3865. [CrossRef]
21. Grimme, S. Semiempirical GGA-type density functional constructed with a long-range dispersion correction. *J. Comput. Chem.* **2006**, *27*, 1787–1799. [CrossRef]
22. Henkelman, G.; Arnaldsson, A.; Jónsson, H. A fast and robust algorithm for Bader decomposition of charge density. *Comput. Mater. Sci.* **2006**, *36*, 354–360. [CrossRef]
23. Tang, W.; Sanville, E.; Henkelman, G. A grid-based Bader analysis algorithm without lattice bias. *J. Phys. Condens. Matter* **2009**, *21*, 084204. [CrossRef]
24. Taylor, J.; Guo, H.; Wang, J. Ab initio modeling of quantum transport properties of molecular electronic devices. *Phys. Rev. B* **2001**, *63*, 245407. [CrossRef]
25. Brandbyge, M.; Mozos, J.L.; Ordejón, P.; Taylor, J.; Stokbro, K. Density-functional method for nonequilibrium electron transport. *Phys. Rev. B* **2002**, *65*, 165401. [CrossRef]
26. Büttiker, M.; Imry, Y.; Landauer, R.; Pinhas, S. Generalized many-channel conductance formula with application to small rings. *Phys. Rev. B* **1985**, *31*, 6207–6215. [CrossRef]
27. Zhou, J.; Wang, Q.; Sun, Q.; Chen, X.S.; Kawazoe, Y.; Jena, P. Ferromagnetism in semihydrogenated graphene sheet. *Nano Lett.* **2009**, *9*, 3867–3870. [CrossRef]
28. Tian, Y.L.; Ren, J.F.; Yue, W.W.; Hu, G.C.; Yuan, X.B. Adsorption of chloroform on N-doped and Al-doped graphene: A first-principle study. *Chem. Phys. Lett.* **2017**, *685*, 344–348. [CrossRef]
29. Tang, Y.; Liu, Z.; Dai, X.; Yang, Z.; Chen, W.; Ma, D.; Lu, Z. Theoretical study on the Si-doped graphene as an efficient metal-free catalyst for CO oxidation. *Appl. Surf. Sci.* **2014**, *308*, 402–407. [CrossRef]
30. Yang, L.; Xiao, W.; Wang, J.; Li, X.; Wang, L. Formaldehyde gas sensing properties of transition metal-doped graphene: A first-principles study. *J. Mater. Sci.* **2021**, *56*, 12256–12269. [CrossRef]
31. Yang, L.; Xiao, W.; Wang, J.; Li, X.; Wang, L. Tunable formaldehyde sensing properties of palladium cluster decorated graphene. *RSC Adv.* **2021**, *11*, 37120–37130. [CrossRef]
32. Chen, X.; Xu, L.; Liu, L.L.; Zhao, L.S.; Chen, C.P.; Zhang, Y.; Wang, X.C. Adsorption of formaldehyde molecule on the pristine and transition metal doped graphene: First-principles study. *Appl. Surf. Sci.* **2017**, *396*, 1020–1025. [CrossRef]
33. Chi, M.; Zhao, Y.P. Adsorption of formaldehyde molecule on the intrinsic and Al-doped graphene: A first principle study. *Comput. Mater. Sci.* **2009**, *46*, 1085–1090. [CrossRef]
34. Topsakal, M.; Bagci, V.M.K.; Ciraci, S. Current-voltage (I−V) characteristics of armchair graphene nanoribbons under uniaxial strain. *Phys. Rev. B* **2010**, *81*, 205437. [CrossRef]

crystals

MDPI

Article

Effect of pH and Concentration of Dopamine Solution on the Deposited Polydopamine Film and the Prepared Graphene-like Material

Dongyang Wang [1], Qiang Wang [1,*], Zongqiong Lin [2,*], Tiantian Pang [1] and Ning Ding [1]

1 School of Electronic Information and Artificial Intelligence, Shaanxi University of Science & Technology, Xi'an 710119, China
2 Institute of Flexible Electronics, Northwestern Polytechnical University, Xi'an 710069, China
* Correspondence: qiang.wang@sust.edu.cn (Q.W.); iamzqlin@nwpu.edu.cn (Z.L.)

Abstract: In this study, according to the acquired polydopamine deposition rates, polydopamine films with equal thickness were prepared under different conditions on SiO_2 substrates. Subsequently, we investigated the influence of dopamine solution pH and concentration on the formation of surface aggregates of the deposited polydopamine films. Assumptions were made to explain how pH and concentration execute their effects. Based on the optimized parameters, a continuous and smooth polydopamine film with a thickness of about 14 nm and a roughness of 1.76 nm was fabricated on a silicon dioxide substrate, through the deposition for 20 minutes in a dopamine solution with a concentration of 1.5 mg/mL and a pH of 8.2. The prepared polydopamine film was then employed as a precursor and subjected to a high-temperature process for the carbonization and graphitization of the film. Raman spectroscopy analysis indicated that the resulting graphene-like film had fewer structural defects in comparison with previous works and the results of XPS indicated that most of the carbon atoms were bound into the cross-linked honeycomb lattice structure. The prepared graphene-like material also exhibited high electrical conductivity and satisfying mechanical elasticity.

Keywords: polydopamine; pH; concentration of dopamine solution; graphene-like film

Citation: Wang, D.; Wang, Q.; Lin, Z.; Pang, T.; Ding, N. Effect of pH and Concentration of Dopamine Solution on the Deposited Polydopamine Film and the Prepared Graphene-like Material. *Crystals* **2023**, *13*, 607. https://doi.org/10.3390/cryst13040607

Academic Editors: Kefayat Ullah, Salah Ud Din and Mustafa Haider

Received: 5 March 2023
Revised: 30 March 2023
Accepted: 31 March 2023
Published: 2 April 2023

1. Introduction

As an emerging surface modification technology, mussel biomimetic chemistry is applied in many scientific research fields such as biology, energy, and environment technology. In this trend, the technology based on dopamine garnered significant attention. In 2007, Lee et al. [1] reported that dopamine can undergo oxidative self-polymerization in a weak alkaline environment, leading to the formation of polydopamine (PDA) coatings on various substrates. Further investigations by Yminu et al. [2] revealed that PDA coatings can be formed on a wide range of materials, including noble metals (Au, Ag, Pt, and Pd), metals with native oxide on the surface (such as Cu, stainless steel, and NiTi shape memory alloys), oxides (TiO_2, non-crystalline SiO_2, quartz, Al_2O_3, and Nb_2O_5), semiconductors (such as GaAs and Si), ceramics, and synthetic polymers, etc.

The self-polymerization reaction of dopamine requires only moderate preconditions, resulting in facile deposition of polydopamine (PDA) layers [3]. Moreover, PDA exhibits good adhesion properties on various surfaces, making it a versatile material for constructing functional surfaces. Ultrathin PDA films can be achieved by regulating the deposition time and through subsequent high-temperature process, the films can be graphitized to graphene-like materials [4,5]. In comparison with other graphene fabrication methods, e.g., chemical vapor deposition (CVD), chemical synthesis, exfoliation of graphite and SiC pyrolysis, the approach with PDA has many advantages. There are no dangerous chemicals, extreme high-temperature processes, or severe environment harmful waste products involved in the PDA method. Moreover, with other approaches, the prepared graphene, in most cases, must be transferred to a suitable substrate, e.g., a substrate with a dielectric layer on the surface, to form the desired devices. With the fabrication method in this study, however, the graphene-like film can be achieved in situ on a SiO_2 substrate [6,7]. Therefore, a transfer-process is not necessary anymore. Additionally, the graphene-like material prepared with this method adhere very well to the underlying SiO_2 film, while transferred graphene supposedly is not so well adherent.

In 2011, Liu et al. [8] reported the successful conversion of solution oxidation-derived PDA to carbon at 800 °C in a nitrogen atmosphere, yielding a carbon material with approximately 7.28% nitrogen doping and nearly 60% carbon yield. Building on this work, J. Kong et al. [9] used various test apparatuses to investigate the structure of the PDA film on a SiO_2 substrate after carbonization, revealing that the resulting carbonized PDA (C-PDA) coating had a graphite-like layered structure. Li et al. [10] deposited PDA on a substrate and obtained a highly ductile conductive carbon film by annealing the PDA coating. Lim et al. [11] adjusted the thickness of the carbon film from a few nanometers to tens of nanometers by varying the number of repetitions of the coating process, and the size of the resulting carbon film could be as large as that of the underlying silicon wafer. While graphene-like carbon films were fabricated through high-temperature carbonization of PDA in these studies, these films exhibited many structural defects. It is believed that the structure of the precursor is a key factor affecting the quality of the carbonized final products. Shin et al. [12] showed that the use of organic matter as a precursor for graphene synthesis results in better quality graphene with thinner films. However, there are few reports on obtaining graphene-like materials with fewer defects by improving the structure of the PDA films.

PDA coatings are commonly obtained through the self-deposition of dopamine under alkaline conditions, and physical measures can be employed to adjust PDA deposition. Wei et al. [13] utilized heating and stirring of the dopamine solution to quickly deposit a thicker PDA film, and Gao's research group [14,15] developed a technology for cyclically extracting and reloading the dopamine deposition solution to prepare PDA coatings with gradient thickness. However, these studies did not pay attention to the problem that many aggregates on the film surface are easy to occur during the PDA self-deposition process., which can negatively affect the performance of the final product when PDA is used as a precursor for carbon-based materials. Although Kim et al. [16] obtained a flat and smooth PDA film with low roughness by passing oxygen into the solution, this preparation method is complex and costly. Ultrasound was also used by Diana et al. [17] to remove aggregates from the surface of the PDA coating, but this approach damaged the film. In this study, we investigated the effects of pH and dopamine concentration on aggregate formation on the PDA surface during the self-deposition process to obtain a PDA film with more ideal structure, so that better graphene-like materials can be achieved employing the improved PDA film as a precursor. In another word, we improved the quality of the precursor to optimize the prepared graphene-like materials and realize the in situ fabrication of the graphene-like film on the substrate. With optimized parameters, we were able to prepare a continuous and flat PDA film with a thickness of about 14 nm and a roughness of 1.76 nm on a SiO_2 substrate. The deposition process lasted 20 min in a dopamine solution with a pH of 8.2 and a dopamine concentration of 1.5 mg/mL. The resulting PDA film was carbonized at 800 °C and 1000 °C with a heating rate of 10 °C/min under a vacuum of 0.1 Pa. The generated graphene-like film had fewer defects and higher conductivity compared to those fabricated with other PDA films.

2. Materials and Methods

2.1. Materials

The dopamine hydrochloride powder and the tris powder were purchased from Shanghai Aladdin Biochemical Technology Co., Ltd., Shanghai, China. The former had a purity of 98%, and was stored in a dry environment at 2–8 °C. The purity of the latter was greater than or equal to 99.8% with 0–0.005% H_2O insoluble matter. Before the experiments, the tris powder was stored in a dry environment at room temperature. The PDA films were deposited on monocrystalline silicon substrates with 300 ± 20 Å silicon dioxide layer on top of them. These substrates were acquired from Tianjin Semiconductor Research Institute, China, and produced in April 2022. They were 10 × 10 mm quadrate samples and were 500 μm thick.

2.2. Preparation

2.2.1. Cleaning of Silicon Dioxide Substrates

At first, the silicon dioxide substrates were flushed 3 times with deionized water, soaked in deionized water, and ultrasonically cleaned for 10 min to remove coarse contaminants. After that, SC1 cleaning for substrates was carried out. A total of 5 parts of deionized water, 1 part of ammonia water (25% by weight of NH_3), and 1 part of H_2O_2 solution (30%) were poured into a beaker. Subsequently, the samples were placed into the beaker and immersed in the solution at 80 °C for 12 min to remove organic residues from the substrates. After the SC1 cleaning was completed, samples were removed from the beaker, and rinsed with deionized water to remove the residual from SC1. Afterwards, the SC2 cleaning was also performed. The cleaning solution for the SC2 was comprised of 6 parts of deionized water, 1 part of hydrochloric acid solution (37% by weight), and 1 part of hydrogen peroxide solution (H_2O_2, 30%). The silicon dioxide substrates were laid into the solution and soaked there at 80 °C for 10 min to remove possible metal residues from the samples. After that, the samples were washed with deionized water again in order to remove the residues from the SC2 cleaning and kept in deionized water waiting for further processing.

2.2.2. Preparation of Dopamine Solution

- Solutions with different pH values:
  Firstly, three portions of dopamine hydrochloride powder were weighed, and were found to be 50 mg each, and they were dissolved each in 50 mL deionized water to form dopamine hydrochloride solution with a concentration of 1 mg/mL. Then, the tris powder was added slowly to set the pH of the three solutions to 7.9, 8.2 and 8.5, respectively.
- Solutions with different dopamine concentrations:
  Three portions of dopamine hydrochloride powder (50 mg, 75 mg, 100 mg) were weighed at first. Subsequently, each of them was dissolved in 50 mL deionized water to prepare the dopamine hydrochloride solution with the concentration of 1 mg/mL, 1.5 mg/mL, 2 mg/mL, respectively. Afterwards, tris powder was added slowly to adjust the pH of the solution to 8.2.

2.2.3. Preparation of Polydopamine Films

The cleaned silicon dioxide substrates were immersed in a dopamine hydrochloride solution. Then, the solution with the samples was placed in an orbital shaker to deposit the PDA film in a dark environment. The rotational speed of the orbital shaker was held at 80 r/min during the deposition. When the deposition was finished, the surface of the silicon dioxide samples was rinsed with deionized water and then dried with N_2.

2.2.4. Carbonization of Polydopamine Films

The N_2 dried samples were divided into 2 groups. They were annealed in a vacuum of 0.1 Pa at 800 °C or 1000 °C for 60 min, respectively. During the annealing processes, the samples were heated from room temperature to the set temperature at a heating rate of 10 °C/min.

2.3. Characterization

The cross section of the silicon dioxide samples with the deposited PDA film were characterized with scanning electron microscope (SEM) with the model of JIB-4700F from

JEOL. The accelerating voltage of electrons were set to be 10 kV or 15 kV. The image magnification at the observation was adjusted between 50,000 and 85,000 to have clear images of the cross sections of the PDA films. For the SEM characterization of the film surface, the accelerating voltage was selected as 10 kV, and the magnifications were set to be 50 and 1000 to obtain the overall and the local images of the surface aggregations on the PDA film.

The surface of the PDA film deposited for a short time was observed with an optical microscope (SMIC) at a magnification of 100. Some sample surfaces were scratched with a tweezer on purpose. With the optical microscope, it can be judged, to some degree, whether a continuous PDA film was formed, based on whether obvious scratches could be made with the tweezer on the film surface. The model of the employed Atomic Force Microscope (AFM) was SPA400-SP13800N, from Seiko Corporation. The surface of the PDA films with different deposition time was characterized with AFM by an image size of 5 μm × 5 μm to investigate the surface roughness and morphology.

The applied X-ray diffraction (XRD) equipment was a product from Rigaku, Japan, with the model of Smart Lab 9 kW. It scanned the flat PDA films and the films with a great quantity of surface aggregations with the 2θ angle varying from 5° to 50° to detect the possible different crystalline structures. A Laser micro-Raman imaging spectrometer from Thermo Fisher (China) with the model of DXRxi was employed for the characterization of the molecular structure of the carbonized PDA films. The measurements were carried out with a spectral resolution <1.5 cm^{-1}. The Raman spectroscopy analysis was performed on the flat films without aggregations, the films with many aggregates on the surface and the films with big thickness, to determine the structural defects in the films.

A pH meter (ST3100E from Ohaus) was deployed to detect the pH value of the reaction solution in order to explore the effect of the pH on the self-deposition of dopamine. The electrical conductivities of the carbonized PDA films were measured using a Hewlett Packard 4140BpA Meter/DC Voltage source.

3. Results and Discussion

Lee's method [1] was employed to prepare PDA films. The elements ratio of the films was analyzed at various positions in the cross section of the prepared samples using EDS (energy dispersive spectroscopy). Figure 1a depicts the detection positions, where point 1 corresponds to the deposited PDA film, and point 2 lies in the SiO_2/Si substrate area. The EDS measurement at point 1 is depicted in Figure 1b, which revealed that the main element in the deposited film was C, accounting for 51.8%, while there were also 5.4% N, 9.7% Si, and 33.1% O in the film. The presence of Si and O may have originated from the environment or some SiO_2 particles present at the cross section. The existence of O may also be attributed to the oxidation of the deposition solution. On the other hand, the EDS result at point 2 shown in Figure 1c was significantly different from that of the deposited film. At point 2, the main elements were O and Si, with proportions of 50.5% and 32%, respectively. Through the EDS measurement, we also detected a certain proportion of C (16.8%) and a very small percentage of N (0.7%). These elements may have originated from the deposited film and might have reached the cross section during the EDS analysis, as the sample was broken. Additionally, C may have also originated from the environment. The observed film structure in Figure 1a and the different proportions of elements at the film cross section and the substrate cross section confirm the formation of PDA films on the surface of the SiO_2 substrate.

During the self-deposition process of dopamine, aggregates are prone to form on the film surface, as illustrated in Figure 1d. The XRD analysis revealed that the molecular chain orientation and crystallinity of the aggregates were distinct from those of the flat film (XRD analysis of the film and the aggregate shown in Figure S1). The aggregates not only affected the smoothness of the PDA film but also influenced the properties of the graphene-like layer produced through carbonization of the former. Therefore, we explored the influencing factors for the formation of the aggregates. Firstly, we explored the effect of different pH values on the formation of aggregates on the surface of PDA films. The pH value of the polymerization system exerted a significant influence on the polymerization reaction, and it also held true for the self-polymerization of dopamine to form PDA under alkaline conditions. Specifically, a lower pH value was detrimental to the formation of PDA on the substrate surface, while an excessively high pH value can lead to the presence of a significant number of aggregates on the surface of the resulting PDA film. Thus, it was necessary to investigate the effects of pH on the formation of PDA in order to obtain a smooth, continuous, and aggregate-free PDA film on the substrate surface [18,19]. In 2009, Zhu et al. [20] discovered that the dopamine composite film, prepared from a buffer solution with a pH of 8.5, exhibited superior adsorption properties. However, this does not necessarily imply that the optimal pH value for the formation of the graphene-like carbonized film will also be 8.5 when the PDA layer serves as the starting material.

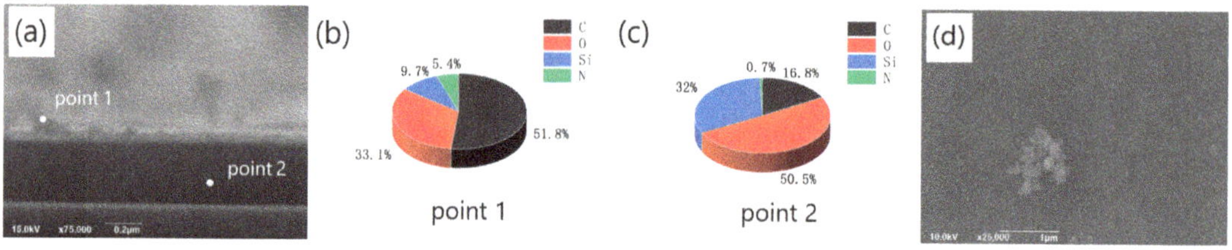

Figure 1. (**a**) SEM image of the cross-section of the film obtained on the silicon dioxide substrate and EDS detection position (point1 corresponds to the cross-section of the film, point2 corresponds to the cross-section of the substrate); (**b**) the proportion of different elements at the film cross-section (point1); (**c**) The proportion of different elements at the cross section of the substrate (point2); (**d**) SEM image of the aggregates on the film surface.

We studied the deposition rate of PDA on SiO_2 substrates with dopamine solutions of different pH values, under the condition that the dopamine hydrochloride concentration of the solutions was 1 mg/mL. Based on the obtained deposition rates, PDA films of identical thickness were prepared to investigate the effects of different pH values on the formation of aggregates on the film surface. Figure 2a–c depict the SEM images of the cross-section of PDA films prepared with a deposition time of 60 min on the surface of the SiO_2 substrate under pH values of 7.9, 8.2, and 8.5, respectively (the test results of the pH values are displayed in Figure S2). The thicknesses of the deposited films were 23.62 nm, 27.06 nm, and 28.09 nm, respectively. Figure 2d–f illustrate the SEM images of the cross-section of PDA films deposited for 90 min under the same conditions as previously mentioned, i.e., pH values of 7.9, 8.2, and 8.5. The thicknesses of the films were 36.32 nm, 43.31 nm, and 45.28 nm, respectively. We utilized the software IrfanView to calculate the width of different positions of the film in the figure, based on the scale in the figure, and then, we calculated the average value.

We conducted a series of SEM measurements to determine the thickness of PDA films under various deposition time conditions, and the results are summarized in Table 1. The relationship between the film thickness and the pH value is shown in Figure 3a. It is evident that, under the same deposition time, the film thickness increased with an increasing pH value. However, this increasing trend weakened as the pH value became higher. The different shades of darkness observed in the deposition solutions with different pH values (Figure S3) were consistent with the film thickness results presented in Figure 3a (the color of the solution serves as an indicator of the deposition rate. Specifically, the solution with a pH of 7.9 appears lighter, suggesting a slower deposition rate and smaller film thickness. On the other hand, solutions with pH values of 8.2 and 8.5 are similar in color, indicating that their deposition rates are also similar and their resulting film thicknesses are close). In Figure 3b, the film thickness is plotted against the deposition time, with the solution pH value being taken as the parameter. Within the range of deposition time between 0 and 90 min and a pH range from 7.9 to 8.5, the thickness of the deposited PDA film varied almost linearly with the deposition time. The deposition rates of PDA under the conditions of pH = 7.9, 8.2, and 8.5 were 0.399 nm/min, 0.466 nm/min, and 0.486 nm/min, respectively.

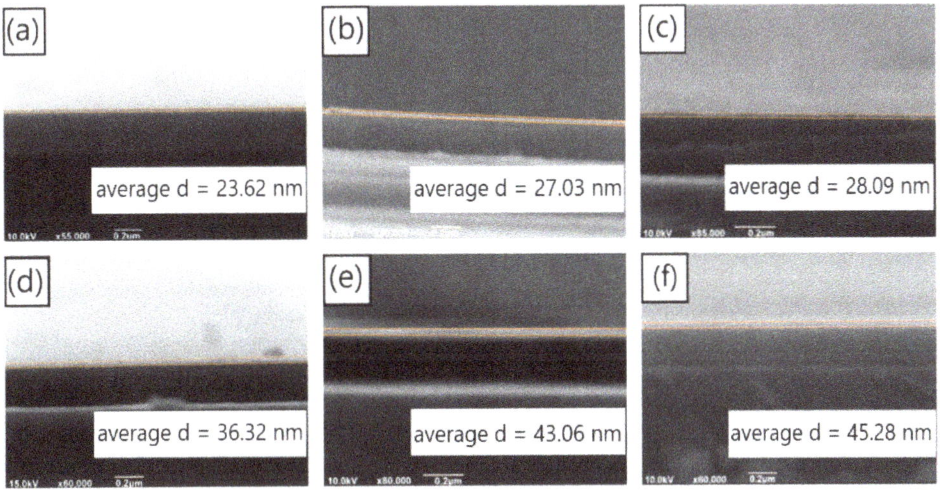

Figure 2. SEM images of the cross section of the PDA films on silicon dioxide substrates deposited for 60 min under the conditions of (**a**) pH = 7.9, (**b**) pH = 8.2 and (**c**) pH = 8.5; SEM images of the cross section of the PDA films on silicon dioxide substrates deposited for 90 min, with (**d**) pH = 7.9, (**e**) pH = 8.2 and (**f**) pH = 8.5.

Table 1. Thickness of the PDA film deposited for different time.

pH	Thickness for Different Deposition Time			
	45 min (nm)	**60 min (nm)**	**75 min (nm)**	**75 min (nm)**
7.9	18.00	23.62	30.03	36.32
8.2	20.11	27.03	35.16	43.06
8.5	20.76	28.09	36.05	45.28

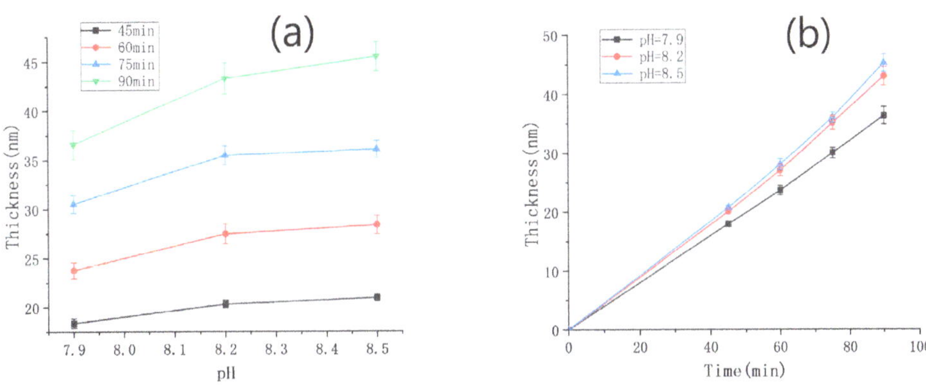

Figure 3. (**a**) PDA film thickness against pH under different time conditions; (**b**) PDA film thickness against deposition time under different pH conditions.

Based on the obtained self-deposition rates of dopamine under different pH conditions, we prepared three sets of PDA film samples with equal thickness on SiO_2 substrates using solutions with pH values of 7.9, 8.2 and 8.5, respectively, deposited for 73 min, 63 min and 60 min, respectively. The density of aggregates on the film surface, film flatness, and continuity were investigated using SEM. Initially, the local morphology of the films was observed using 1000× magnified images. Figure 4a–c correspond to pH values of 7.9, 8.2, and 8.5, respectively. It was observed that the film surface was smooth with almost no visible aggregates at pH values of 7.9 and 8.2, while numerous aggregates appeared on the surface of the film deposited at pH = 8.5. From the curve in Figure 3a, it can be inferred that the effect of pH on the PDA film primarily concerns the establishment of film thickness between pH values of 7.9 and 8.2, whereas within the range of pH 8.2–8.5, the effect on film thickness was insignificant, and the increase in pH significantly influenced the formation of aggregates. This phenomenon could be attributed to the gradual increase in the binding rate of dopamine molecules on the silica surface as the concentration of OH- increased under weaker alkaline conditions, resulting in the rapid formation of dopamine molecules on the surface of silica. When the concentration of OH- reaches a certain value, the increase in the binding rate of dopamine molecules on the SiO_2 surface was no longer significant. Instead, it influenced the self-polymerization of freely moving dopamine molecules in the solution, resulting in more aggregates on the film surface. Another possible explanation is that a weaker alkaline solution tends to donate more protons to saturate the dangling bonds of dopamine molecules, reducing their chemical activity. Therefore, less self-polymerization will occur under weaker alkaline conditions, leading to smooth films without aggregates.

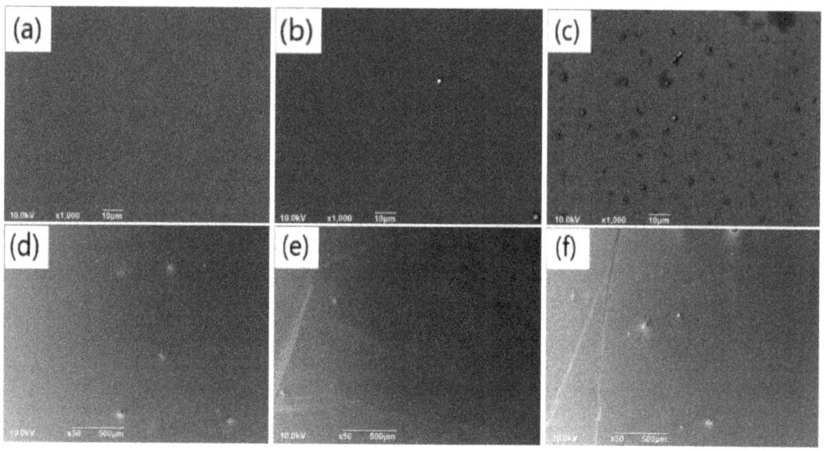

Figure 4. SEM images (1000×) showing the surface of the PDA films with the same thickness deposited for different duration on silicon dioxide substrates under the conditions of (**a**) pH = 7.9, (**b**) pH = 8.2 and (**c**) pH = 8.5; 50× SEM images of the surface of the above-mentioned films corresponding to (**d**) pH = 7.9, (**e**) pH = 8.2, and (**f**) pH = 8.5, respectively.

SEM images with 50× magnification of the surface of the films deposited at pH values of 7.9, 8.2, and 8.5 are presented in Figure 4d–f. The particle size and quantity of surface aggregates are illustrated in Figure S4a–c. The results indicate that, under each pH condition, aggregates appeared on the film surface. However, the number of aggregates on the film surface was relatively small with pH = 8.2, and the average particle size of the aggregates was also small (12.48 μm) compared to the other films. On the other hand, higher numbers of aggregates (8 and 15) arose on the film surface under the condition of pH = 7.9 and 8.5, respectively. Moreover, the average size of the aggregates increased to 13.28 μm and 12.91 μm. Combined with the previous observations on subtle aggregates (Figure 4a–c), it can be concluded that under the condition of pH = 7.9, there were a few large aggregates, but almost no fine aggregates on the film surface. In contrast, under the condition of pH = 8.2, the surface of the film was relatively smooth, with few large or subtle aggregates. However, under the condition of pH = 8.5, there were not only some large aggregates but also a great quantity of subtle aggregates on the film surface. As -mentioned above, the smooth surface of the film deposited with the solution of pH = 8.2 can be attributed to the plenty of protons in the solution, which can saturate the dangling bonds of the dopamine molecules. When the pH decreased further, it is supposed that the film that was built beginning with discrete aggregates before a continuous layer was constructed. This may lead to the final film with large aggregates on the surface, just like in the case of the pH of 7.9.

In chemical reactions, the rate of reaction under some conditions is influenced by the concentration of the reactants [21,22], thereby exerting an effect on the final product. In light of the impact of pH values on the formation of surface aggregates in the PDA film, we selected pH 8.2 with fewer aggregates and further investigated the influence of dopamine concentration in the deposition solution on aggregate formation. This investigation aimed to minimize the number of aggregates and achieve a smooth film.

Similar to the exploration of the impact of pH value, we investigated the effects of different dopamine solution concentrations on the deposition rates of PDA films on SiO_2 substrates, with the aim of identifying conditions that lead to reduced aggregates and a smoother film. Specifically, we prepared PDA films of the same thickness under various concentration conditions and studied the aggregates on their surfaces. SEM images of the cross-sectional views of PDA films deposited for different time periods with varying dopamine concentrations are shown in Figure 5a–e, which represent film thicknesses of 41.34 nm, 53.15 nm, 64.95 nm, and 79.16 nm, respectively. We also examined a low dopamine concentration of 0.5 mg/mL, as shown in Figure S5a, where only a few aggregates were observed on the SiO_2 substrate, and no continuous film was formed. To confirm the formation of a continuous PDA film, EDS analysis was performed on the film surface, a relatively high proportion of C, O and N element was detected only at the aggregates, and there was nearly no C or N element observed in the non-aggregated areas. It could be verified that a PDA film with irregular surface was formed on the SiO_2 substrate.

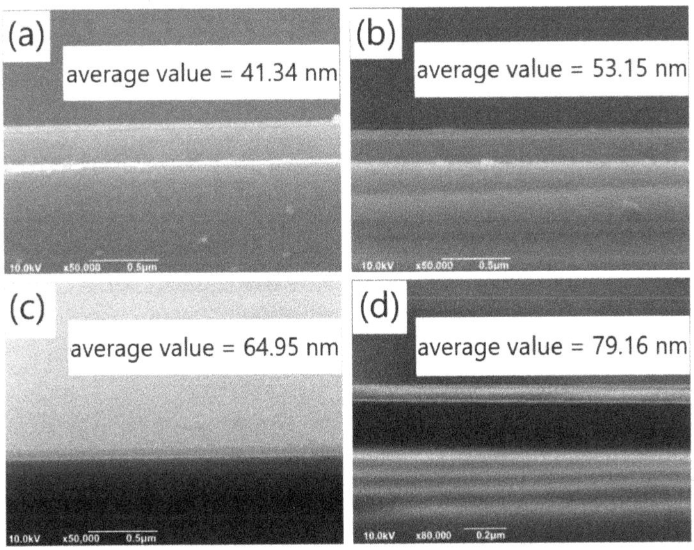

Figure 5. Cross-sectional SEM images of PDA films obtained on silicon dioxide substrates under the concentration conditions of (**a**) 1.5 mg/mL and (**b**) 2 mg/mL with a deposition time of 60 min; Cross-sectional SEM images of the PDA film obtained on the silicon dioxide substrates under the concentration conditions of (**c**) 1.5 mg/mL and (**d**) 2 mg/mL with a deposition time of 90 min.

The thickness of the PDA films was measured using SEM under various deposition time conditions, and the results are presented in Table 2. By combining these results with the thickness of the PDA film deposited with a solution concentration of 1 mg/mL (Table 1), the relationship between the film thickness and the concentration was plotted in Figure 6a. As shown in the figure, the film thickness gradually increased with the concentration for all deposition times below 90 min, and the relationship between the two variables was linear. This finding is consistent with the chemical reaction rate formula $\Delta v(A) = \Delta c(A)/\Delta t$, where v represents the reaction rate, c is the reactant concentration, and t denotes the reaction time. Figure 6b illustrates the relationship between the film thickness and deposition time for different solution concentrations. It can be observed that the film thickness increased linearly with deposition time, when the concentration was between 1 and 2 mg/mL and the deposition time was below 90 min. Specifically, at solution concentrations of 1.5 mg/mL and 2 mg/mL, the deposition rates of dopamine were 0.705 nm/min and 0.883 nm/min, respectively.

Table 2. Thickness of the PDA film deposited for different time.

Concentration	Thickness for Different Deposition Time			
	45 min (nm)	60 min (nm)	75 min (nm)	75 min (nm)
1.5 mg/mL	30.89	41.34	52.55	64.95
2 mg/mL	40.73	53.15	65.91	79.19

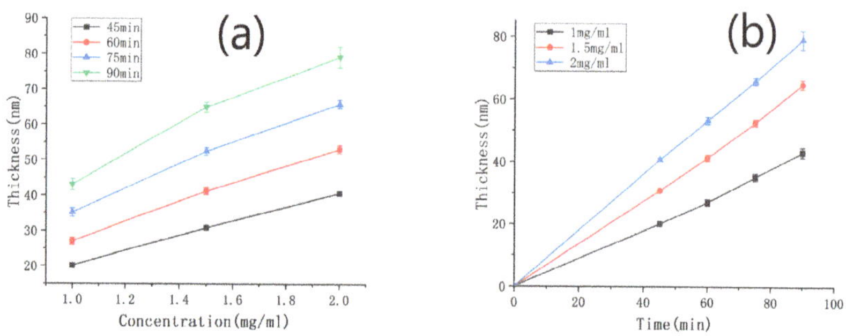

Figure 6. (**a**) Curve of PDA film thickness against the concentration of the dopamine solution, with the deposition time as a parameter; (**b**) curve of PDA film thickness with the deposition time, under different concentrations.

According to the deposition rates obtained from our experiments, we fabricated two sets of samples with SiO_2 substrates. The first set underwent PDA deposition for 42 min in a solution with a concentration of 1.5 mg/mL, while the second set underwent deposition for 32 min in a solution with a concentration of 2 mg/mL. Both groups were prepared using a solution with a pH value of 8.2, resulting in PDA films with the same thickness under different concentration conditions. The density of aggregates on the film surface, as well as the flatness and continuity of the film, were investigated using SEM. Similar to our exploration of the effect of pH value, we examined the subtle particles on the film surface using 1000× magnified images. Figure 7a,b correspond to the concentrations of 1.5 mg/mL and 2 mg/mL, respectively. It was observed that the film surface corresponding to 1.5 mg/mL was smooth, without any subtle aggregated particles. In contrast, a great quantity of aggregates was present on the surface of the film prepared with a concentration of 2 mg/mL. SEM images with a magnification of 50×, depicting the surface of the film having the concentrations of 1.5 mg/mL and 2 mg/mL, are shown in Figure 7c,d, respectively. The particle size and quantity of surface aggregates are summarized in Figure S6a,b. It is noted that, for the concentration of 1.5 mg/mL, the number of aggregates on the film surface was very small, and the average particle size of the aggregates was 9.36 μm. In contrast, when the concentration increased to 2 mg/mL, the number of aggregates on the film surface increased significantly compared to the concentration of 1.5 mg/mL, exceeding 50, and the average particle size was 10.41 μm.

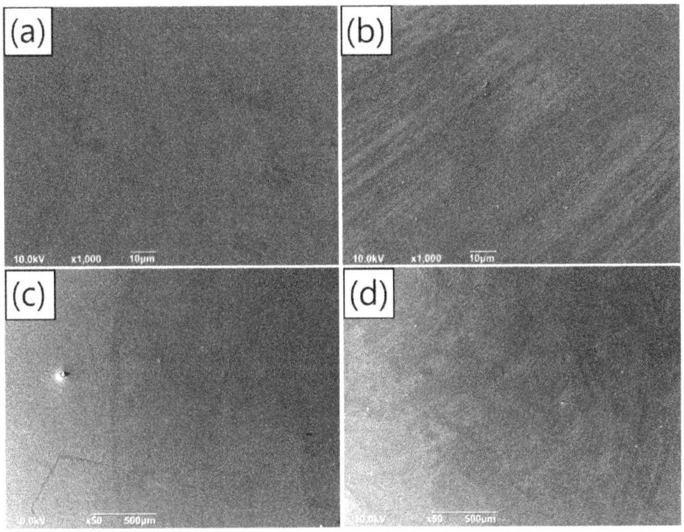

Figure 7. Surface SEM images (1000×) of the same thickness PDA films prepared with different deposition time on silicon dioxide substrates under the concentrations of (**a**) 1.5 mg/mL and (**b**) 2 mg/mL; surface SEM images (50×) of the same thickness PDA films fabricated with different deposition time on silicon dioxide substrates under the concentrations of (**c**) 1.5 mg/mL and (**d**) 2 mg/mL.

By observing the surface aggregates on the PDA film under two concentration conditions and combining with the previous inspection of the PDA film formed at the concentration of 1 mg/mL (Figure 4b,e), we can conclude that the number of large particles and tiny aggregates on the film of 2 mg/mL was greater than that of the films deposited under the other two concentrations. In the cases of concentrations of 1 mg/mL and 1.5 mg/mL, the film surface was relatively flat, and there were few large particles on the surface. Compared with the film of 1 mg/mL, the PDA film deposited at the concentration of 1.5 mg/mL was smoother. Furthermore, no subtle aggregates were observed on the film surface in the latter case, and the average particle size of the large particles was also smaller. This is probably because the concentration of 1 mg/mL was insufficient to make dopamine molecules form film structure rapidly. At this concentration, the film can only be formed through continuous accumulation of dot aggregates, leading to aggregates arising on the surface. The concentration condition of 1.5 mg/mL satisfied the requirement for dopamine molecules to build a film structure rapidly. As a result, no subtle aggregates are formed on the film surface. However, the continuous increase in concentration also promoted the self-polymerization of the freely moving dopamine molecules in the solution, resulting in more aggregates on the surface of the deposited film. This may be the case for the film of 2 mg/mL [23,24].

Based on the investigation of the impact of varying pH levels and concentrations on the deposition of PDA film, a pH level of 8.2 and a concentration of 1.5 mg/mL were selected to deposit the PDA film with fewer surface aggregates. To determine the appropriate parameters for a continuous and thin PDA film, the deposition time was gradually increased. PDA films with deposition times of 2 min, 10 min, and 20 min were prepared. However, when the film thickness was very small, it was challenging to ascertain whether a continuous film structure was formed from the cross-sectional SEM images. According to this situation, we tried to make use of the optical microscope (OM) and scratches made in the film, assisted with the SEM inspection, to assess whether a continuous film was formed on the sample surface. The surface of the PDA film was investigated with OM, and the sample surface of the film was scraped to make some scratches. The OM observations are presented in Figure 6, where Figure 6a–c illustrate the OM images of the PDA film surface deposited for 2 min, 10 min, and 20 min, respectively. It can be observed that the sample surface characterized by OM was blue-purple under a deposition time of 2 min, which corresponds to the color of SiO_2. This suggests that at the deposition time of 2 min, there was either no PDA film formed on top of the SiO_2 layer or only sporadic dotted PDA accumulations. In contrast, under a deposition time of 10 min, the OM image of the sample surface appeared light yellow, and no obvious scratches could be observed after the surface was lightly scratched. It is supposed, under the deposition time of 10 min, layered PDA may be formed in some areas on the surface of the sample, but it lacks continuity and integrity. When the deposition time reached 20 min, the light-yellow color became more intense than that of the PDA film deposited for 10 min. Obvious scratches were found on the sample surface when it was lightly scraped. Thus, we conclude that a continuous and complete PDA film was formed on the sample with a deposition time of 20 min.

Atomic force microscopy (AFM) was utilized to investigate the surface roughness of the PDA film and to validate the observations obtained from OM. AFM images of the PDA film deposited on a SiO_2 substrate are shown in Figure 8d–f, corresponding to deposition times of 2 min, 10 min, and 20 min, respectively. The surface roughness of the PDA film deposited for 2 min was found to be 0.6458 nm, which is comparable to that of the SiO_2 substrate. The presence of abundant white spots in the AFM image suggests the absence of a continuous PDA layer on the sample surface. When the deposition time was increased to 10 min, the number of white spots decreased, and the surface roughness increased to 3.185 nm. These observations imply that a layered PDA film was formed in some regions on the sample surface, but the film structure was non-uniform, and some point-like aggregates still existed on the substrate surface. Comparatively, the PDA film deposited for 20 min exhibited a significant change in the surface morphology. The absence of white spots in the AFM image suggests the presence of a uniform and continuous PDA film on the SiO_2 substrate, and the surface roughness decreased to 1.763 nm as compared to the 10 min deposition. The cross-sectional view of the AFM scanning of the PDA film deposited for 20 min is presented in Figure 8g (its corresponding surface AFM image and test position are shown in Figure S7), where a depression ten of nanometers wide is observed. The absence of the PDA film at this location may be due to direct contact of the AFM probe with the SiO_2 layer at the bottom of the depression. The thickness of the PDA film, estimated from the difference between the lowest part of the depression and the film surface, was about 14 nm, which is consistent with the PDA deposition rate deduced from Figure 6.

The PDA film was synthesized using optimal parameters (pH = 8.2, solution concentration of 1.5 mg/mL, and deposition for 20 min), followed by carbonization to obtain graphene-like material through annealing at 800 °C in a vacuum (0.1 Pa) for 60 min at a heating rate of 10 °C/min. Raman spectroscopy was employed to analyze the carbon chain structure of the carbonized PDA film and compare it with that of films with the same thickness but different surface characteristics. Specifically, we compared the Raman spectra of the carbonization products of the PDA film with a smooth surface and no surface aggregates prepared with the optimized parameters (red curve), the carbonization products of the thicker PDA film (green curve), and the carbonization products of the PDA film with surface aggregates (blue curve). In Figure 9a, the Raman spectra of all three curves show obvious D peaks, indicating the presence of structural defects. Additionally, all three curves display 2D peaks, indicating the presence of multi-layered graphene. The I_D/I_G and I_{2D}/I_G ratios were compared and are summarized in Figure 7b. The I_D/I_G ratio of the carbonization products of the PDA film with a smooth surface and no surface aggregates prepared with optimal parameters was lower than that of the thicker PDA film and the PDA film with surface aggregates, suggesting fewer structural defects. Moreover, the half-peak width of the red curve was comparatively narrow, indicating a higher degree of graphitization. The I_{2D}/I_G ratios in all three cases were significantly less than 1, indicating the carbonized films consist of multi-layered graphene rather than one-layer graphene.

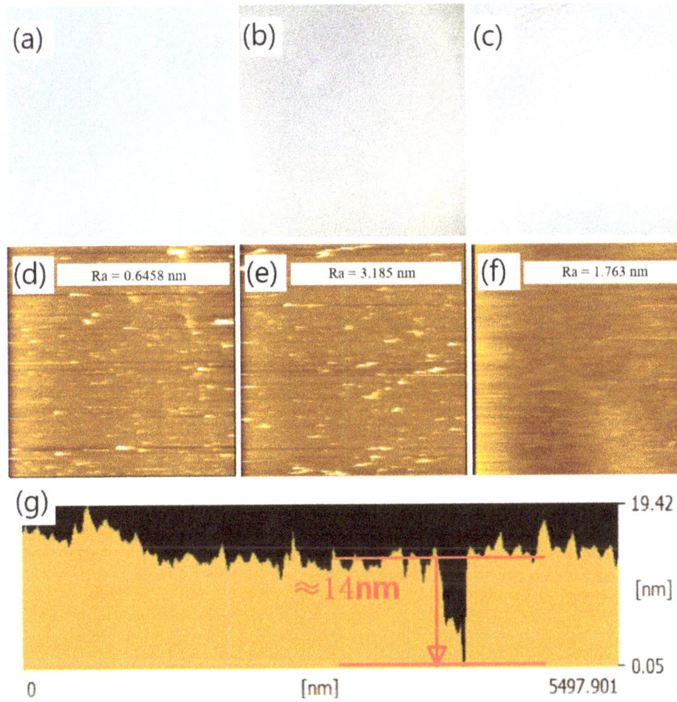

Figure 8. OM image of the surface of the PDA film deposited on silicon dioxide substrate under the pH of 8.2 and the concentration of 1.5 mg/mL with a deposition time of (**a**) 2 min, (**b**) 10 min, and (**c**) 20 min; the AFM image of the surface of the PDA film deposited on silicon dioxide substrate under the pH of 8.2 and the concentration of 1.5 mg/mL with a deposition time of (**d**) 2 min, (**e**) 10 min, and (**f**) 20 min; (**g**) the cross-sectional view of the AFM scanning of the surface of the PDA film under the deposition time of 20 min.

The PDA films were carbonized at 1000 °C under vacuum conditions (0.1 Pa) while maintaining the same heating rate. The resulting Raman spectra are presented in Figure 9c. It was evident from the spectra that the D peaks of the three curves weaken significantly with increasing annealing temperature, while the 2D peaks also exhibited a marked reduction. This behavior may be attributed to the accelerated decomposition of PDA at high temperature, promoting the rapid and thorough combination of carbon atoms and improved crystallinity of the carbonized layer. The ratios of I_D/I_G and I_{2D}/I_G in the three cases were compared, as depicted in Figure 9d. Consistent with the previous case of 800 °C, the I_D/I_G ratio of the carbonization products of the PDA film with a smooth surface and without surface aggregates is comparatively lower than the other two cases. Furthermore, the I_{2D}/I_G ratio of the carbonization products of the optimally prepared film was higher than that of the other two films, indicating that the number of stacked graphene layers was smaller. Notably, the I_D/I_G ratios of all three films carbonized at 1000 °C were evidently lower than those at 800 °C, suggesting a substantial reduction in structural defects. The Raman spectrum of the C-PDA acquired by Yu [2] also showed the D peaks and G peaks, confirming the existence of the graphite-like nanostructures. Compared with research results of Yu, our study presented the improved uniformity and the reduced thickness of the PDA film. In our study, the G peak of the Raman spectrum was more distinct, and at

2700 cm^{-1}, a 2D peak appeared, indicating the formation of the graphene-like structure. This is also consistent with the research results from Li [25], who used neutron diffraction, Raman spectroscopy, and other techniques to verify that C-PDA has some graphite-like structures as the annealing temperature increased.

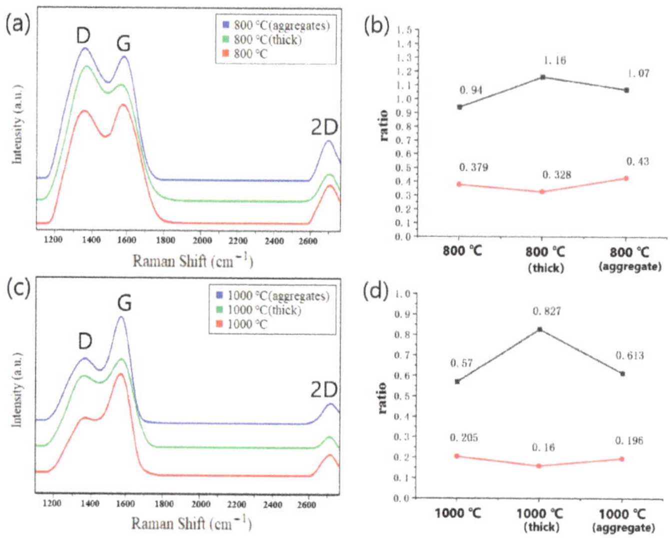

Figure 9. (**a**) Comparison of Raman spectra of the at 800 °C carbonized PDA films with different morphologies (red curve corresponds to PDA film prepared according to the achieved optimal parameters, green curve corresponds to PDA film with smooth surface but bigger thickness, the blue curve corresponds to PDA film with the same thickness as the red curve, but aggregates on the surface). (**b**) The ratios of I_D/I_G and I_{2D}/I_G in three cases after carbonization at 800 °C; (**c**) Raman spectrum comparison after carbonization at 1000 °C. (**d**) The ratios of I_D/I_G and I_{2D}/I_G in three cases after carbonization at 1000 °C.

The electrical conductivity of the carbonization products of the PDA film with a smooth surface and without surface aggregates was approximately 670 S·m^{-1}, as illustrated in Figure 10a. This value is close to that of multi-layer graphene. The attained high electrical conductivity may be attributed to the modified molecular charge transfer behavior induced by nitrogen doping or the improved crystallinity of the carbonized film resulting from the optimized deposition parameters [26,27]. Compared to the thicker film (550 S·m^{-1}) and the film with numerous aggregates (605 S·m^{-1}) (Figure 10b,c), the carbonized PDA film with a smooth surface and no aggregates exhibited superior electrical conductivity.

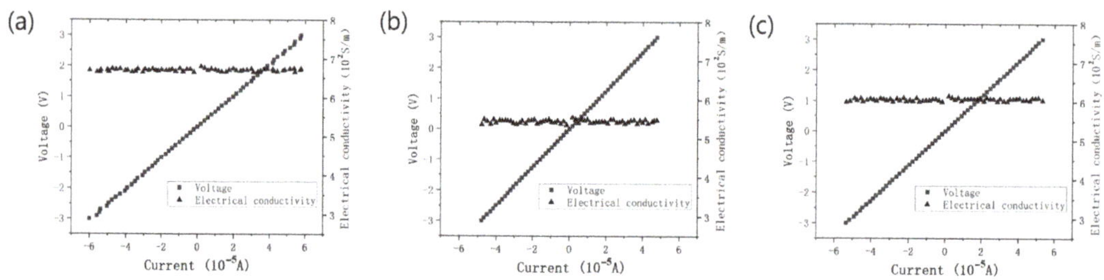

Figure 10. The I–V curves and electrical conductivity of the carbonization product of (**a**) the smooth and no aggregates PDA film, (**b**) the thicker PDA film, and (**c**) the PDA film with many aggregates.

Through the Raman spectrum analysis and electrical conductivity measurement of the at different temperatures carbonized films, it can be shown that the carbonization product of the PDA film with a smooth surface and without surface aggregates, prepared with the optimized parameters, had fewer structural defects and a higher electrical conductivity.

Figure 11a displays the XPS spectra of the PDA film with a smooth surface and without surface aggregates carbonized at 1000 °C. The strong O1s peaks and the appearance of Si2s and Si2p peaks suggest that the thickness of the carbonized film was remarkably thin, likely below the detection depth of the XPS (≈10 nm). Upon closer inspection of the C1s peak (Figure 11b), it can be observed that the peak at a binding energy of 288 eV was almost absent, indicating the absence of the C = O structure. In contrast, the peak at a binding energy of 286 eV corresponds to the C-O/C-N structure, indicating the presence of a small amount of O and N doping in the film. The peak with a binding energy of 285 eV corresponds to the C-C/C = C structure, and the large peak indicates that most of the carbon atoms are bound into the cross-linked honeycomb lattice structure.

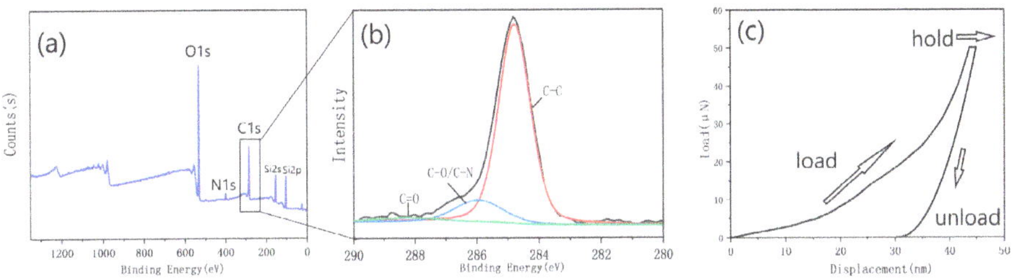

Figure 11. (**a**) XPS spectrum of the film with the optimized parameters; (**b**) the C1s peak of the XPS spectrum; (**c**) a representative load vs. displacement curve obtained from the thick sample.

Due to the excessively small thickness of the prepared graphene-like material, it was not suitable for testing using the nanoindentation instrument. Therefore, a thick PDA film (with a thickness much greater than that shown in Figure 9) was fabricated and carbonized under 1000 °C. Based on the force-indentation depth curve of the carbonized thick film (Figure 11c), the Young's modulus of the material was calculated to be approximately 30 GPa. A certain degree of creep could be observed. This was mainly due to the weak van der Waals forces between the atom layers in the graphite structure in the carbonized film. When subjected to external force, the carbon atom planes would bend and deform, maintaining the lateral structural stability without requiring the in-plane rearrangement of the carbon atoms to adapt to the external force.

Based on the mechanical properties of the carbonized product of the thick film, those of the prepared thin graphene-like material were inferred. Comparatively, the graphene-like thin film contained fewer structural defects (according to the results in Figure 9, the thicker the PDA film, the more defects generated after carbonization). For materials with graphene-like structures, an increase in the concentration of point defects and single vacancy defects will lead to a significant decrease in their Young's modulus [28]. Furthermore, as the thickness decreased, the Young's modulus within the graphene plane significantly increased [29]. This all indicates that the prepared thin graphene-like material had a higher Young's modulus. Additionally, its multi-layered structure will partially retain its creep properties.

4. Conclusions

By investigating the impact of varying pH values on the formation of aggregates on the surface of deposited PDA film, distinct effects were observed across different pH value intervals. Within the range of 7.9 to 8.2, pH predominantly influences film thickness, while in the interval of 8.2 to 8.5, pH primarily affects the aggregation of particles on the film surface. For the deposition time below 90 min and the pH values between 7.9 and 8.5, the thickness of the PDA film exhibits a linear relationship with time. The deposition of a PDA film with minimal aggregates was achieved at pH 8.2. This was attributed to the various concentrations of the protons or OH− in the solutions. Solutions with a deficiency in the amount of OH− tend to donate more protons to saturate the dangling bonds of dopamine molecules, reducing their chemical activity and inhibiting their aggregation, while excessive OH− can promote the self-polymerization of dopamine molecules that move freely in the solution, resulting in more aggregates on the film surface [30–32]. To investigate the effects of different concentrations on the formation of aggregates on the PDA film surface, a pH of 8.2 was maintained. For the deposition time below 90 min, the film thickness exhibited a linear relationship with solution concentration and deposition time. A PDA film with negligible subtle aggregates and only a minimal quantity of large aggregates was formed at a solution concentration of 1.5 mg/mL.

Finally, to produce a thin and smooth film without surface aggregates, the deposition time was gradually increased under the conditions of pH 8.2 and the concentration of 1.5 mg/mL. It was observed that at a deposition time of 20 min, a continuous, smooth, and uniform PDA film with a thickness of 14 nm was successfully deposited on the SiO_2 substrate. Utilizing the prepared PDA as a precursor, a graphene-like film was obtained through a high-temperature carbonization process. According to the Raman spectrum and the four-terminal sensing analysis, the multilayered graphene derived from the film deposited with the optimized parameters showed fewer structural defects and higher electrical conductivity compared to the carbonized products obtained from thicker PDA films and the films with surface aggregates. With optimized pH value and dopamine solution concentration, the structure of the deposited PDA film was improved, so the finally prepared graphene-like product possesses better structural properties. The XPS results also indicate that the carbonization products of the PDA film with a smooth surface and without surface aggregates are primarily composed of carbon atoms bonded in a cross-linked honeycomb lattice structure. Moreover, the nanoindentation analysis verified that the fabricated graphene-like film had good mechanical properties. Therefore, it is quite suitable for the application in flexible electronics.

Supplementary Materials: The following supporting information can be downloaded at: https://www.mdpi.com/article/10.3390/cryst13040607/s1, Figure S1: XRD curves of the smooth areas of the film and the surface aggregates; Figure S2: pH value detection of each reaction solution, (a) pH = 7.9, (b) pH = 8.2 and (c) pH = 8.5; Figure S3: Color comparison of the deposition solutions after 90 min in dark environment; Figure S4: The size and quantity of the aggregates calculated based on the surface SEM images (50×) of the same thickness PDA films deposited for different time length on SiO_2 substrates under (a) pH = 7.9, (b) pH = 8.2 and (c) pH = 8.5; Figure S5: (a) SEM images of cross-sections of the PDA films prepared on SiO_2 substrates at a concentration of 0.5 mg/mL; EDS analysis on (b) the aggregated particle and (c) on the smooth area of the film (without aggregates); Figure S6: Statistical graph of particle size and quantity of surface aggregates calculated through the 50× magnified SEM images of films deposited under (a) c = 1.5 mg/mL and (b) c = 2 mg/mL; Figure S7: The surface AFM image and the test position for Figure 8g.

Author Contributions: Conceptualization, D.W. and Q.W.; methodology, D.W. and T.P.; validation, D.W. and N.D.; preparation, D.W.; writing—review and editing, Q.W. and Z.L. All authors have read and agreed to the published version of the manuscript.

Funding: This research was funded by Key R&D Project in Shaanxi Province "Brain-inspired computing device based on two-dimensional material/polymer semiconductor heterojunction" (No.: 2020GXLH-Z-012).

Institutional Review Board Statement: Not applicable.

Informed Consent Statement: Not applicable.

Data Availability Statement: Not applicable.

Conflicts of Interest: The authors declare no conflict of interest.

References

1. Lee, H.; Dellatore, S.M.; Miller, W.M.; Messersmith, P.B. Mussel-inspired surface chemistry for multifunctional coatings. *Science* **2007**, *318*, 426–430. [CrossRef] [PubMed]
2. Yu, X.; Fan, H.; Liu, Y.; Shi, Z.; Jin, Z. Characterization of carbonized polydopamine nanoparticles suggests ordered supramolecular structure of polydopamine. *Langmuir* **2014**, *30*, 5497–5505. [CrossRef] [PubMed]
3. Yang, H.-C.; Pi, J.-K.; Liao, K.-J.; Huang, H.; Wu, Q.-Y.; Huang, X.-J.; Xu, Z.-K. Silica-decorated polypropylene microfiltration membranes with a mussel-inspired intermediate layer for oil-in-water emulsion separation. *ACS Appl. Mater. Interfaces* **2014**, *6*, 12566–12572. [CrossRef]
4. Lee, K.; Park, M.; Malollari, K.G.; Shin, J.; Winkler, S.M.; Zheng, Y.; Park, J.H.; Grigoropoulos, C.P.; Messersmith, P.B. Laser-induced graphitization of polydopamine leads to enhanced mechanical performance while preserving multifunctionality. *Nat. Commun.* **2020**, *11*, 4848. [CrossRef]
5. Xiao, C.; Chu, X.; Yang, Y.; Li, X.; Zhang, X.; Chen, J. Hollow nitrogen-doped carbon microspheres pyrolyzed from self-polymerized dopamine and its application in simultaneous electrochemical determination of uric acid, ascorbic acid and dopamine. *Biosens. Bioelectron.* **2011**, *26*, 2934–2939. [CrossRef]
6. Awasthi, A.K.; Gupta, S.; Thakur, J.; Gupta, S.; Pal, S.; Bajaj, A.; Srivastava, A. Polydopamine-on-liposomes: Stable nanoformulations, uniform coatings and superior antifouling performance. *Nanoscale* **2020**, *12*, 5021–5030. [CrossRef]
7. Kang, S.M.; Ryou, M.-H.; Choi, J.W.; Lee, H. Mussel- and diatom-inspired silica coating on separators yields improved power and safety in Li-Ion Batteries. *Chem. Mater.* **2012**, *24*, 3481–3485. [CrossRef]
8. Liu, R.; Mahurin, S.M.; Li, C.; Unocic, R.R.; Idrobo, J.C.; Gao, H.; Pennycook, S.J.; Dai, S. Dopamine as a carbon source: The controlled synthesis of hollow carbon spheres and yolk-structured carbon nanocomposites. *Angew. Chem.* **2011**, *123*, 6931–6934. [CrossRef]
9. Kong, J.; Yee, W.A.; Yang, L.; Wei, Y.; Phua, S.L.; Ong, H.G.; Ang, J.M.; Li, X.; Lu, X. Highly electrically conductive layered carbon derived from polydopamine and its functions in SNO2-based lithium-ion battery anodes. *Chem. Commun.* **2012**, *48*, 10316. [CrossRef]
10. Li, R.; Parvez, K.; Hinkel, F.; Feng, X.; Müllen, K. Bioinspired wafer-scale production of highly stretchable carbon films for transparent conductive electrodes. *Angew. Chem.* **2013**, *125*, 5645–5648. [CrossRef]
11. Lim, K.H.; Kweon, H.; Kim, H. Polydopamine-derived nitrogen-doped graphitic carbon for a bifunctional oxygen electrode in a non-aqueous li-O2battery. *J. Electrochem. Soc.* **2017**, *164*, A1595–A1600. [CrossRef]
12. Byun, S.-J.; Lim, H.; Shin, G.-Y.; Han, T.-H.; Oh, S.H.; Ahn, J.-H.; Choi, H.C.; Lee, T.-W. Graphenes converted from polymers. *J. Phys. Chem. Lett.* **2011**, *2*, 493–497. [CrossRef]
13. Zhou, P.; Deng, Y.; Lyu, B.; Zhang, R.; Zhang, H.; Ma, H.; Lyu, Y.; Wei, S. Rapidly-deposited polydopamine coating via high temperature and vigorous stirring: Formation, characterization and Biofunctional Evaluation. *PLoS ONE* **2014**, *9*, e113087. [CrossRef] [PubMed]
14. Zhao, M.-X.; Li, J.; Gao, X. Eliminating diffusion limitations at the solid–liquid interface for rapid polymer deposition. *ACS Biomater. Sci. Eng.* **2017**, *3*, 782–786. [CrossRef]
15. Zhao, M.-X.; Li, J.; Gao, X. Gradient coating of polydopamine via CDR. *Langmuir* **2017**, *33*, 6727–6731. [CrossRef]
16. Kim, H.W.; McCloskey, B.D.; Choi, T.H.; Lee, C.; Kim, M.-J.; Freeman, B.D.; Park, H.B. Oxygen concentration control of dopamine-induced high uniformity surface coating chemistry. *ACS Appl. Mater. Interfaces* **2013**, *5*, 233–238. [CrossRef]
17. Li, Y.; Choi, J.; Sun, Z.; Russell, T.P.; Carter, K.R. Fabrication of sub-20 nm patterns using dopamine chemistry in self-aligned double patterning. *Nanoscale* **2018**, *10*, 20779–20784. [CrossRef]
18. Du, X.; Li, L.; Behboodi-Sadabad, F.; Welle, A.; Li, J.; Heissler, S.; Zhang, H.; Plumeré, N.; Levkin, P.A. Bio-inspired strategy for controlled dopamine polymerization in basic solutions. *Polym. Chem.* **2017**, *8*, 2145–2151. [CrossRef]
19. Tien, H.-W.; Huang, Y.-L.; Yang, S.-Y.; Hsiao, S.-T.; Liao, W.-H.; Li, H.-M.; Wang, Y.-S.; Wang, J.-Y.; Ma, C.-C.M. Preparation of Transparent, Conductive Films by Graphenenanosheet Deposition on Hydrophilic or Hydrophobic Surfaces through Control of the ph Value. *J. Mater. Chem.* **2012**, *22*, 2545–2552. [CrossRef]
20. Xi, Z.-Y.; Xu, Y.-Y.; Zhu, L.-P.; Wang, Y.; Zhu, B.-K. A facile method of surface modification for hydrophobic polymer membranes based on the adhesive behavior of poly(dopa) and poly(dopamine). *J. Membr. Sci.* **2009**, *327*, 244–253. [CrossRef]
21. Ball, V.; Frari, D.D.; Toniazzo, V.; Ruch, D. Kinetics of polydopamine film deposition as a function of ph and dopamine concentration: Insights in the polydopamine deposition mechanism. *J. Colloid Interface Sci.* **2012**, *386*, 366–372. [CrossRef] [PubMed]
22. Wang, Z.; Xu, C.; Lu, Y.; Wei, G.; Ye, G.; Sun, T.; Chen, J. Microplasma-Assisted Rapid, chemical oxidant-free and controllable polymerization of dopamine for surface modification. *Polym. Chem.* **2017**, *8*, 4388–4392. [CrossRef]
23. Bernsmann, F.; Ponche, A.; Ringwald, C.; Hemmerlé, J.; Raya, J.; Bechinger, B.; Voegel, J.-C.; Schaaf, P.; Ball, V. Characterization of dopamine−melanin growth on silicon oxide. *J. Phys. Chem. C* **2009**, *113*, 8234–8242. [CrossRef]
24. Lee, M.; Lee, S.-H.; Oh, I.-K.; Lee, H. Microwave-Accelerated Rapid, chemical oxidant-free, material-independent surface chemistry of poly(dopamine). *Small* **2016**, *13*, 1600443. [CrossRef]
25. Gong, W.; Chen, W.; He, J.; Tong, Y.; Liu, C.; Su, L.; Gao, B.; Yang, H.; Zhang, Y.; Zhang, X. Substrate-independent and large-area synthesis of carbon nanotube thin films using zno nanorods as template and dopamine as carbon precursor. *Carbon* **2015**, *83*, 275–281. [CrossRef]

26. Klosterman, L.; Riley, J.K.; Bettinger, C.J. Control of heterogeneous nucleation and growth kinetics of dopamine-melanin by altering substrate chemistry. *Langmuir* **2015**, *31*, 3451–3458. [CrossRef]
27. Huang, X.; Rao, W.; Chen, Y.; Ding, W.; Zhu, H.; Yu, M.; Chen, J.; Zhang, Q. Infrared Emitting Properties and environmental stability performance of aluminum/polymer composite coating. *J. Mater. Sci. Mater. Electron.* **2016**, *27*, 5543–5548. [CrossRef]
28. Hao, F.; Fang, D.; Xu, Z. Mechanical and Thermal Transport Properties of Graphene with Defects. *Appl. Phys. Lett.* **2011**, *99*, 041901. [CrossRef]
29. Lee, C.; Wei, X.; Kysar, J.W.; Hone, J. Measurement of the Elastic Properties and Intrinsic Strength of Monolayer Graphene. *Science* **2008**, *321*, 385–388. [CrossRef] [PubMed]
30. Yang, H.-C.; Wu, Q.-Y.; Wan, L.-S.; Xu, Z.-K. Polydopamine gradients by oxygen diffusion controlled autoxidation. *Chem. Commun.* **2013**, *49*, 10522. [CrossRef]
31. Madhurakkat Perikamana, S.K.; Lee, J.K.; Shin, Y.M.; Ahmad, T.; Kim, S.-jeong; Park, K.M.; Shin, H. Oxygen-dependent generation of a graded polydopamine coating on nanofibrous materials for controlling stem cell functions. *J. Mater. Chem. B* **2017**, *5*, 8865–8878. [CrossRef] [PubMed]
32. Liebscher, J.; Mrówczyński, R.; Scheidt, H.A.; Filip, C.; Hădade, N.D.; Turcu, R.; Bende, A.; Beck, S. Structure of Polydopamine: A Never-Ending Story? *Langmuir* **2013**, *29*, 10539–10548. [CrossRef] [PubMed]

crystals

MDPI

Brief Report

Laser-Induced Graphene in Polyimide for Antenna Applications

Aivaras Sartanavičius, Justina Žemgulytė, Paulius Ragulis, Karolis Ratautas and Romualdas Trusovas *

Center for Physical Sciences and Technology, Savanoriu Avenue 231, LT-02300 Vilnius, Lithuania
* Correspondence: romualdas.trusovas@ftmc.lt

Abstract: Laser-induced graphene (LIG) has gained considerable attention recently due to its unique properties and potential applications. In this study, we investigated using LIG in polyimide (PI) as a material for antenna applications. The LIG-–PI composite material was prepared by a facile picosecond laser (1064 nm) irradiation process, which resulted in a conductive graphene network within the PI matrix. Furthermore, LIG formation was confirmed by Raman spectroscopy and sheet resistance measurements. Finally, a patch antenna from LIG with 2.45 GHz microwaves was simulated, produced and tested. These findings suggest that LIG–PI composites have great potential for use in high-frequency electronic devices and can provide a new avenue for the development of flexible and wearable electronics.

Keywords: picosecond laser; graphene; LIG; antenna

1. Introduction

Laser-induced graphene (LIG) is a graphene material synthesized using lasers. Laser irradiation heats and transforms the material into a three-dimensional carbon structure during this process. Mainly being a one-step process, LIG formation gathers considerable attention due to its unique properties, such as mechanical flexibility, biocompatibility, and large electrical conductivity. These properties make this material for various wearable and flexible electronic device applications [1–8].

LIG can be produced, when the graphene oxide (GO) reduction process induces the removal of oxygen-containing groups from the GO matrix, the resulting conductive multilayer graphene matrix [9–14]. Furthermore, LIG formation can be performed on organic materials such as paper, wood, and food products and various polymers such as PDMS, polyimide, and PEI [5,6,15–20]. In addition, various groups have already applied the LIG process on various materials using different laser sources, such as pulsed 1064 nm, 532 nm, and 355 nm laser irradiation and CO_2 lasers [20–24]. When pulsed lasers and shorter wavelengths are applied, the photochemical effect is more dominant during the LIG formation as photons, which possess higher energy and directly excite molecules of precursor materials. As a result, chemical bonds are broken, when high photon energy is absorbed. On the other hand, while 1064 nm of CO_2 laser irradiation is applied, the photothermal effect, resulting in thermal energy production, is more prominent. In this case, laser irradiation causes lattice vibrations, increasing thermal energy and breaking chemical bonds [3,25].

One key aspect that affects manufacturing consistency is the laser parameters used during the synthesis process. Factors such as laser power, wavelength, duration, scanning speed, and scanning time [26] influence the quality and properties of the resulting LIG. Therefore, controlling these parameters precisely and consistently is crucial to obtaining uniform LIG samples while maintaining high production throughput [27].

Different polymers can yield varying results in terms of LIG quality and properties. Therefore, the choice of precursor materials plays a significant role in manufacturing

Citation: Sartanavičius, A.; Žemgulytė, J.; Ragulis, P.; Ratautas, K.; Trusovas, R. Laser-Induced Graphene in Polyimide for Antenna Applications. *Crystals* **2023**, *13*, 1003. https://doi.org/10.3390/cryst13071003

Academic Editor: Conrad Becker

Received: 20 May 2023
Revised: 13 June 2023
Accepted: 21 June 2023
Published: 23 June 2023

consistency. Ensuring a consistent and reliable supply of high-quality precursor materials is essential for manufacturing consistency in LIG production.

Cao et al. [28] demonstrated that LIG could be stable and used in severe environmental conditions. In addition, different research groups demonstrated scalability options for LIG production [29–32]. Achieving large-scale, uniform, and high-quality LIG production remains a challenge that requires advancements in production techniques, equipment, and process optimization. The cost of laser equipment can be significant for industrial-scale production. On the other hand, the laser process offers flexibility for switching shapes and types of LIG-based electronic devices. Furthermore, choosing precursor materials for LIG production can impact the overall cost. Some precursors, such as polyimide, are more expensive than others. Recent results of LIG formation in paper or biomass products [6,26,32,33] may significantly decrease production costs and contribute to the circular economy.

It is worth noting that the research on laser-induced graphene is ongoing, and advancements are being made to improve manufacturing consistency [34]. As the understanding of the synthesis process deepens and standardized protocols are developed [27], it is expected that the manufacturing consistency of laser-induced graphene will continue to improve over time.

The application of LIG has already been demonstrated for sensors and IoT devices [8,35–39]. Some research groups have shown that LIG structures can be used as antennae [1,33,40]. LIG is lightweight and flexible. This flexibility allows for integrating LIG antennae into curved surfaces or wearable electronics, expanding their potential applications. The lightweight nature of LIG also minimizes the overall weight of devices, which is crucial for portable and lightweight electronics. This flexibility in LIG integration enables the development of innovative antenna designs and complex electronic systems. These benefits make LIG a promising material for advancing the performance and functionality of antennae and electronic devices. Therefore, this research presents our initial study of the LIG−PI application for a 2.45 GHz patch antenna working in the WiFi range.

2. Materials and Methods

2.1. Samples

Our experiments utilized Polyimide Kapton® films manufactured by DuPont (Wilmington, DE, USA) with a thickness of 127 μm.

2.2. LIG Formation

A picosecond solid-state laser Atlantic (Ekspla, Vilnius, Lithuania), operating at 1064 nm wavelength, was used for LIG formation experiments. The pulse duration was 10 ps, and the pulse repetition rate was 100 kHz. A SCANgine (ScanLab, Munich, Germany) galvanometer scanner with a telecentric F-theta objective of a 160 mm focal length was used for laser beam positioning. The beam scanning speed was 200 mm/s. The laser irradiation dose was changed by varying laser average power (4 and 5 W), defocusing and changing hatch distances between 20 μm and 30 μm. The sample was put under the focusing objective, and LIG patterns were formed by scanning the laser beam over the PI surface (Figure 1). In initial tests for selecting optimal parameters, the squares (10 × 10 mm) of LIG were formed on the PI surface. Raman measurements and sheet resistance measurements were performed on these samples. A patch antenna was produced using optimal laser process parameters (lowest sheet resistance to LIG).

2.3. Optical Microscopy

The optical microscope ECLIPSE LV100D (Nikon, Tokyo, Japan) was used to evaluate the thickness of the formed LIG layer.

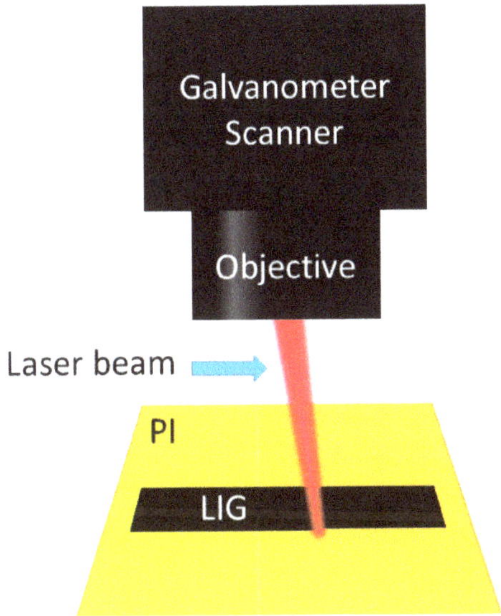

Figure 1. Experimental setup of LIG formation.

2.4. Resistance Measurement

Sheet resistance measurements were performed using a source meter (Keithley 2602A, Keithley, Cleveland, OH, USA) with the measurement software (TSP® Express, Keithley, Cleveland, OH, USA). Measurements were performed using the 4-probe method [41]. Four co-linear electrical probes were identically spaced form a four-point probe. To operate, a DC current (I) was applied between the outer two probes, and the voltage drop between the inner two probes was then measured. The following equation can then be used to obtain the sheet resistance:

$$R_s = \frac{\pi}{\ln(2)} \frac{\Delta V}{I} \tag{1}$$

where R_s is the sheet resistance, ΔV is the change in voltage measured between the inner probes, and I is the current applied between the outer probes.

2.5. Raman Measurements

A Raman spectrometer/microscope inVia (Renishaw, Wotton-under-Edge, UK) was utilized for conducting Raman spectroscopic measurements. Equipped with a thermoelectrically cooled CCD detector at −70 °C, the spectrometer measured the spectra at a 532 nm laser irradiation dispersed by a 1800 grooves/mm grating. The laser power at the sample was 0.5 mW. The measurements were performed using a 50×/0.75 NA objective lens with a 100 s accumulation time, and the resulting spectral peaks were processed using OriginPro 8.5 software from Microlab (Northampton, MA, USA).

2.6. Antenna Modelling

CST Studio suite (Simulia, Johnston, RI, USA) software was employed to simulate LIG antennae.

2.7. Antenna Measurement

The following parameters were measured to evaluate the manufactured antennae: reflection coefficient, operating frequency, maximum gain, half-power beam width, and surface conductivity. Gain and directivity parameters were measured in an anechoic chamber. The measurement setup is explained in Figure 2. The antenna under investigation was placed on a rotation stand that could be rotated 360° by a stepping motor. The excitation port of an antenna was connected to a calibrated NRZP-Z24 (Rhode & Schwartz, Munich, Germany) power meter. The power meter was connected to a PC via a USB cable. A broadband horn antenna with known specifications was positioned 2 m away from the transmitting antenna (far-field). The horn antenna was connected to an E8257D (Agilent technologies, Santa Clara, CA, USA) programmable radio signal generator through a coaxial connection. The stand, the power meter, and the signal generator were all controlled by a LabVIEW (NI, Austin, TX, USA) program.

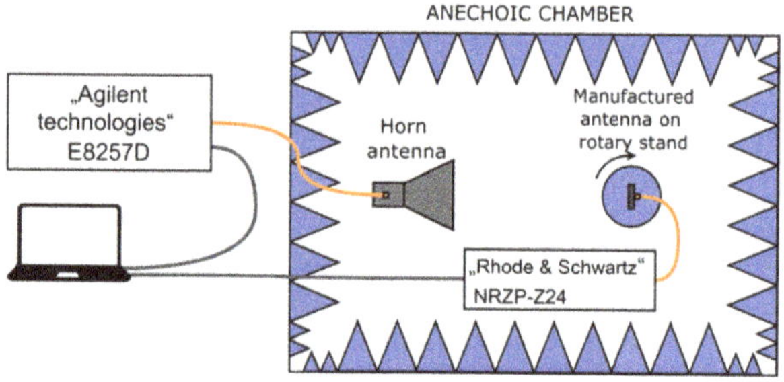

Figure 2. Measurement setup in an anechoic chamber.

The frequency response was measured by keeping the power of the signal fed to the horn antenna constant (20 dBm) and changing the frequency from 2 GHz to 3 GHz in 0.01 GHz increments. The gain dependence on an angle was measured by setting a constant power (20 dBm), locking the frequency at 2.45 GHz and rotating the stand from 0° to 360° in 3° increments. Then, using Equation (2) derived from telegraph equations, the gain was calculated as follows:

$$G = \frac{P(4 \cdot \pi \cdot R)^2}{(P_{TR} \cdot G_{TR} \cdot \lambda)^2} \tag{2}$$

where P is the measured power, R is the distance between antennae, P_{TR} is the transmitted power, G_{TR} is the transmitting antenna gain, and λ is the wavelength.

For more straightforward data interpretation, we normalized the gain measurements using the following equation:

$$G_N = 10log\left(\frac{G}{G_{MAX}}\right); \tag{3}$$

where G_{MAX} is the highest gain measured and G is the measured gain at angles or frequencies.

The antenna was also characterized by the frequency dependence of the reflection, which described the matching quality between the antenna and its excitation port (which was 50 Ω). For this purpose, a vector network analyzer was used.

3. Results and Discussion

3.1. Influence of the Laser Process Parameters on LIG Sheet Resistance and Raman Spectra

Our initial tests showed that varying laser process parameters could control sheet resistance. Therefore, we chose the 1064 nm picosecond laser irradiation, as it showed a relatively wide range of parameters, by using which high-quality LIG can be formed. Furthermore, Chyan et al. [26] showed that LIG formation multiple lasing—repeating irradiation—is crucial to achieving high-quality LIG materials. In other words, specific fluence and sufficient laser irradiation dose are required to form LIG.

During our laser tests, the laser scanning speed was kept at 200 mm/s, and the irradiation dose was varied by changing the average laser power (4 and 5 W), by varying the sample position according to the focal plane of the focusing objective—at the focus and 1 mm above and finally by the number of scanning times (2 and 3). The radiation dose assessment involved the multiplication of the laser fluence with the total number of pulses within the beam spot area.

Figure 3 shows the Raman spectrum of the LIG with the smallest resistance of 36.6 Ω/sq. This LIG structure was formed using a 148 J/cm^2·N irradiation dose. The main peaks prominent for the graphene Raman spectra were visible. $I(D)/I(G)$ was 0.61, and $I(2D)/I(G)$ was 0.48. Further investigations of LIG antennae were performed using the laser process parameters from Figure 2 with the lowest sheet resistance.

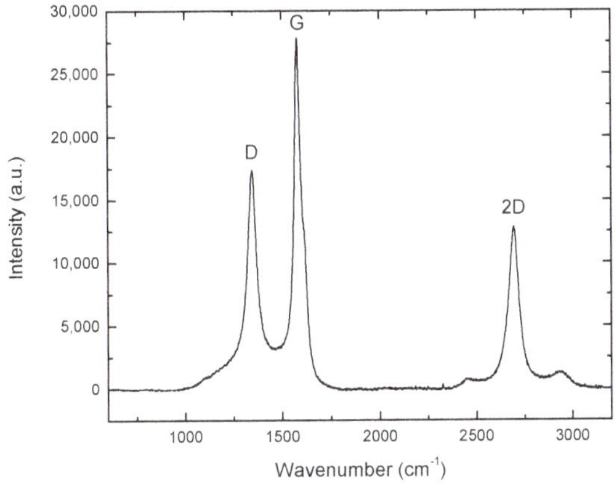

Figure 3. Raman spectrum of LIG, formed using a 148 J/cm^2·N irradiation dose and possessing the lowest sheet resistance of 36.6 Ω/sq.

Figure 4 shows the dependence of the sheet resistance from this accumulated irradiation dose. It is worth noticing that when the irradiation dose was in the range of 90–150 J/cm^2·N, the sheet resistance was around 100 Ω/sq and significantly smaller compared to the ones measured at irradiation doses greater than 150–150 J/cm^2·N. It is also worth noticing that all the lowest LIG sheet resistance values were achieved while LIG was processed with a defocused beam (1 mm above the focal plane of the focusing objective).

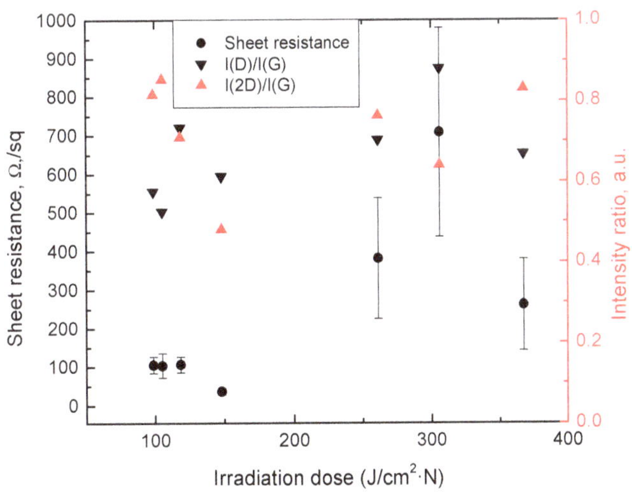

Figure 4. Dependence of LIG sheet resistance and $I(2D)/I(G)$ and $I(2D)/I(G)$ intensity ratios of Raman spectra band intensities from the laser irradiation dose applied on PI.

The $I(D)/I(G)$ ratio is the intensity ratio between the D and G peaks in graphene Raman spectra. The D peak represents defects or disorders in the graphene lattice, while the G peak corresponds to graphite-like lattice vibrations. Therefore, a higher $I(D)/I(G)$ ratio indicates a higher degree of defects or disorder in the graphene structure.

On the other hand, the $I(2D)/I(G)$ ratio represents the intensity ratio between the 2D peak and the G peak. The 2D peak is associated with the second-order Raman scattering and is a characteristic feature of graphene. The $I(2D)/I(G)$ ratio determines the number of graphene layers. A higher $I(2D)/I(G)$ ratio suggests an enhanced graphene quality: High-quality graphene with a well-ordered lattice structure typically displays a higher $I(2D)/I(G)$) ratio. Conversely, lower-quality graphene samples, such as those with a higher degree of disorder, defects, or impurities, tend to have lower $I(2D)/I(G)$ ratios. These ratios provide insights into a sample's quality, structural integrity, and number of graphene layers. Figure 4 presents the distributions of Raman spectra intensity ratios, i.e., $I(D)/I(G)$ and $I(2D)/I(G)$, and their values at different irradiation doses are plotted.

It is worth noticing that through the whole range of tested irradiation doses, we succeeded in achieving high-quality LIG formation, which was confirmed by the $I(2D)/I(G)$ ratio higher than 0.5. The LIG samples formed with a low irradiation dose exhibited the highest $I(2D)/I(G)$ values, indicating a high-quality graphene structure. Additionally, these samples demonstrated the lowest $I(D)/I(G)$ values, suggesting a lower number of LIG structural defects. As mentioned before, this particular processing regime included a defocusing the laser beam. Therefore, with a higher beam diameter, the multiple lasing effect was more prominent, resulting in the formation of high-quality LIG; furthermore, the defocused laser beam induced the lowest resistance due to the lower laser beam intensity. Therefore, lower ablation rates and nonlinear effects such as multiple photon absorptions occur rarely during the process. The process thus generates more thermal energy, leading to graphene formation.

3.2. LIG Antenna Measurements

In order to demonstrate the possibility of using laser-induced graphene for microwave antenna application, we modelled a microwave patch antenna shown in Figure 5. The antenna was made of an FR-4 substrate ($\tan\delta = 0.03$) with a thickness of 1.45 mm and $\varepsilon = 4.2$. One side of the FR4 substrate was completely covered with a thin copper layer and formed the ground plane of the antenna. The dimensions of the FR4 substrate were FR4x = 110 mm and FR4y = 54 mm. The polyimide was deposited on the FR-4 substrate using a commercially available adhesive designed for bonding polyimide films to various substrates. The adhesive was applied uniformly on the backside of the polyimide film, ensuring complete coverage. The use of the adhesive facilitated a strong and reliable bond between the polyimide film and the FR-4 substrate, ensuring mechanical stability and secure attachment. After successfully attaching the polyimide film to the substrate, we proceeded with the laser-induced graphene (LIG) formation on the polyimide film. The thickness of the polyimide was approximately 130 μm), and the LIG layer's thickness was around 10 μm (evaluated using an optical microscope). The dimensions of the LIG patch were GRx = 20 mm and GRy = 29 mm. The LIG patch was placed in the middle of the FR4 substrate. During the antenna modelling, the best patch size and the place of the SMA connector were found to have the smallest S11 and the highest antenna gain.

Reflection and normalized directivity measurements were used to characterize a graphene antenna, which are shown in Figure 6. Looking at the measured reflection dependence on frequency, the antenna appeared to have a reasonably broad operating frequency band. The reflection coefficient was below −10 dB from 2.1 GHz and continued to decrease to −13 dB at 5 GHz. This did not match the typical patch antenna behavior. LIG's low conductivity might cause a mismatch compared to traditionally used copper. This phenomenon was investigated by changing the patch sheet resistance in the antenna model using CST Studio suite for antenna simulation (Figure 6a). Simulation results showed that when the LIG sheet resistance was low, the antenna acted like a regular patch antenna and had a reflection minimum at the intended frequency (2.45 GHz). When the sheet resistance of the LIG layer increased, the resonance disappeared and the reflections in the whole frequency range decreased. This could be explained due to internal losses in the LIG layer. The simulation of the results coincided best with the measured S11, when the sheet resistance was around 30–40 Ω/sq. A similar sheet resistance was obtained by performing four-probe measurements. Considering the high internal losses in the LIG layer, it was required to develop a higher-conductivity LIG layer to achieve better impedance matching. Enhancing the conductivity of the LIG material is an area for future research. Directivity measurements showed that the radiation pattern resembled a traditional patch antenna. The antenna beamwidths were 63° in the vertical position and 80° in the horizontal position.

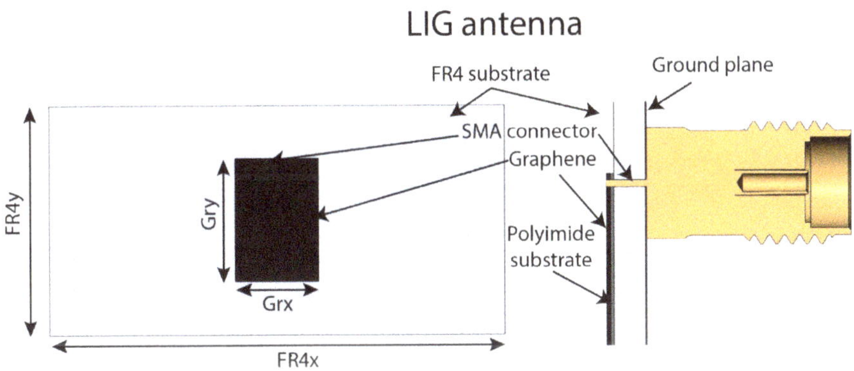

Figure 5. Proposed patch antenna with LIG graphene. **Left**—patch antenna; **right**—antenna cross-section.

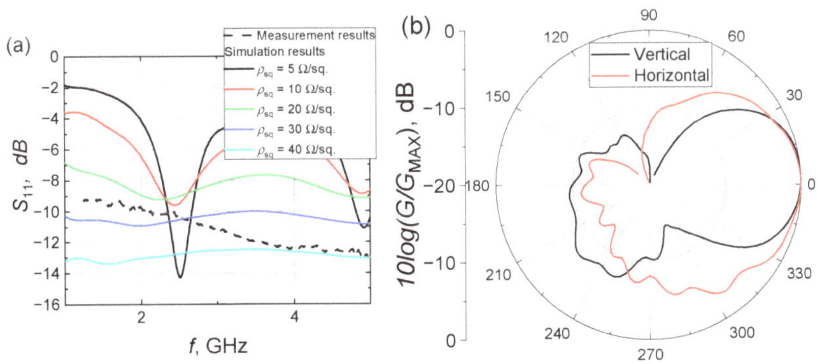

Figure 6. Antenna simulation and measurement results: (**a**) reflection coefficient S_{11}. Solid lines represent simulation results, and the dashed line represents measurement results; (**b**) antenna directivity.

Table 1 presents simulation results for the patch antennae with both laser-induced graphene (LIG) and copper patches at 2.45 GHz. Comparing the results, it is evident that the LIG patch antennae exhibited lower performance in terms of S_{11} and directivity compared to the copper patch antenna with a sheet resistance of 0.01 Ω/sq. However, the LIG patch antennae still demonstrate a remarkable potential. As the sheet resistance of the LIG patch decreased, there were noticeable improvements in S11, directivity, beamwidth, and efficiency. The results indicate that if the sheet resistance of the LIG patch can be further reduced, it has the potential to compete with the copper patch antenna in terms of performance metrics. This suggests a promising future for LIG-based patch antennae, with the possibility of achieving higher efficiency and comparable or even better overall performance.

Table 1. Comparison of performance parameters of LIG and copper patch antennae.

Antenna Type	Sheet Resistance (Ω/sq)	S11 (dB) at 2.45 GHz	Directivity (dBi) at 2.45 GHz	Beamwidth (°) at 2.45 GHz	Efficiency (%)
LIG patch	40	−12.8	6.3	96.8	0.3
LIG patch	10	−9.6	7.4	92.5	2
LIG patch	5	−13.3	7.4	92.3	4
Copper patch	0.01	−30	7.5	85	40

Table 2 compares the results of our proposed laser-induced graphene (LIG) patch antenna with the results from other sources in the literature. It highlights the advantages of our proposed antenna design. Regarding the reflection coefficient, our LIG patch antenna demonstrated a favorable performance with a value of −13 dB, indicating lower signal reflection compared to the other antenna types in the table, suggesting improved impedance matching and reduced signal loss. Additionally, our antenna exhibited a sheet resistance ranging from 30 to 40 Ω/sq, which was relatively low compared to some other antennae [42]. This lower sheet resistance contributed to enhanced conductivity and improved signal transmission efficiency.

Table 2. Comparison of performance parameters of reported graphene-based antennae.

Article	Antenna Type	Reflection Coefficient (dB)	Sheet Resistance (Ω/sq)	Gain (dBi)	Radiation Efficiency (%)
This paper	Patch	−13	30–40	−8	0.3
[40]	Monopole	−25	6–15	0	No data
[39]	Patch	−14	No data	1.82	50
[42]	Trapezoidal Meandered	−5.52	50	No data	No data
[33]	Patch	−25	No data	No data	No data
[43]	Nano dipole	−45.5	No data	−10.9	3.34

4. Conclusions

Successful LIG formation was performed with a wide range of laser irradiation doses and was confirmed by Raman spectroscopy—$I(2D)/I(G)$ was higher than 0.5. The sheet resistance was smallest at irradiation doses from 98 to 148 $J/cm^2 \cdot N$, when the PI sample was irradiated at the position 1 mm above the focal plane of the focusing objective. The minimum sheet resistance of LIG was 36.6 Ω/sq. Using optimal process parameters, a patch antenna at 2.45 GHz (WiFi) was produced and investigated.

The results of the patch antenna with an LIG layer show promising results as one of the alternative methods as green and sustainable solutions to manufacturing removable and flexible antennae. However, additional investigations still need to be performed to reduce the LIG layer sheet resistance and increase the antenna efficiency. For this purpose, further optimization of the LIG process by tuning laser process parameters and investigating other substrates will be performed.

Author Contributions: Conceptualization, P.R., K.R. and R.T.; methodology, A.S., J.Ž., P.R. and R.T.; software, P.R.; validation, A.S., J.Ž., P.R., K.R. and R.T.; formal analysis, A.S. and J.Ž.; investigation, P.R., K.R. and R.T.; writing—original draft preparation, P.R. and R.T.; writing—review and editing, R.T.; visualization, P.R. and R.T.; supervision, R.T.; project administration, K.R. and R.T. All authors have read and agreed to the published version of the manuscript.

Funding: This project has received funding from European Regional Development Fund (project BEGAMA No 01.2.2-LMT-K-718-03-0038) under grant agreement with the Research Council of Lithuania (LMTLT).

Data Availability Statement: The data supporting this study's findings are available from the corresponding author upon reasonable request.

Acknowledgments: Authors are grateful to Martynas Talaikis for accessing Raman spectroscopy at the Laboratory of Spectroelectrochemistry of the Department of Organic Chemistry (FTMC, Lithuania).

Conflicts of Interest: The authors declare no conflict of interest.

References

1. Wanjari, V.P.; Reddy, A.S.; Duttagupta, S.P.; Singh, S.P. Laser-Induced Graphene-Based Electrochemical Biosensors for Environmental Applications: A Perspective. *Environ. Sci. Pollut. Res.* **2023**, *30*, 42643–42657. [CrossRef] [PubMed]
2. Liu, H.; Chen, Y. Laser-Induced Graphene Film and Its Applications in Flexible Electronics. *Appl. Sci.* **2022**, *12*, 1233. [CrossRef]
3. Vivaldi, F.M.; Dallinger, A.; Bonini, A.; Poma, N.; Sembranti, L.; Biagini, D.; Salvo, P.; Greco, F.; Di Francesco, F. Three-Dimensional (3D) Laser-Induced Graphene: Structure, Properties, and Application to Chemical Sensing. *ACS Appl. Mater. Interfaces* **2021**, *13*, 30245–30260. [CrossRef] [PubMed]
4. Barber, R.; Cameron, S.; Devine, A.; McCombe, A.; Kirsty Pourshahidi, L.; Cundell, J.; Roy, S.; Mathur, A.; Casimero, C.; Papakonstantinou, P.; et al. Laser Induced Graphene Sensors for Assessing PH: Application to Wound Management. *Electrochem. Commun.* **2021**, *123*, 106914. [CrossRef]
5. Peng, Z.; Lin, J.; Ye, R.; Samuel, E.L.G.; Tour, J.M. Flexible and Stackable Laser-Induced Graphene Supercapacitors. *ACS Appl. Mater. Interfaces* **2015**, *7*, 3414–3419. [CrossRef]
6. Kulyk, B.; Silva, B.F.R.; Carvalho, A.F.; Silvestre, S.; Fernandes, A.J.S.; Martins, R.; Fortunato, E.; Costa, F.M. Laser-Induced Graphene from Paper for Mechanical Sensing. *ACS Appl. Mater. Interfaces* **2021**, *13*, 10210–10221. [CrossRef] [PubMed]
7. Stanford, M.G.; Zhang, C.; Fowlkes, J.D.; Hoffman, A.; Ivanov, I.N.; Rack, P.D.; Tour, J.M. High-Resolution Laser-Induced Graphene. Flexible Electronics beyond the Visible Limit. *ACS Appl. Mater. Interfaces* **2020**, *12*, 10902–10907. [CrossRef]
8. Žutautas, V.; Trusovas, R.; Sartanavičius, A.; Ratautas, K.; Selskis, A.; Pauliukaite, R. A Sensor for Electrochemical PH Monitoring Based on Laser-Induced Graphene Modified with Polyfolate. *Chemosensors* **2023**, *11*, 329. [CrossRef]
9. Trusovas, R.; Ratautas, K.; Račiukaitis, G.; Barkauskas, J.; Stankevičiene, I.; Niaura, G.; Mažeikiene, R. Reduction of Graphite Oxide to Graphene with Laser Irradiation. *Carbon N. Y.* **2013**, *52*, 574–582. [CrossRef]
10. Wei, Z.; Wang, D.; Kim, S.; Kim, S.Y.; Hu, Y.; Yakes, M.K.; Laracuente, A.R.; Dai, Z.; Marder, S.R.; Berger, C.; et al. Nanoscale Tunable Reduction of Graphene Oxide for Graphene Electronics. *Science* **2010**, *328*, 1373–1376. [CrossRef]
11. Kumar, P.; Subrahmanyam, K.S.; Rao, C.N.R. Graphene Produced by Radiation-Induced Reduction of Graphene Oxide. *Int. J. Nanosci.* **2011**, *10*, 559–566. [CrossRef]
12. Sokolov, D.A.; Shepperd, K.R.; Orlando, T.M. Formation of Graphene Features from Direct Laser-Induced Reduction of Graphite Oxide. *J. Phys. Chem. Lett.* **2010**, *1*, 2633–2636. [CrossRef]
13. Sokolov, D.A.; Rouleau, C.M.; Geohegan, D.B.; Orlando, T.M. Excimer Laser Reduction and Patterning of Graphite Oxide. *Carbon N. Y.* **2013**, *53*, 81–89. [CrossRef]
14. Trusovas, R.; Račiukaitis, G.; Barkauskas, J.; Mažeikienė, R. Laser Induced Graphite Oxide/Graphene Transformation. *J. Laser Micro Nanoeng.* **2012**, *7*, 49. [CrossRef]
15. Ye, R.; James, D.K.; Tour, J.M. Laser-Induced Graphene. *Acc. Chem. Res.* **2018**, *51*, 1609–1620. [CrossRef]
16. Luong, D.X.; Yang, K.; Yoon, J.; Singh, S.P.; Wang, T.; Arnusch, C.J.; Tour, J.M. Laser-Induced Graphene Composites as Multifunctional Surfaces. *ACS Nano* **2019**, *13*, 2579–2586. [CrossRef] [PubMed]
17. Lin, J.; Peng, Z.; Liu, Y.; Ruiz-Zepeda, F.; Ye, R.; Samuel, E.L.G.; Yacaman, M.J.; Yakobson, B.I.; Tour, J.M. Laser-Induced Porous Graphene Films from Commercial Polymers. *Nat. Commun.* **2014**, *5*, 5714. [CrossRef] [PubMed]
18. Lopes, D.V.; Santos, N.F.; Moura, J.P.; Fernandes, A.J.S.; Costa, F.M.; Kovalevsky, A.V. Design of Laser-Induced Graphene Electrodes for Water Splitting. *Int. J. Hydrogen Energy* **2023**, *48*, 4158–4172. [CrossRef]
19. Trusovas, R.; Ratautas, K.; Račiukaitis, G.; Niaura, G. Graphene Layer Formation in Pinewood by Nanosecond and Picosecond Laser Irradiation. *Appl. Surf. Sci.* **2019**, *471*, 154–161. [CrossRef]
20. Fiodorov, V.; Ratautas, K.; Mockus, Z.; Trusovas, R.; Mikoliūnaitė, L.; Račiukaitis, G. Laser-Assisted Selective Fabrication of Copper Traces on Polymers by Electroplating. *Polymers* **2022**, *14*, 781. [CrossRef]
21. Ruan, X.; Wang, R.; Luo, J.; Yao, Y.; Liu, T. Experimental and Modeling Study of CO2 Laser Writing Induced Polyimide Carbonization Process. *Mater. Des.* **2018**, *160*, 1168–1177. [CrossRef]
22. Ma, L.; Wang, Y.; Wang, Y.; Wang, C.; Gao, X. Graphene Induced Carbonization of Polyimide Films to Prepared Flexible Carbon Films with Improving-Thermal Conductivity. *Ceram. Int.* **2020**, *46*, 3332–3338. [CrossRef]
23. In, J.B.; Hsia, B.; Yoo, J.H.; Hyun, S.; Carraro, C.; Maboudian, R.; Grigoropoulos, C.P. Facile Fabrication of Flexible All Solid-State Micro-Supercapacitor by Direct Laser Writing of Porous Carbon in Polyimide. *Carbon N. Y.* **2015**, *83*, 144–151. [CrossRef]
24. Wang, S.; Yu, Y.; Li, R.; Feng, G.; Wu, Z.; Compagnini, G.; Gulino, A.; Feng, Z.; Hu, A. High-Performance Stacked in-Plane Supercapacitors and Supercapacitor Array Fabricated by Femtosecond Laser 3D Direct Writing on Polyimide Sheets. *Electrochim. Acta* **2017**, *241*, 153–161. [CrossRef]
25. Wang, L.; Wang, Z.; Bakhtiyari, A.N.; Zheng, H. A Comparative Study of Laser-Induced Graphene by CO_2 Infrared Laser and 355 nm Ultraviolet (UV) Laser. *Micromachines* **2020**, *11*, 1094. [CrossRef] [PubMed]
26. Chyan, Y.; Ye, R.; Li, Y.; Singh, S.P.; Arnusch, C.J.; Tour, J.M. Laser-Induced Graphene by Multiple Lasing: Toward Electronics on Cloth, Paper, and Food. *ACS Nano* **2018**, *12*, 2176–2183. [CrossRef]

27. Murray, R.; Burke, M.; Iacopino, D.; Quinn, A.J. Design of Experiments and Optimization of Laser-Induced Graphene. *ACS Omega* **2021**, *6*, 16736–16743. [CrossRef]
28. Cao, L.; Zhu, S.; Pan, B.; Dai, X.; Zhao, W.; Liu, Y.; Xie, W.; Kuang, Y.; Liu, X. Stable and Durable Laser-Induced Graphene Patterns Embedded in Polymer Substrates. *Carbon N. Y.* **2020**, *163*, 85–94. [CrossRef]
29. Gupta, A.; Sharma, C.P.; Arnusch, C.J. Simple Scalable Fabrication of Laser-Induced Graphene Composite Membranes for Water Treatment. *ACS ES&T Water* **2021**, *1*, 881–887. [CrossRef]
30. Lee, Y.A.; Jang, K.; Yoon, H. A Facile and Scalable Fabrication of High-Performance Flexible Laser-Induced-Graphene Micro-Supercapacitors Using Ultrafast Pulse Laser. *ECS Meet. Abstr.* **2022**, *MA2022-02*, 2562. [CrossRef]
31. Liu, F.; Gao, Y.; Wang, G.; Wang, D.; Wang, Y.; He, M.; Ding, X.; Duan, H.; Luo, S. Laser-Induced Graphene Enabled Additive Manufacturing of Multifunctional 3D Architectures with Freeform Structures. *Adv. Sci.* **2023**, *10*, 2204990. [CrossRef]
32. Wang, Y.; Wang, Y.; Zhang, P.; Liu, F.; Luo, S. Laser-Induced Freestanding Graphene Papers: A New Route of Scalable Fabrication with Tunable Morphologies and Properties for Multifunctional Devices and Structures. *Small* **2018**, *14*, 1802350. [CrossRef] [PubMed]
33. Ahmad, M.; Cantarella, G.; Angeli, M.A.C.; Madagalam, M.; Ebner, C.; Ciocca, M.; Riaz, R.; Ibba, P.; Petrelli, M.; Merino, I.; et al. 2.4 GHz Microstrip Patch Antenna Fabricated by Means of Laser Induced Graphitization of a Cellulose-Based Paper Substrate. In Proceedings of the 2021 IEEE International Flexible Electronics Technology Conference (IFETC), Columbus, OH, USA, 8–11 August 2021; pp. 44–46.
34. Le, T.-S.D.; Phan, H.-P.; Kwon, S.; Park, S.; Jung, Y.; Min, J.; Chun, B.J.; Yoon, H.; Ko, S.H.; Kim, S.-W.; et al. Recent Advances in Laser-Induced Graphene: Mechanism, Fabrication, Properties, and Applications in Flexible Electronics. *Adv. Funct. Mater.* **2022**, *32*, 2205158. [CrossRef]
35. Mostaccio, A.; Antonelli, G.; Bragaglia, M.; Martinelli, E.; Marrocco, G. Comparative Evaluation of Laser Induced Graphene (LIG) Traces on Polyimide under Soft and Hard Stress for IoT Applications. *IEEE J. Flex. Electron.* **2023**, 1. [CrossRef]
36. Liao, J.; Zhang, X.; Sun, Z.; Chen, H.; Fu, J.; Si, H.; Ge, C.; Lin, S. Laser-Induced Graphene-Based Wearable Epidermal Ion-Selective Sensors for Noninvasive Multiplexed Sweat Analysis. *Biosensors* **2022**, *12*, 397. [CrossRef]
37. Wan, Z.; Nguyen, N.-T.; Gao, Y.; Li, Q. Laser Induced Graphene for Biosensors. *Sustain. Mater. Technol.* **2020**, *25*, e00205. [CrossRef]
38. Zhu, J.; Guo, X.; Wang, H.; Song, W. Cost-Effective Fabrication and High-Frequency Response of Non-Ideal RC Application Based on 3D Porous Laser-Induced Graphene. *J. Mater. Sci.* **2018**, *53*, 12413–12420. [CrossRef]
39. Sindhu, B.; Kothuru, A.; Sahatiya, P.; Goel, S.; Nandi, S. Laser-Induced Graphene Printed Wearable Flexible Antenna-Based Strain Sensor for Wireless Human Motion Monitoring. *IEEE Trans Electron. Devices* **2021**, *68*, 3189–3194. [CrossRef]
40. Abdul-Aziz, M.R.R.; Mohassieb, S.A.; Eltresy, N.A.; Yousef, M.M.K.; Anis, B.; Abdellatif, S.O.; Khalil, A.S.G. Enhancing the Performance of Polygon Monopole Antenna Using Graphene/TMDCs Heterostructures. *IEEE Trans. Nanotechnol.* **2020**, *19*, 269–273. [CrossRef]
41. Smits, F.M. Measurement of Sheet Resistivities with the Four-Point Probe. *Bell Syst. Tech. J.* **1958**, *37*, 711–718. [CrossRef]
42. Rivadeneyra, A.; Salmeron, J.F.; Rodriguez, N.; Morales, D.P.; Colella, R.; Chietera, F.P.; Catarinucci, L. Laser-Fabricated Antennas for RFID Applications. In Proceedings of the 2020 50th European Microwave Conference (EuMC), Utrecht, The Netherlands, 12–14 January 2021; pp. 812–815.
43. Inum, R.; Rana, M.M.; Shushama, K.N. Performance Analysis of Graphene Based Nano Dipole Antenna on Stacked Substrate. In Proceedings of the 2016 2nd International Conference on Electrical, Computer & Telecommunication Engineering (ICECTE), Rajshahi, Bangladesh, 8–10 December 2016; pp. 1–4.